Soils and Sediments

Springer

Berlin
Heidelberg
New York
Barcelona
Budapest
Hong Kong
London
Milan
Paris
Santa Clara
Singapore
Tokyo

Hélène Paquet, Norbert Clauer

Soils and Sediments

Mineralogy and Geochemistry

With 138 figures and 18 tables

Dr. Hélène Paquet

Dr. Norbert Clauer

Centre de Géochimie de la Surface (CNRS)
I rue Blessig
67084 Strasbourg
France

ISBN 3-540-61599-7 Springer-Verlag Berlin Heidelberg New York

Die Deutsche Bibliothek - CIP-Einheitsaufnahme

Soils and sediments, mineralogy and geochemistry: a book dedicated to Georges Millot; with 18 tables / Hélène Paquet; Norbert Clauer (ed.) - Berlin; Heidelberg; New York; Barcelona; Budapest; Hong Kong; London; Milano; Paris; Santa Clara; Singapore; Tokyo: Springer 1997
ISBN 3-540-61599-7
NE: Paquet, Hélène [Hrsg.]; Millot, Georges: Festschrift

Coverdesign: design & production, Heidelberg
Production editor: Stefan Pauli, Heidelberg

SPIN: 10541503 VA 32/3136 - 5 4 3 2 1 0 - printed on acid free paper

A BOOK DEDICATED TO PROFESSOR GEORGES MILLOT

Georges MILLOT
(1917-1991)

Dr. Georges Millot received his PhD in 1949 from the University of Nancy. He began his career as Assistant Professor at the same university in 1942 and then as Professor at the "Ecole Nationale Supérieure des Mines" and at the "Ecole Nationale Supérieure de Géologie" of Nancy. In 1954, he joined the University of Strasbourg and became successively or concomitantly Head of the Geological Institute (1954-1981), Founder and Head of the "Centre de Sédimentologie et Géochimie de la Surface" (1963-1981) and Dean (1962-1965). He retired as Emeritus Professor in 1981, but remained in charge of many activities at the French Academy of Sciences where he was elected in 1977 a permanent member. In his personal and professional life he was awarded many honorific distinctions, such as Officier de la Légion d'Honneur and Officier de l'Ordre du Mérite National. He was also elected an associate member of the Belgian Academy of Sciences, Letters and Arts (1975), and he was nominated Doctor Honoris Causa at the Universities of Madrid (Spain, 1979), Neuchâtel (Switzerland, 1981), Pavia (Italy, 1981) and Ankara (Turkey, 1991).

Georges Millot's research career, extending over 50 years, almost always involved "Clay Geology", which is the title of the book that he published in French in 1964, in Russian in 1968 and in English in 1970. His research may be qualified by the following: "birth of clay minerals in weathering profiles, evolution of clay minerals during sediment deposition and death of clay minerals during sediment burial". He was among the pioneers identifying clay minerals, defining their formational processes and their relationships with their environment and reporting about their potential as markers. In 1963, he created a research center in Strasbourg which became one of the "clay research centers" known world-wide in the Earth-Science community. He also assumed many responsibilities in the Groupe Français des Argiles and in the European Clay Group and was one of the initiators of the Association Internationale pour l'Etude des Argiles (AIPEA).

Foreword

It was a great honour for me to become acquainted with Georges Millot 40 years ago at the admission committee for teaching positions in junior high schools (Lycées); he was on the jury and I was a candidate. We recognized each other immediately. He was a spiritual guide for me, becoming, year after year, a beloved friend.

I would like to acknowledge this friend, on behalf of my colleagues in Geology and my fellow members of the Académie des Sciences, who were all very attached to him. We all admired the scientist who was always searching for the truth. From an illness at the beginning of his scientific career, which refrained him from fieldwork that he liked so much, Georges Millot made a strength. Due to his action, the Earth Sciences proceeded from a natural to an exact science. It should be recalled that Georges Millot was one of the founders of geochemistry, with which so many are acquainted with today. What an opening in a period mainly characterized by sadness in between the two World Wars! This was the beginning of a new concept of the Earth Sciences which developed more and more over the years.

Behind the scientist was the professor who ensured ceaseless exchanges between research and teaching. For him, teaching was the most noble profession, whatever the level. Georges Millot prepared generations of professors who hold a lasting remembrance of his lectures. He respected and admired the teachers of the primary school who taught the Parisian boy that he was.

Later came the period of responsibility. Georges Millot was a remarkable scientific manager who took up the French geological profession at a very difficult period. He succeeded in installing peace among the specialists of the different disciplines in the Earth Sciences, allowing each to act according to his own capabilities. A fortunate period began for French geology, as many recall nostalgically.

Georges Millot had, in fact, a very high opinion about State service, which he felt with passion and expressed with kind equability. I remember an open letter that he addressed to talented and impatient young colleagues and signed «Georges Millot, servant of the Republic». This reminder of duty impressed many at a moment when rights are claimed permanently!

Erudite professor and upright higher official are two facets of this rich personality dedicated to serving others. Anyone who was in difficulty or in need knew he could rely on Georges Millot; and this extended far beyond material problems, since care of souls was also among his preoccupations. He was obviously among those who pondered and changed the direction of thoughts and things; he was strongly involved in the development of contemporary ideas.

Then came the period of honours, that of membership to the Académie des Sciences. Georges Millot entered the Académie des Sciences in order to serve; once again! Another of his great successes, one of which he was very proud, was to have reenlivened the periodic journal, *Comptes Rendus de l'Académie des Sciences,* in the domain of Earth Sciences and to make it into an up-to-date review, which was impatiently expected by many.

What an influence in the academic assemblies, where his kind common sense, as well as his smiling but corrosive humour did wonders, like elsewhere! His memory will be present for a long time in our debates and many of us will think of him before making decisions in complicated situations.

Indeed, now the time of remembrance has come. The purpose of this book is to reenliven one of George Millot's facets, i.e. the eminent scientist, through the works of some of his close colleagues and disciples, in homage to his illuminating personality.

Jean Aubouin

Preface

After Georges Millot's death in September 1991, we felt that not only his professional career, but also his personal way of life deserved a tribute by at least some of his close collaborators. As a professor, Georges Millot contributed greatly to the formation of generations of earth scientists and, among all these colleagues, some had the privilege to further their own research due to his influence and advice. We were among those who were lucky to be part of his "working community". We learned so much from him, even beyond professional contact. This textbook, which is dedicated to his memory, is our way of thanking him and reviving our attachment to him. The main idea for this volume was twofold:

- a collection of up-to-date reviews by scientists, who were among Georges Millot's closest collaborators and who wanted to remember these years with this dedication, from research fields in which he was a pioneer or in which he initiated new research avenues;
- an English presentation of an up-graded synthetic view of the "French School" in surficial geochemistry.

In fact, as for many colleagues of his generation, Georges Millot's knowledge and practice of foreign languages was such that he could not present his scientific message as easily as he would have wished. We are therefore proud, through this editorial work, to contribute to a presentation of part of the knowledge acquired through his work.

Last, but not at least, we would like to thank sincerely the authors of the various chapters which cover most of the scientific topics that were of major interest to Georges Millot, the reviewers of the contributions who helped to improve the drafts and the publisher of this textbook.

Strasbourg, 12 February 1996
Hélène Paquet
Norbert Clauer

Table of contents

List of Contributors

Jean-Paul Ambrosi and
Daniel Nahon
Laboratoire de Géosciences de
l'Environnement, UM GECO, CEREGE
Université d'Aix-Marseille III
B.P. 80
13545 Aix en Provence, France

Bruno Boulangé,
Instituto de Geociências
USP, ORSTOM
C.P. 11.348
05422-970 São Paulo (S.P.), Brazil

René Boulet
Instituto de Geociências
USP ORSTOM
C.P. 11.348
05422-970 São Paulo (S.P.), Brazil

Hervé Chamley
URA 719, SN 5
University de Lille 1
59655 Villeneuve d'Ascq, France

Sam Chaudhuri
Department of Geology, Kansas
State University
Manhattan, Kansas 66506, USA

Norbert Clauer
Centre de Géochimie de la Surface
1 Rue Blessig
67084 Strasbourg, France

Fabrice Colin
Laboratoire de Géosciences de
l'Environnement, UM GECO, CEREGE
Université d'Aix-Marseille III
B.P. 80,
13545 Aix en Provence, France
and ORSTOM, 213 Rue La Fayette
75480 Paris, Cedex 05, France

Emmanuel Fritsch
Instituto de Geociências
USP, ORSTOM
C.P. 11.348
05422-970 São Paulo (S.P.), Brazil

Daniel Jeannette
Centre de Géochimie de la Surface
1 Rue Blessig
67084 Strasbourg, France

Bernard Kübler
Institute de Géologie
Université de Neuchâtel
11 Rue Emile-Argand
2007 Neuchâtel, Suisse

Jean-Claude Leprun
Mission ORSTOM
C.P. 4010
Boa Viagem
51020 Recife (PE), Brazil

Jacques Lucas
Centre de Géochimie de la Surface
1 Rue Blessig
67084 Strasbourg, France

Yves Lucas
Université de Toulon
B.P. 132
83957 La Garde, France

Hélène Paquet
Centre de Géochimie de la Surface
1 Rue Blessig
67084 Strasbourg, France

Georges Pédro
Académie d'Agriculture
18 Rue de Bellechasse
75007 Paris, France

L. Prévôt-Lucas
Centre de Géochimie de la Surface
1 Rue Blessig
67084 Strasbourg, France

Médard Thiry
Ecole des Mines de Paris
Centre d'Informatique Géologique
35 Rue St Honoré
77305 Fontainebleau, France

Jean-Jacques Trescases
Laboratoire de Pétrologie
de la Surface
Université de Poitiers
40 Avenue du recteur Pineau
86022 Poitiers, France

Marie-Madeleine Blanc Valleron
Laboratoire de Géologie
Muséum National d'Histoire
Naturelle
43 Rue Buffon
75005 Paris, France

Francis Weber
Centre de Géochimie de la Surface
1 Rue Blessig
67084 Strasbourg, France

1 Clay Minerals in Weathered Rock Materials and in Soils

Georges Pédro

Since Georges Millot's initial contribution to clays and his thesis on the relationship between composition and development of sedimentary clayey rocks defended in 1949, half a century has passed by. So much has been discovered on clays during that period that it would almost be impossible either to list the work undertaken or to give a comprehensive account of the present situation in a concise presentation. I have, therefore, selected some topics in which Georges Millot was mainly interested and I will proceed according to his methods, i.e. I will present facts observed in nature, examine them with the best-suited techniques available and finally organize the data into a coherent picture, even though I am not as bright as he was. I will try and aim for clarity rather than for the enlightenment which was his.

Georges Millot soon realized, after his thesis, that studying soils and pedological profiles covering the continents was essential if research on the origins of clays was to progress any further. Thus, he soon became interested in clays in weathered material and in soils (cf. Millot 1964, 1970), and I will review the current research on this topic in his memory.

1 Introduction

Soils or pedological covers are, according to a now classical viewpoint, the outermost thin layers of loose material of the geosphere which support life on earth. A soil, which appears as a natural entity interfacing the lithosphere with the atmosphere and the biosphere, is like the skin of our geoid. Actually, it results from two types of processes which are quite distinct and do not develop at the same rate (Pédro 1992):

1. The first type of process is related to interactions between the mineral world (the superficial geosphere) and the atmosphere. These processes are inert and azoic, and could be considered as some sorts of surface metamorphism or «atmospheric metamorphism». This type is subdivided into two subclasses in which water plays a major role: (1) changes in the chemical and mineralogical composition of the superficial geosphere (weathering = alteration) during which water acts as a solvent and a reagent; and (2) struc-

tural changes with plasma formation (clay plasmation) during which water acts mainly as a lubricant.

2. The second type of process, which is of biotic and organic nature, is related to the biological invasion of the surface of our planet and consequent development of the biosphere. The involved biochemical reactions include synthesis reactions followed by degradation when organisms die, leading to organic matter which is incorporated into the soil as humic plasma.

To what extent these processes affect any region on earth depends on the processes involved, on their intensity and on their duration. They always seem to be vertically stratified, especially in well developed situations where the altered layers are some 10 m thick. Three levels are organized in a more or less parallel way above the compact unweathered rock (Fig. 1). The lower level (III) is only affected by geochemical weathering (without any pedoplasmation and biological activity whatsoever). The maturity of the resulting mineral material varies but the original rock frame can always be recognized. Water acts both as a reagent and as a lubricant in the middle level (II), leading to reorganization and pedoplasmation processes which erase the initial rock structure. Both geochemical weathering and clay plasmation strongly affect this mineral level which often corresponds to the pedological horizon B. The upper level (I) is affected by all three types of processes but is mainly characterized by biological activity. Organo-clay plasmation takes over from initial weathering and is specific to the pedological horizon A.

The outlines of this chapter having been set up, we can now be more specific. The full stratification, which has just been described, is characteristic of deep soils typical – but not exclusively – of intertropical regions where soil formation has been intense over long periods. Usually, only the upper two levels (A/B) are present and the soils are termed normal. Also, superficial soils comprise only the top level (horizon A) and are affected by all three types of proc-

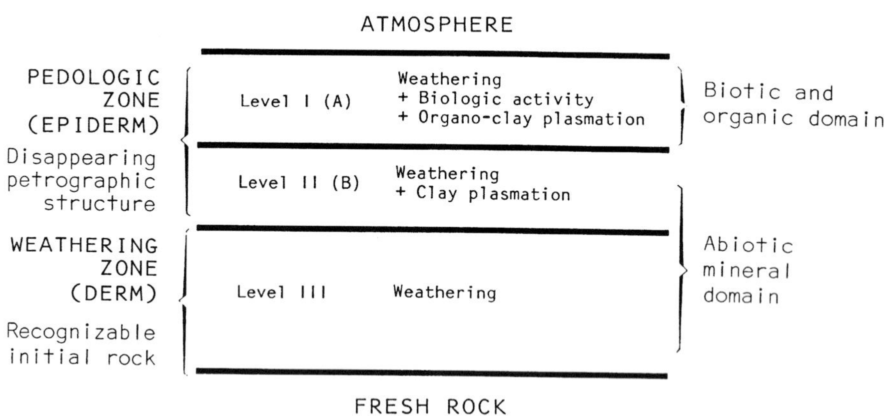

Fig. 1. Stratification of superficial processes

esses – i.e. weathering, clay plasmation and biological activity – which may occur simultaneously. The three levels can be grouped into two petrological zones: (1) the weathering zone which includes the deeper zone of level III where the original framework of the rocks can still be recognized; and (2) the pedological zone "s.s." or the soil in its initial sense, which includes levels II and I where the original material can no longer be recognized because of changes and rearrangements.

Similarly to skin, the pedological cover can thus be divided into dermis and epidermis, and two main domains can be distinguished on biological and organic grounds: the incorporation of organic matter is the hallmark of the superficial biotic domain, and the abiotic domain includes levels II and III. His extensive studies in this domain led Georges Millot to emphasize the importance of hydrolysis.

The present chapter reviews what is currently known about the abiotic domain, successively dealing with the weathering zone (level III in which the original rock can be recognized), its various states and its relationship with the type of secondary minerals obtained, especially the phyllosilicates, and the submicrometric clay particles which are typical of the mostly mineral level II, where the original rock can no longer be recognized.

2 Hydrolytic Weathering Processes and Genesis of Secondary Constituents

2.1 The Concept of «Intensity of Hydrolysis»: Certainties and Limitations

Hydrolytic weathering of the earth's surface always results, from a geochemical standpoint, in at least some loss of silica and alkalies from the initial rock, whereas Al (and Fe) are maintained in situ. Lamellar secondary constituents, which are usually Al phyllosilicates (siallitization) but which can also be Al hydroxides under extreme conditions (allitization), appear in these conditions.

The degree of hydrolysis and the type of newly formed minerals are, of course, related as shown by field work (e.g. Lacroix 1913; Harrasowitz 1926; Tardy 1969; Quantin 1993), experimental investigations (Pédro 1964, 1966, 1978; Robert 1970) and theoretical considerations (Fritz and Tardy 1973; Fritz 1981; Bourrié 1990). Three prime scenarios have long been recognized (Pédro 1966, 1978, 1984).

1. Total hydrolysis leads to complete loss of silica and alkalies inducing formation of gibbsite, $Al(OH)_3$. Allitization is on the way.
2. If hydrolysis is less pervasive, loss of silica might be only partial, whereas loss of alkalies is complete. Kaolin-type phyllosilicates with a 1/1 layering are formed, which is recognized as the monosiallitization process.
3. Hydrolysis can even be weak with incomplete loss of silica and alkalies. Low charge 2/1 phyllosilicates of smectite type (beidellite) form and bisiallitization is in progress.

For a long time, the intensity of hydrolysis and the degree of weathering were considered to be directly related. However, facts accumulated from major surveys following World War II until the 1960s, have provided evidence for a revision of this view. Thus in some very slightly weathered formations, where small amounts of secondary minerals are formed, these secondary minerals are of the allitic type (gibbsite), whereas other massively weathered formations correspond to a very low level of hydolysis and hence contain very large amounts of smectite-type clay.

These facts led to a different approach , in which two factors were considered simultaneously (Pédro et al. 1975; Sequeira Braga et al. 1989): (1) the degree of weathering measured as the amount of weatherable primary minerals that have actually disappeared; and (2) the type of weathering, which is mainly based on the geochemical nature of the authogenic components. Since the type of weathering has just been briefly described, a closer look might be needed to evaluate the concept of degree of weathering.

2.2 Introduction to the Concept of "Degree of Weathering". Major Types of Weathering

Epigenetic replacement («épigénie») which is a specific form of weathering, will not be considered here, and is discussed in the contribution of H. Paquet and A. Ruellan.

The main parameter with respect to the degree of weathering is the ratio between secondary and primary components in weathered rock units. Two cases may occur.

The first case can be considered as massive weathering which develops when the conditions are such that chemical decomposition affects synchronously, and in the same way, all the weatherable minerals of the parent rock, including the most stable aluminosilicates (i.e. K-feldspars and muscovite). Weathering is considered to be massive when chemical weathering is more important than physical weathering, and when all weatherable minerals quickly disappear, replaced by large amounts of identical secondary minerals, formed regardless of the constitution of the initial primary minerals. These secondary components are either gibbsite, kaolinite or smectites (low charge 2/1 clay minerals). This case is known as a stage of alterite formation, i.e. loose material in which the original petrographic structure of the parent rock is preserved – although the volume may change – and which always contains a large amount of secondary phyllosilicates (typical clay weathering). These changes occur at level III which was defined above.

In the second case, weathering acts progressively in a limited way. This kind of weathering induces differential leaching of the primary framework minerals, starting with the most vulnerable. The resulting material tends to be loose and contains a large amount of coarse grains (by the release of primary minerals) and few secondary components, because the reaction kinetics for crystallization and neoformation are very slow. The clay minerals are often very diverse and the types of minerals obtained depend much more on the initial primary minerals than on the general bioclimatic conditions. The most typical minerals are high-charged 2/1 phyllosilicates resulting from changes in the trioctahedral structures of the micas. A small amount of neoformed minerals (gibbsite, kaolinite and smectites) also result from weathering of other weatherable minerals such as plagioclases, amphiboles and others. Physical weathering is always more intense than chemical weathering in this case; the degree of weathering – measured by the secondary constituents/primary constituents ratio – being variable. This form of weathering usually develops in very superficial soils (A) and in normally developed soils (A/B). Sometimes, especially on coarse-grained rocks, a very thick unit III called arena can develop (e.g. Collier 1961; Tardy 1969; Seddoh 1973; Dejou et al. 1977; Meunier 1980; Sequeira Braga et al. 1989). Arenization designates the kind of weathering affecting a large thickness of material, while leaving the lithological pattern of the original rock unchanged, although the total volume changes because of swelling.

2.3 Conditions for the Formation of Alterite and Arena

Generally speaking, alterization typically occurs in wet and semi-wet climates, the type of alterization depending mainly on the length of the warm rainy season. The longer the rainy season, as in equatorial areas, the more pronounced is the silica loss (kaolinite–gibbsite); the shorter the rainy season (dry tropical areas), the more smectitic is the type of weathering. On the other hand, arenization which develops on coarse-grained rocks requires mild geochemical weathering which occurs either under temperate or cold wet climates and low temperatures (<10 °C), where the kinetics of neoformation reactions are reduced,

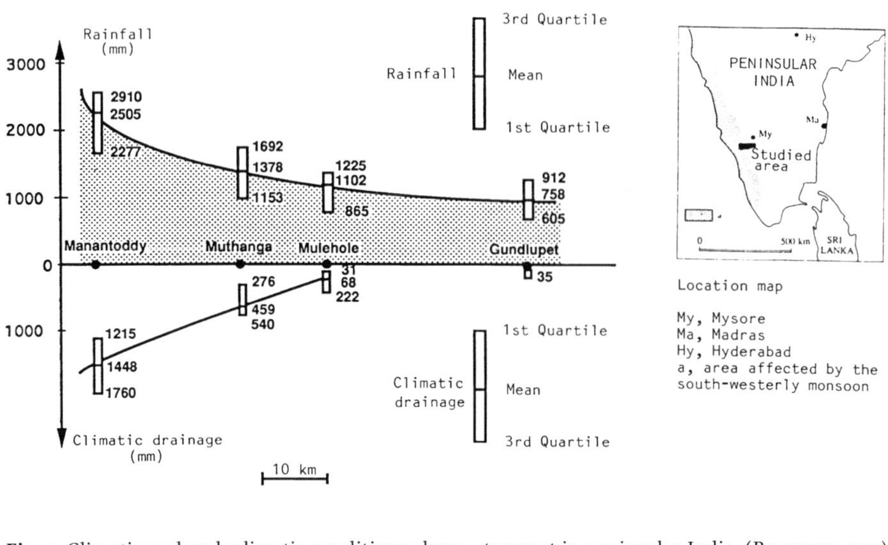

Fig. 2. Climatic and pedoclimatic conditions along a transect in peninsular India. (Bourgeon 1991)

or under warm dry climates, where deep drainage is insufficient for subsurface material to be affected.

The transition from arena to alterite under a warm climate has been studied and the key role of drainage demonstrated (Bourgeon 1991; Bourgeon and Pédro 1992) owing to the recent discovery of a transect in southern India, which bears all the steps of a full-scale natural experiment. The parent material (gneiss) is the same throughout the transect and the temperature is high and constant throughout the year (around 24 °C in Mysore). Mean annual rainfall increases over 60 km from 720 mm in the dry zone to 2650 mm in the wet zone because of the monsoon climate (Fig. 2). Mean climatic drainage [Dcm] which represents the amount of water that actually percolates yearly the profile beyond a depth of 1 m, could be evaluated for the various sites over several years. Present-day drainage values were then used for estimating real climatic drainage per year [Dc] and its variability (1st and 3rd quartiles). The results showed a transition from a dry zone where Dcm is near zero – present-day drainage being sporadic – to a zone where Dcm is small but occasionally large every fourth year, to a wet zone where Dcm is systematically positive every year, increasing regularly towards the foothills of the Ghâts (Dcm>1500 mm). All weathering profiles were caused by primary weathering and a simple relationship could be identified between the pedoclimatic conditions and the main changes affecting the soil, especially those affecting level III (Fig. 3).

Arenization develops in the driest part of the transect, where the mean drainage is close to zero. Weathering is of very limited extent leading to arenas that are always cohesive in semi-arid zones (i.e. where drainage is sporadic) and increases wherever deep drainage is effective some years, leading to less

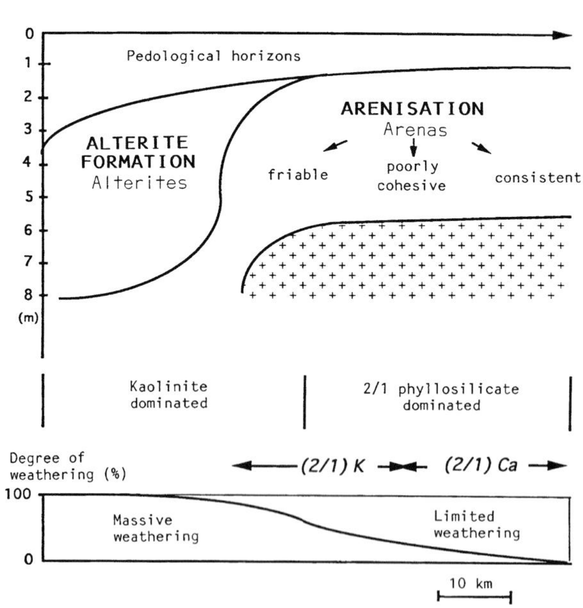

Fig. 3. Geochemical and mineralogic characteristics of the weathering zone along the transect in Fig. 2. (Bourgeon 1991)

compact arenas. Formation of alterites is characteristic of the zone where drainage is regularly positive: as long as Dcm remains low, the degree of weathering is less than 100, which is the domain of very friable arenas where weathering just about penetrates the core of feldspar grains. As soon as Dcm increases, the degree of weathering becomes equal to 100 (all weatherable minerals are decomposed) and the resulting material is an alterite in the strict sense of the term. The amount of drainage now affects the whole altered level.

The type of secondary component appearing along this series changes in parallel with the drainage (Fig. 3). When arenization is weak (consistent and poorly cohesive arenas), plagioclase remains intact whereas biotite and amphibole are affected and replaced by 2/1 low-charge mineral structures which are swelling (10–14 Å/1.0–1.4 nm) mixed-layer minerals. Geochemically speaking, Ca is entirely maintained in the profile. When arenization intensifies (friable arenas), the plagioclase grains are also affected, yielding high-charge K 2/1 phyllosilicates (sericite and illite). Export of Ca becomes considerable, whereas K is retained. Finally, when weathering gets intense, removal of alkalies significantly increases. Kaolinite appears together with illite in the very friable arena domain. As the degree of weathering reaches 100, the 1/1 kaolin phyllosilicates invade the alterites.

The transect clearly shows that there is a domain in which high-charge K 2/1 phyllosilicates appear, which corresponds to advanced arenization. This domain lies between that of low-charge Ca phyllosilicates (smectites) characteristic of hot and dry tropical regions, where arenization is restricted, and that of 1/1 alkali-poor phyllosilicates characteristic of wet tropical regions (formation of alterites). This shows that there is some coherence between the type and the degree of weathering under warm climates. This no longer holds for arenas in temperate regions.

3 Pedological Clay Minerals: Nature and Characterization

For a long time, standard investigations on secondary components (clays, oxides, hydroxides) were based only on X-ray diffraction data – especially on the (001) reflections – and on chemical analyses. In other words, one did not take totally into account either crystallography or crystal chemistry. Likewise, shape, size (lateral extension and thickness) and morphology of clay crystallites, or the way the particles are organized with respect to each other were ignored.

Secondary components of alterites (level III) are often microcrystalline – usually (2–10 µm), sometimes even 100 µm – and slightly hydrated. At the other end of the scale, the components of the pedological horizons (levels II and III) are cryptocrystalline, of submicronic size (ca. 0.1 µm), usually hydrated with well-marked colloidal characteristics. In between is a vast world which had to be considered, before better understanding of pedological clays could be gained.

In order to make things clear and synthetic and for demonstration purposes, let us, as a first approximation, consider the following two major divisions:

1. Phyllosilicates with *neutral and slightly hydrated layers* characteristic of extremely alkali-depleted environments (cations of alkali and alkaline-earth metals are virtually removed), are typical of monosiallitization (1/1 phyllosilicates) where kaolinite appears, and of allitization in which gibbsite neoformation occurs. Talc, pyrophyllite, serpentines, etc, obviously belong to this type.
2. Phyllosilicates with *charged and more hydrated layers* making up the bulk of the 2/1 phyllosilicates, indicate that bisiallitization has occurred, regardless of whether this resulted from «transformations» (apparent bisiallitization) or neoformations (true bisiallitization) (Paquet 1970).

This distinction corresponds roughly to the separation between low-activity clays and high-activity clays, which is the nomenclature commonly used now.

3.1 Clays of the Kaolin Group (Low-Activity Clays)

Clay minerals of this group have a uniform layer which is neutrally charged and has a well-defined composition, i.e. $Si_2O_5Al_2(OH)_4$. However, the shape, size and «texture» of the minerals vary considerably. For instance, size and morphology differ markedly for kaolinite particles from ferrallitic alterites and for those from micro-aggregated ferrallitic horizons of the same profile (e.g. Robain et al. 1990). Also, normal kaolinite is poorly hydrated and has a very low cation exchange capacity (CEC), whereas halloysite is hydrated and was thought to have a higher CEC (e.g. Delvaux et al. 1990).

Current data show that these differences are mainly due to the presence of lattice defects caused by trace amounts of elementary impurities. These defects may be within the layers, which introduced concepts such as domain of coherence, intergrowth and segregation (e.g. Mestdagh et al. 1980; Manceau and Calas 1985; Muller and Bocquier 1987; Calas et al. 1990), or due to the presence of different types of layers in the crystallite, as seen in hybrid structures such as kaolin/smectite mixed-layer minerals and which now account for the large cation-exchange capacity of what was once believed to be halloysite in many soils (e.g. Delvaux et al. 1990; Gaboriau 1991).

These discoveries are due to the use of direct analytical methods (combined spectroscopy and microscopy) involving physical methods such as infrared (IR) spectroscopy, extended X-ray absorption fine structure spectroscopy (EXAFS), nuclear magnetic resonance (NMR), electron spin resonance (ESR), Mössbauer effect, lattice imaging, high resolution transmission electron microscopy (HRTEM) and scanning transmission electron microscopy (STEM) etc. This topic will not be further considered here because it seems premature to give a synopsis, as the cases studied so far do not provide a broad basis for a general model.

3.2 Charged Clays of the 2/1 Group (High-Activity Clays)

This group of clay minerals is very different from the previous one, inasmuch as it includes a wide range of minerals whose 10-Å-silicate layers present a wide range of compositions (as determined with classical means) and whose interlayer spaces are charged and hydrated to various extents. This provides for a wide diversity of behaviours.

The first analytical step was based on X-ray diffraction patterns (001) and on bulk chemistry. This allowed us to distinguish among species such as smectites, illites, vermiculites etc. The main criterion used was the charge of the layers, the value 0.6 being considered to set a major discontinuity in the *z* values ranging from 0 to 1 (Mering and Pédro 1969). The *z* value is relative to the formula Si_4O_{10} for phyllosilicates. However, it turned out that results based on an assumption of strict ordering between X-ray and chemical data, were not as relevant as expected. Several concepts were then introduced in order to clear things up: (1) smectites derived from transformation with charges ranging from 0.4 to 0.6 have an atypical crystal chemistry; (2) random mixed layers are responsible for a series of irrational (001) reflections; and (3) crystallinity, used mainly in sedimentary sequences containing illite (e.g. Kubler 1990), is estimated from the width of the (001) reflection.

Unfortunately, the introduction of these concepts was not very satisfactory. The next step went beyond characterization of the layers and the stacking of layers within the crystallites, as analysed in mineral powders which were not or only slightly hydrated. One had to take into account parameters relating to the size of the particles (extension and thickness/number of layers), their shape , their degree of flexibility or rigidity and the way they were arranged with respect to each other. These parameters have to be determined on hydrated and undisturbed natural samples. This was possible due to: (1) the availability of special techniques for preparing wet samples (Tessier and Berrier 1979; Tessier 1984); (2) the application of electron miscroscopy to this material (SEM and HRTM) and, more importantly, of small angle scattering (SAS; Pons 1980; Ben Rhaim et al. 1986; Gaboriau 1991); and (3) the introduction of various parameters such as the hydration level, the exchange capacity, the total surface area, etc.

On the basis of the currently available data and by using the layer charge *z* as an index, the crystal chemistry of 2/1 phyllosilicates can be divided into three categories: the high-charged species (z=0.8–0.9), the low-charged species (z<0.45) and the intermediate category (where 0.45<z<0.8).

3.2.1 The High-Charged Category (z=0.8–0.9)

This category includes species in which the layer stacking is of an ordered or a semi-ordered type (corresponding to a translational disorder). The species of this category correspond (1) to illites and micaceous clays when the charges are compensated by *non-hydrated fixed cations* such as K (anhydrous interlayer

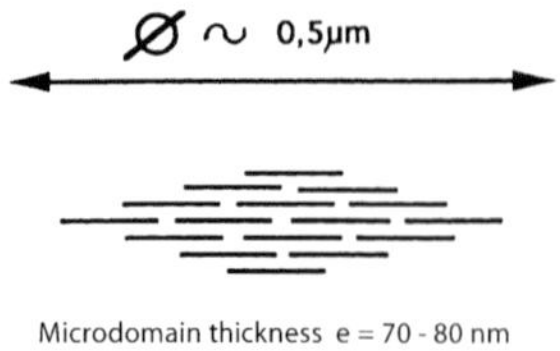

Fig. 4. Diagram of a polycrystalline microdomain. (Tessier 1984)

space); (2) to vermiculites when the charges are compensated by *hydrated exchangeable cations* such as Mg – hydrated interlayers appearing although hydration is limited (two structured sheets of water molecules); and (3) to aluminous intergrades of soils and chlorite-like minerals when charge compensation involves *hydroxy cations* of the $[Al_{3-x}(OH)_6]3x^+$ type.

In species belonging to this group, the term «elementary particle» (well-delimited rigid platelets) makes sense and condensation leads to structures combining a large number of particles of the same type with diverse orientations. Platelets are joined by discontinuous arrangements of the assembled and disconnected type which leads to a structure similar to a house of playing cards. The total surface area in contact is very small, which explains why stability is so low. This type of arrangement is obviously found in the phyllosilicates with neutral charged layers mentioned previously.

The main issue at this point is to find out what is the basic particle in such a structure. The basic particles can be monocrystalline units, i.e. elementary crystallites. This occurs in micromicas with large lateral extensions (ca. 1 μm) and a thickness greater than 10 nm, in (true) vermiculites and occasionally in chlorites. The basic particles can also be polycrystalline, which is characteristic of illites (*sensu* Grim) in which the small crystallites (lateral extension ca. 5 nm) join face to face into stable lenticular units. These units are the true units called microdomains (diameter 0.5 nm; thickness<100 nm; Aylmore and Quirk 1971; Fig. 4). These microdomains then aggregate into structures similar to houses of playing cards, the packing density varying according to the physical and hydric state of the system.

Because of this, it has been suggested that clay minerals with rigid platelets be characterized relative to a nested model, similar to that for proteins. The various levels that have been suggested are the primary structure

Scale (order of magnitude)	Levels of organization			Clay constituents	
				micromica vermiculite	illite
1 nm	primary structure	layer + interlayer		structural unit	structural unit
100 nm	secondary str.	"texture"	monocrystalline entity	crystallite	crystallite
	tertiary str.		polycrystalline entity	–	microdomain
10 μm	quaternary str.	fabric	arrangement of particles (house of cards)	crystallite aggregate	microdomain aggregate

Table 1. Polycrystalline arrangements of clay

(layer=structural unit), the secondary structure (monocrystalline particle= elementary crystallite), the tertiary structure (polycrystalline particle) and the quaternary structure (type of fabric; Table 1; Pédro and Tessier 1987).

3.2.2 The Low-Charged Category ($z<0.45$)

The low-charged category includes species without ordering whatsoever (turbostratic structure), in which there is no limit to interlayer hydration, leading to the formation of true diffuse double layers. Smectites, especially montmorillonites, are the species of this category, in which all interlayer cations are exchangeable. No discrete elementary particles with well-defined dimensions are known in this category. Condensation always leads to loose sponge-like continuous frameworks, such as in polymers. The walls of these continuous networks are perfectly flexible like a film and the mesh size is determined by the thickness, lateral extension, etc. These parameters are functions of the type of smectite: the weaker the charge and, for a given charge, the more octahedral the charge deficit, the thinner is the wall. They are also functions of the geochemical conditions determining the type of exchangeable cations and the concentration of the interstitial solutions, and the physical and hydric conditions (water potential).

The organization of the walls of the networks consists of a face-to-face subparallel disposition with overlapping layers or groups of layers facing each other. Large surfaces of contact are created which allow easy sliding, resulting in broad units connected together. The final structures are termed delaminated and connected and have a fractal organization (Ben Ohoud and Van Damme 1990).

Figure 5 outlines the arrangement in a Na-montmorillonite and in a Ca-montmoillonite under identical conditions (dilution=10^{-3} M; 3 kPa=pF 1.5). The Na samples exhibit adjustments with few layers (M=8), but the water interlayer

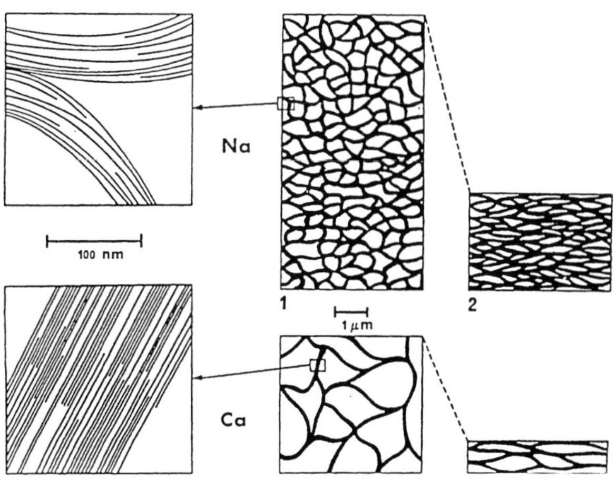

Fig. 5. Network arrangement of a montmorillonite sample in Na and Ca states. Details of the framework and appearance of the network (under well-hydrated and poorly hydrated conditions; Ben Ohoud and Van Damme 1990).

(diffuse layer) is very thick (d001=85 Å) with a small lateral extension; this structure is known as a tactoid and is extremely flexible. In the Ca samples, the adjustments are broader and thicker (M=55 layers, ca. 100 nm) and the interlayer water is organized into a monomolecular sheet (d001=20 Å – 3 layers), this structure having been called a quasi-crystal (Aylmore and Quirk 1971).

It is obvious from these observations that there is a strict relation between the wall thickness (number of layers) and the total surface area of the sample. Surface area measurements can, therefore, provide valuable information on the «texture» of the material. The nested structures defined in the first category become imprecise. The layer (primary structure) and framework (quaternary structure) can even be directly related (as in Li smectites of the hectorite type). Usually, however, intermediate structures combining levels III and IV exist: tactoids and quasi-crystals.

3.2.3 The Intermediate Category

The intermediate category is the most commonly found among clay minerals of soils, especially in temperate ones. It includes all intermediate species with a layer charge z between 0.45 and 0.7. This domain is by far the most difficult to characterize because neither the chemical nor the X-ray diffraction data are discriminating enough. Progress could only be obtained using a more comprehensive approach, i.e. going beyond characterizing single layers by including the other nesting levels involved in the textural fabric. This is why measuring parameters such as hydrability, cation-exchange capacity, total surface area and volume per unit mass were needed, using undisturbed and hydrated samples (Tessier 1984).

Several relationships were inferred from a series of 2/1 clays (true vermiculites not included) covering a broad range of crystal chemistry, the layer charge ranging from 0.3 to 1.

The first relationship was between charge and hydration. Hydration is measured as the amount of water contained under well-defined conditions. Figure 6 (Tessier and Pédro 1987) was obtained from Na clays in contact with diluted solutions (10^{-3} M NaCl) at a water potential of 3 kPa (pF=1.5). Two domains appear in this figure: (1) the material is well hydrated and organized as a continuous framework when $z<0.6$, which is typical for smectites in reference to the AIPEA International Nomenclature Committee (Brindley and Pédro 1972): the weaker the charge, the greater the hydration; and (2) hydration is weak and is more or less independent of the charge when $z>0.6$, the presence of discrete particles and houses of playing-card structures being easily demonstrated.

The second relationship is between charge and cation-exchange capacity (CEC; Fig. 7; Tessier and Pédro 1987). Two cases can occur: (1) the CEC=z and all interlayer cations are exchangeable when $z<0.45$; and (2) the CEC is always less than z and some of the interlayer cations are no longer exchangeable when $z>0.45$. Moreover, the closer the charge to 1, the greater is the non-exchangeable fraction, two explanations being possible for this (Fig. 8). Either (1) «structural

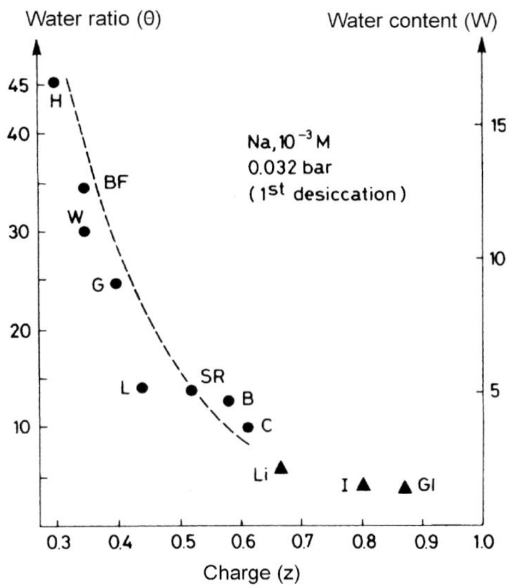

Fig. 6. Sample hydration of several 2/1 clays as a function of layer charge *z*. (Tessier and Pédro 1987)

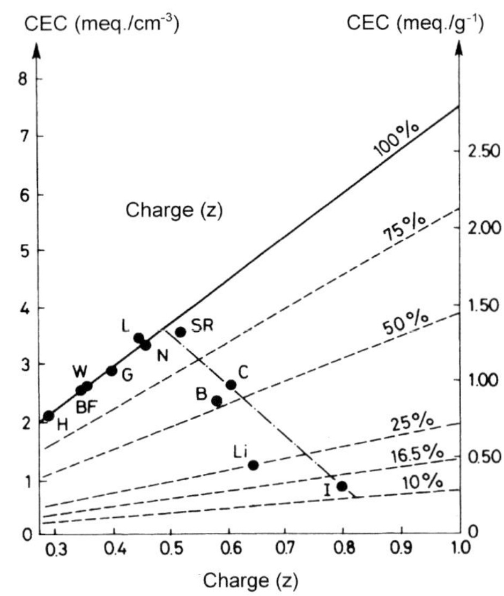

Fig. 7. Cation-exchange capacity of several 2/1 clays as a function of layer charge *z*. (Tessier and Pédro 1987)

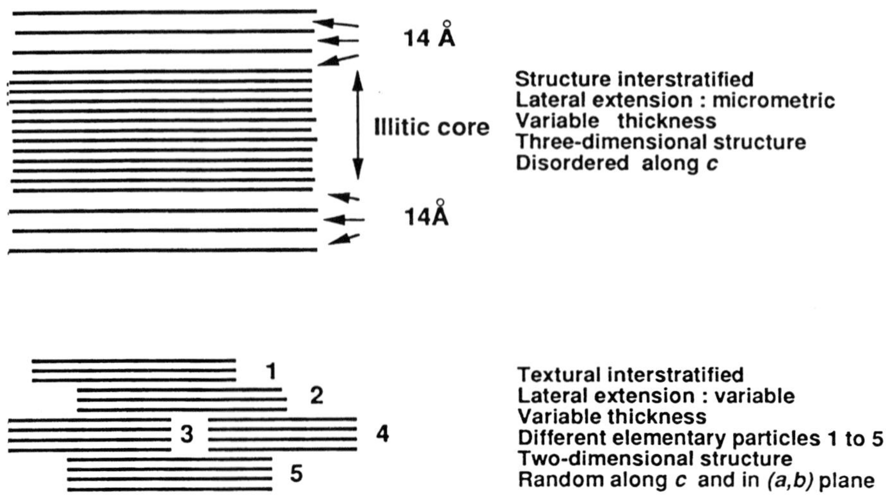

Fig. 8. Interpretation of structural and textural interstratification of 2/1 phyllosilicates. (Robert et al. 1991)

interstratification» is present with alternating anhydrous and hydrated interlayer spaces, which seems to occur in crystallites with a relatively large lateral extension (micromica-like minerals); or (2) the thickness (and the number of layers) of crystallites decreases as the charge decreases and the CEC increases. This occurs in clay particles with a small lateral extension for which ordering between the specific surface and the CEC is strict (Fig. 9). This arrangement, in which small particles are piled up with some overlap into large flexible lamellae has been called «textural interstratification» by Nadeau et al. (1984) and Robert et al. (1991).

3.3 Summary

Several levels must be taken into account when characterizing 2/1 clay minerals in the intermediate domain:

1. The structural level relates to the crystal chemistry of the layers and can be investigated using the *z* charge which ranges from 0.45 to 0.9, depending on the type and extent of isomorphous substitutions. When the charge increases, the thickness of the monocrystalline units increases, the CEC decreases, the selectivity towards K cations increases and the rigidity increases.
2. The textural level relates to the type of polycrystalline units and to the way they are assembled with respect to each other.

A discontinuity occurs at a *z* value of about 0.6 within the 0.45–0.9 domain. When $0.6<z<0.9$, the monocrystalline particles, which are relatively thick and somewhat heavily charged, tend to join their faces in a quite orderly fashion. Discrete, stable and rigid polycrystalline units are formed. These microdomains, as they are called, build up structures like houses of playing cards.

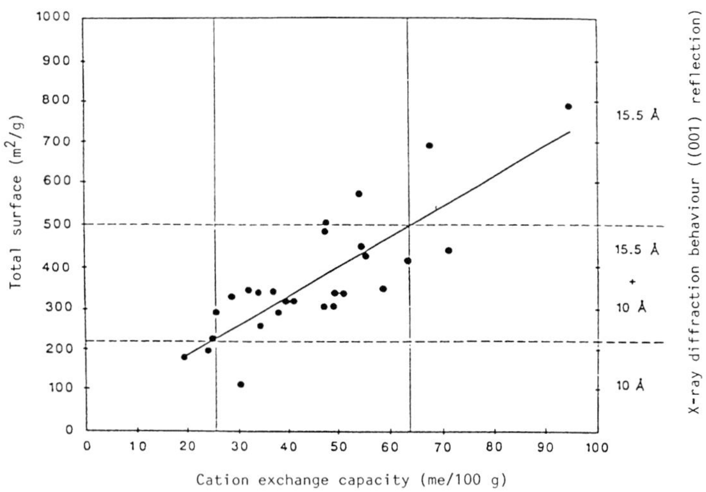

Fig. 9. Relationship between specific surface, exchange capacity and X-ray diffraction patterns of 2/1 clays in soils. (D. Tessier, pers. comm.)

Broad 10 Å reflections are observed, sometimes associated with 15.5 Å reflections. Some 50–60% of the charge is compensated by cations with exchangeable properties. This type of material was previously known as «open illite». When $0.45<z<0.6$, the monocrystalline particles, which are very thin, have a weak charge and vary according to the type of the charge of the interfoliar cation (Na, Ca). They can join turbostratically by overlapping laterally into arrangements which are flexible and join up to form a continuous network. As in smectites, 15.5 Å reflections are observed. Over 70% of the interlayer cations are exchangeable. This type of material corresponds to «transformation clays» with a well-marked beidellite characteristic.

Figure 10 summarizes the characteristics of 2/1 high-activity clays as a function of layer charge z. Two main cases appear:

1. The parameter z is meaningful in terms of crystal chemistry to characterize the different layers within the sample. This is observed first, when the charge is weak ($z<0.45$), as all cations in the interlayer are exchangeable, such as in smectites (i.e. clay material with a turbostratic structure forming a continuous framework); and second, when the charge is very high (z»0.9 to 1), as most cations in the interlayer site are fixed, such as in micaceous clays with a well-ordered structure, rigid crystallites and a discontinuous framework. The z value, which depends on the initial conditions during genesis, determines the type of mineral and its properties. The traditional approach used in mineralogy, which is based on structural parameters, is therefore usually quite efficient.
2. The second case includes all the species of the intermediate category ($0.45<z<0.8$). The z value, which is an average value, results from simultaneous presence of layers and crystallites belonging to both end-member

groups mentioned above, the percentage of each of which depending on the conditions during evolution. The traditional mineralogical approach no longer holds for such mixed structures insofar as fabric aspects take precedence over the other aspects. This is why hydration and other related parameters are so useful. It has also been observed that when the exchange capacity is less than half the charge, as when $z>0.6$, clays have an illite texture and fabric. But when the exchange capacity is over half the interlayer charge ($z<0.6$), the typical fabric is a smectite framework.

The main lesson to be remembered from this review on 2/1 clay minerals in soils is that although the concept of clay species still holds, one should not forget to extend it by including textural aspects. This line of thought agrees with what Georges Millot (1970) felt 30 years ago. He wrote in 1970 in his book, *Geology of Clays* (p. 24): "In fact, we should be aware of two dangers: that of drying up a natural evolution by means of jerky mechanisms going from one species to the next, and that of diluting the concept of a mineral species into a continuous succession in which no distinct steps are discernible." Nowadays, the textural approach can provide a coherent link between these two seemingly contradictory viewpoints.

4 Conclusion

Important facts regarding phyllosilicate clays found in soil and weathered material now seem to emerge from the data presented. The main factor from which everything more or less derives, seems to be the layer charge z.

Clays with neutrally charged layers ($z=0$) belong to low-activity clays. Their characterization depends primarily on the type of layer and to a lesser extent on their crystal chemistry, which usually turns out to be constant, at least for the

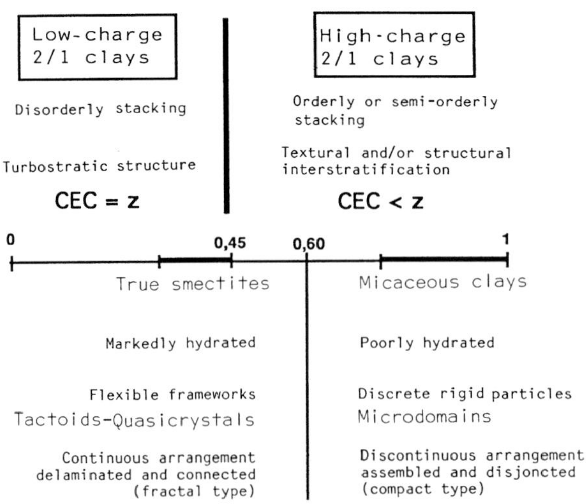

Fig. 10. Structural and textural characteristics of 2/1 clays (excluding vermiculites)

major elements. This characterization is not very thorough by modern standards and the size – and to a lesser extent the shape – of particles (lateral extension, number of layers in each crystallite) should also be taken into account. Size is directly determined by conditions during evolution, as these are responsible for the presence of minor elements within the lattice. Kaolinites, for instance, may reach thicknesses far greater than 2 µm. However, even when crystallites with sizes less than 1 µm are considered, a kaolinite from a ferrallitic alterite (e>1000 nm) is completely different from a kaolinite of the B horizon of an oxisol (e~10–100 nm; e.g. Muller and Bocquier 1987; Robain et al. 1990), with clear implications regarding the understanding of the genesis conditions and the behaviour of the constituents.

Clays with charged layers (z>0) form high-activity clays, illustrating the colloidal tendency of this type of mineral. The layer which forms the basic unit is always the same (2/1 layer) but, because the heterovalent isomorphic substitutions are so diverse, charge z seems to extend over a wide range: 0 to 2 in theory, 0.3 to 0.9 in nature. These variations lead to frameworks whose atomic structure (tetrahedral or octahedral substitution), texture (type and size of crystallites) and type of particles (continuous or discontinuous) vary considerably. Two groups can now be clearly distinguished:

1. The low charge 2/1 clays (z<0.6) behaving basically like smectites – although their domain is somewhat broader than that of «true» smectites – are well hydrated and markedly selective for divalent cations, especially for Ca.
2. The high charge 2/1 clays (z>0.6) typically occurring as discrete particles – generally polycrystalline – are weakly hydrated and are highly selective towards anhydrous monovalent cations, especially K.

This sheds some light on the vast world of clay minerals which are tiny but essential components of the earth's surface, lying somewhere between «crystal and smoke» to take a favourite quotation of Georges Millot from the title of a book by H. Atlan. However, much remains to be done bearing in mind that any clay – even a monomineral clay – represents a population of particles with diverse compositions and/or sizes (Paquet et al. 1983; Duplay 1984). It would be appropriate to quote here one of the last scientific messages of Georges Millot taken from the opening address of the IXth International Clay Conference in September 1989, which incidentally was held in Strasbourg: «Any clay, even a monomineral clay is a population of different particles. Each particle is itself a population of micro-domains. When the environment changes, each micro-domain and each particle starts changing. Each of them shifts towards a new thermodynamic equilibrium according to its own speed: population dynamics are going on. Population dynamics apply to clay mineralogy today.»

References

Aylmore LAG, Quirk JP (1971) Domains and quasi-crystalline regions in clay systems. Soil Sci Soc Am Proc 35:652–654

Ben Ohoud M, Van Damme H (1990) La texture fractale des argiles gonflantes. CR Acad Sci Paris 311 (II):665–670

Ben Rhaiem H, Tessier D, Pons CH (1986) Comportement hydrique et évolution structurale des montmorillonites au cours d'un cycle de dessiccation-humectation. Partie 1: cas des montmorillonites Ca. Clay Min 21:9–19

Bourgeon G (1991) Les «sols rouges» de l'Inde péninsulaire méridionale: pédogenèse fersiallitique sur socle cristallin en milieu tropical. Thèse, Univ Paris VI, 269 p

Bourgeon G, Pédro G (1992) Rôle majeur du drainage climatique dans la différenciation altéritique et pédologique des sols des régions chaudes. CR Acad Sci Paris 314 (II):717–725

Bourrié G (1990) Conséquences de l'existence de complexes polynucléaires Al(III) sur la stabilité des minéraux alumineux: une reconsidération du système gibbsite-kaolinite-quartz. CR Acad Sci Paris 310 (II):765–770

Brindley GW, Pédro G (1972) Report of the AIPEA Nomenclature Committee (Madrid). AIPEA Newslett 7:8–13

Calas G, Ildefonse P, Manceau A, Muller JP (1990) Crystal chemistry of clays and associated oxides: constraints for element transfer and mineral formation processes at the earth's surface. Chem Geol 84:253–254

Collier D (1961) Mise au point sur les processus de l'altération des granites en pays tempérés. Ann Agron 12:273–331

Dejou J, Guyot J, Robert M (1977) Evolution superficielle des roches cristallines et cristallophylliennes dans les régions tempérées, vol 1. INRA, Paris, 464 pp

Delvaux B, Herbillon AJ, Dufey JE, Vielvoye L (1990) Surface properties and clay mineralogy of hydrated halloysitic clays I. Existence of interlayer K+ specific sites. Clay Min 25:129–139

Duplay J (1984) Analyses chimiques ponctuelles d'argiles. Sci Géol Bull (Strasb) 37 (4):307–317

Fritz B (1981) Etude thermodynamique et modélisation des réactions hydrothermales et diagénétiques. Sci Géol Mém (Strasb) 65:197

Fritz B, Tardy Y (1973) Etude thermodynamique du système gibbsite-quartz-kaolinite-gaz carbonique. Sci Géol Bull (Strasb) 26 (4):339–367

Gaboriau H (1991) Interstratifiés smectites-kaolinite de l'Eure Relations entre la structure et la texture. Thèse, Univ Orléans, 274 pp

Harrassowitz H (1926) Laterit. Fortschr Geol Paläontologie (Berl) 4:253–565

Kubler B (1990) «Cristallinité» de l'illite et mixed-layers. Brève révision. Schweiz Mineral Petrogr Mitt 70:89–93

Lacroix A (1913) Les latérites de la Guinée et les produits d'altération qui leur sont associés. Nouv Arch Mus Hist Nat 5 (5):255–356

Manceau A, Calas G (1985) Heterogeneous distribution of nickel in hydrous silicates from New-Caledonia ore deposits. Am Mineral 70:549–558

Mering J, Pédro G (1969) Discussion à propos des critères de classification des phyllosilicates 2/1. Bull Gr Fr Argiles XXI:1–30

Mestdagh MM, Vielvoye L, Herbillon AJ (1980) Iron in kaolinite: II The relationship between kaolinite crystallinity and iron content. Clay Min 15:1–13

Meunier A (1980) Les mécanismes de l'altération des granites et le rôle des microsystèmes.Thèse, Univ Poitiers 1977, 248 pp, Mém Soc Géol Fr, 1980, 40, 80 pp

Millot G (1949) Relations entre la constitution et la genèse des roches sédimentaires argileuses. Rev Géol Appl Prosp Min (Nancy) 2:1–352

Millot G (1964) Géologie des argiles. Masson, Paris, 499 pp

Millot G (1970) Geology of clays: weathering, sedimentology, geochemistry. Springer, Berlin Heidelberg New York, 429 pp

Muller JP, Bocquier G (1987) Textural and mineralogical relationships between ferruginous nodules and surrounding clayey matrices in a laterite from Cameroon. In: Schultz LG, Van Olphen H, Mumpton FA (eds) Proc VIIIth Int Clay Conf, Denver 1985. Clay Minerals Society, Bloomington, pp 186–194

Nadeau P, Wilson MJ, Mc Hardy WJ, Tait JM (1984) Interparticle diffraction: a new concept for interstratified clays. Clay Min 19:757–769

Paquet H (1970) Evolution géochimique des minéraux argileux dans les altérations et les sols des climats méditerranéens et tropicaux à saisons contrastées. Mém Serv Carte Géol Als Lorr (Strasb) 30:212

Paquet H, Duplay J, Nahon D, Tardy Y, Millot G (1983) Analyse chimique de particules isolées dans les populations de minéraux argileux. CR Acad Sci Paris 296 (II):699–704

Pédro G (1964) Contribution à l'étude expérimentale de l'altération géochimique des roches cristallines. Thèse Paris, Ann Agron 15:85–191, 243–333, 339–456

Pédro G (1966) Essai sur la caractérisation géochimique des différents processus zonaux résultant de l'altération des roches superficielles (cycle aluminosilicique). CR Acad Sci Paris 262 (D):1828–1831

Pédro G (1978) Caractérisation générale des processus de l'altération hydrolytique. Base des méthodes géochimiques et thermodynamiques. Séminaire AFES, Sci du Sol, 1979, 2/3:93–105

Pédro G (1981) Les grands traits de l'évolution cristallochimique des minéraux au cours de l'altération superficielle des roches. Rend Soc Ital Mineral Petrol (Jubilé du 40e anniversaire) 37 (2):633–666

Pédro G (1984) La genèse des argiles pédologiques Ses implications minéralogiques, physico-chimiques et hydriques. Sci Géol Bull (Strasb) 37 (4):333–347

Pédro G (1992) La pédogenèse et ses relations avec les phénomènes et les enveloppes biogéosphériques. Sciences 92 (2/3):37–72 (109e Congrès Association Française pour l'Avancement des Sciences: «L'homme et le sol», Orléans, 1990)

Pédro G, Delmas AB (1980) Regards actuels sur les phénomènes d'altération hydrolytique Cah. ORSTOM Pédol XVIII (3–4):217–234

Pédro G, Tessier D (1985) Importance de la prise en compte des paramètres texturaux dans la caractérisation des argiles. Transactions 5th Meet European Clay Groups, Prague 1983, pp 417–428

Pédro G, Delmas AB, Seddoh FK (1975) Sur la nécessité et l'importance d'une distinction fondamentale entre type et degré d'altération. CR Acad Sci Paris 280 (D):825–828

Quantin P (1993) Les sols de l'archipel volcanique des Nouvelles Hébrides (Vanuatu). Trav Doc ORSTOM, 494 pp

Pons CH (1980) Mise en évidence des relations entre la texture et la structure dans les systèmes eau-smectites par diffusion aux petits angles du rayonnement synchrotron. Thèse, Univ Orléans, 175 p

Robain H, Tessier D, Grimaldi M, Elsass F (1990) Importance de la texture des kaolinites dans la caractérisation des couvertures ferrallitiques. CR Acad Sci Paris 311 (II):239–246

Robert M (1970) Etude expérimentale de la désagrégation du granite et de l'évolution des micas. Extrait d'une thèse de doctorat, 21 mai 1970. In: Ann Agron 21 (1970):777–817; 22 (1971):43–93, 155–181

Robert M, Hardy M, Elsass F (1991) Crystallochemistry, properties and organization of soil clays derived from major sedimentary rocks in France. Clay Min 26:409–420

Seddoh FK (1973) Altération des roches cristallines du Morvan Etude minéralogique, géochimique et micromorphologique. Thèse Sci, Univ Dijon, 377 pp

Sequeira Braga MA, Lopes Nunes J, Paquet H, Millot G (1989) Essai sur les arènes de l'Europe Atlantique. Place de l'arénisation parmi les grands systèmes mondiaux d'altération météorique. CR Acad Sci Paris 309 (II):1955–1962

Sequeira Braga MA, Lopes Nunes J, Paquet H, Millot G (1990) Climatic zonality of coarse granitic saprolites («arenes») in Atlantic Europe from Scandinavia to Portugal. In: Proc IXth Int Clay Conf Strasbourg 1989. Sci Géol Mém (Strasb) 85:99–108

Tardy Y (1969) Géochimie des altérations. Etude des arènes et des eaux de quelques massifs cristallins d'Europe et d'Afrique. Mém Serv Carte Géol Als Lorr (Strasb) 31:199 pp

Tessier D (1984) Etude expérimentale de l'organisation des matériaux argileux. Hydratation, gonflement et structuration au cours de la dessiccation et de la réhumectation. Thèse Sci, Paris VII, INRA, Paris, 359 pp

Tessier D, Berrier J (1979) Utilisation de la microscopie électronique à balayage dans l'étude des sols. Observation de sols humides à différents pF. Sci Sol 1:67–82

Tessier D, Pédro G (1987) Characterization of 2/1 clays in soils. Importance of the clay texture. In: Schultz LG, Van Olphen H, Mumpton FA (eds) Proc VIIIth Int Clay Conf Denver 1985. Clay Minerals Society, Bloomington, pp 78–84

Tessier D, Bruand A, Beaumont A (1989) Relationship between clay mineralogy and soil behaviour in Paris Basin clayey soils. Abstr IXth Int Clay Conf Strasbourg, p 400

2 Calcareous Epigenetic Replacement («Epigénie») in Soils and Calcrete Formation

Hélène Paquet and Alain Ruellan

1 Introduction

This contribution is a review offering a new interpretation for the genesis of calcretes as defined by Goudie (1973). Calcretes are calcareous crusts which develop within weathering or soil profiles formed under warm climates with a marked dry season, excluding travertine or spring-, river- and water table cementations of various origins.

In Mediterranean and dry tropical areas, calcareous profiles usually display three horizons, as follows: (1) in the middle of the profile, a horizon characterized by calcareous accumulation and referred to as Bca or Cca; (2) above, a less calcareous or even non-calcareous loose horizon which is the A horizon; and (3) below the Bca or Cca horizon, a less calcareous horizon which may also be non-calcareous (parent rock and/or weathered products) and which is the C horizon. The systematics of the structure of these calcareous accumulations are given by Gile et al. (1966), Ruellan (1967, 1970) and Netterberg (1980).

Calcareous concentration always proceeds in the same way:

- In the beginning, a diffuse calcareous accumulation forms, but is difficult to observe.
- A discontinuous calcareous accumulation then becomes visible, with pseudo-myceliums outlining the porosity of the horizons and friable nodules with more or less gradational boundaries that become progressively indurated (hard nodules).
- Finally a continuous calcareous accumulation occurs with the hard nodules becoming more numerous and coalescing into massive, nodular or laminar crusts and eventually into a more compact calcitic slab whose superficial part is extremely hard and platy (ribbon pellicle of Ruellan 1970). In weathered profiles and parent rocks characterized by schistosity planes or joints, discontinuous calcareous accumulation can appear as grids, the bars of which are calcareous (Millot et al. 1977).

2 Vertical and Lateral Development of Calcareous Accumulation

2.1 Vertical Organization of Calcareous Accumulation

The study of a profile with a highly differentiated calcareous accumulation allows following succession to be observed from top to bottom (Fig. 1):

- A loose horizon referred as A which is slightly calcareous (20 to 30%) or even non-calcareous. Its thickness (often between 30 and 70 cm) varies and depends on the geographic location, the parent rocks and the climate. This horizon rests unconformably on the underlying horizon without any transitional feature, but with signs of penetration of the horizon A into the lime crusting horizon which is characteristic of dissolution.
- A lime crusting horizon in which the amount of carbonate is highest near the top. In complete profiles, one can observe a compact slab overlying a crust which, in turn, overlies a non-layered (massive or nodular) lime crusting horizon. Some units may, however, be lacking, such as the compact slab at the top of the profile, or the compact slab and the crust. There is always a progressive transition between the top of the lime crusting horizon and the underlying horizons. This transition affects the carbonate content, and the continuity of the structures and of the components.
- The transition continues through the underlying discontinuous accumulations consisting of nodules or grids, down to the diffuse accumulations at the bottom of the profile.

No relation is observed between the thickness and carbonate amount of the horizon A and thickness and carbonate amount of the lime crusting horizon. Likewise, there is no relation between the thickness and carbonate amount of the lime crusting horizon and the Ca content of the parent rock.

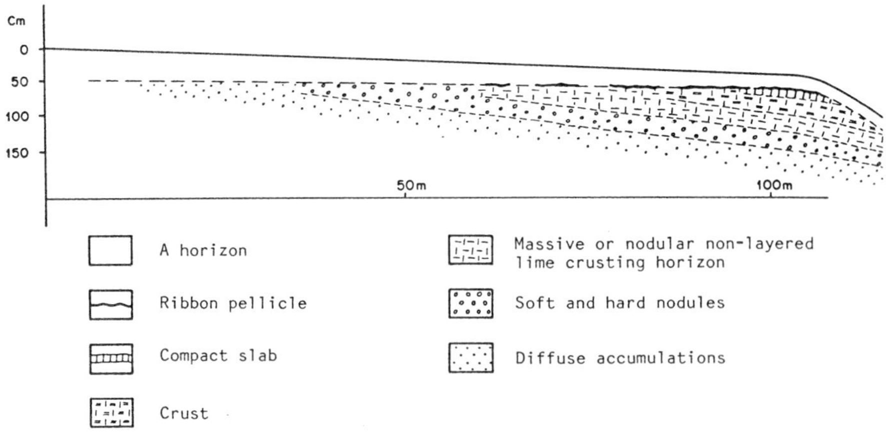

Fig. 1. Schematic pattern of the vertical and lateral distribution of calcareous accumulations. (After Ruellan 1970)

The relationships observed in vertical organization can also be discerned in the lateral variation of calcareous accumulation.

2.2 Lateral Variation of Calcareous Accumulation

In a slope developed on a homogeneous parent rock, the different types of calcareous accumulation can be followed progressively from upslope to downslope (Fig. 1). There is a progressive transition from diffuse and discontinuous accumulations, to continuous accumulation, and finally to most differentiated indurated horizons. Thus, at the bottom of horizon A at a depth of about 50 cm, diffuse accumulations are first observed, followed by a lateral transition to friable nodules, to hard granules and nodules, to a massive or nodular lime crusting horizon, and to a crust and/or a compact slab. The ribbon pellicles are discordant on this sequence; they are facultative and discontinuous. If the observation is carried out at greather depth, the same sequence is found again, but it is progressively displaced dowslope, eventually without the end members.

Thus, the same fundamental sequence is observed vertically and laterally. It consists of diffuse accumulations, soft and hard nodules (or grids), a massive or nodular lime crusting horizon, a crust, a compact slab and eventually ribbon pellicles.

2.3 Dynamics of Calcareous Accumulation

The vertical and lateral distribution of calcareous accumulation allows us to understand its dynamics:

- The lime crusting horizon is not a sedimentary horizon, but a pedologic one developing within a pedologic sequence: the genesis of this horizon can be followed downslope beneath the lixiviated and continuous horizon A.
- The horizon A and the lime crusting horizon belong to the same pedologic system. As long as calcareous accumulation is only slightly differentiated, there is an insensible gradation from horizon A to the discontinuous accumulations. But as calcareous accumulation becomes continuous, giving way to lime crusting horizons, the contrast between the horizons A and B_{ca} progressively increases. This is a pedologic differentiation.
- The thickness and carbonate amount of horizon A may not be related to those of horizons B_{ca} or C_{ca} and the thickness and carbonate amount of the horizons B_{ca} or C_{ca} cannot be related to those of the parent rock. The reason is that the thickness and carbonate amount of the accumulation horizons depend on the Ca lixiviation from horizon A, on the Ca content of the parent rock and on the lateral redistribution of the carbonates, the calcareous accumulation becoming progressively more intense and thicker downslope.
- Within a pedologic system, progressively increasing calcareous accumulation results in a crust consisting in up to 90% carbonates. This means that

most of the parental material has disappeared and that calcareous accumulation can develop if the pedologic system has operated long enough for lateral migrations to occur, but also if the parental material contained minerals whose dissolution could provide Ca.

3 Isovolumetric Replacement by Calcite

As early as 1977, Georges Millot proposed the hypothesis of isovolumetric replacement of silicate parental material by calcite in order to explain the genesis of calcretes in weathering or soil profiles. Occurrence of calcareous epigenesis (or isovolumetric replacement) can be demonstrated in soil profiles containing calcareous accumulation developed on crystalline rocks and on their weathering products, since geometric relationships show that volume remains unchanged during epigenesis. These geometric criteria apply at all scales of observation, i.e. to sedimentary, tectonic and petrographic structures, as well as to textures, fossils, alignment of minerals, macles, cleavages, inclusions, etc. Permanence of structures and volumes is thus real.

3.1 Isovolumetric Replacement in Non-Carbonate Rocks

With reference to Boulet (1974), who described the replacement of plagioclase by calcite in weathering profiles developed on the migmatitic basement of Haute-Volta, Africa, calcareous isovolumetric replacement was demonstrated to be a very general process in subarid and semiarid areas (Nahon et al. 1975; Millot et al. 1977; Ruellan et al. 1979; Bech et al. 1980).

This was first shown in calcareous profiles developed on granites, migmatites, gneiss, schists and quartzites, all these rocks having been selected because they were poor in Ca or devoid of this element and because they exhibited recognizable geometric structures. The most significant example presented here is that of a calcrete occurring in a weathering profile developed on Precambrian greenschists of the Tiznit area in the Anti-Atlas, southern Morocco (Millot et al. 1977).

Plate I, 1-8. Epigenetic replacements by calcite in calcretes developed from non-carbonate rocks. **1** Outcrop of the Precambrian greenschist of the Anzi series (30 km east of Tiznit, Morocco) affected by calcareous epigenetic replacement in the form of a calcareous grid (× 1.5); **2** greenschist, incipient weathering shows palygorskite (*p*) replacing a schist polyhedron (*S*) crossed by microfissures (× 40); **3** greenschist, showing advanced weathering with palygorskite (*p*) replacing a schist polyhedron and calcareous epigenetic replacement (c) after palygorskite from microfissures (× 40); **4** detail: calcitization (*c*) follows palygorskitization (*p*) of greenschist (*S*) (× 80); **5** epigenetic replacement of a quartz grain (*Q*) by calcite (*c*) (× 100) in a granite; **6** epigenetic replacement of plagioclase by calcite in a granite. Calcitic micrite (*c*) replaces plagioclase (*F*) whose macle systems are preserved (× 100); **7** epigenetic replacement of biotite by calcite in a granite. In the biotite crystal (*B*) ferruginous segregations perpendicular to cleavage are preserved in the calcitized zones (*c*) (× 100); **8** epigenetic replacement of biotite (*B*) by calcite (*c*) developing from biotite cleavage (× 100). (After Millot et al, 1977)

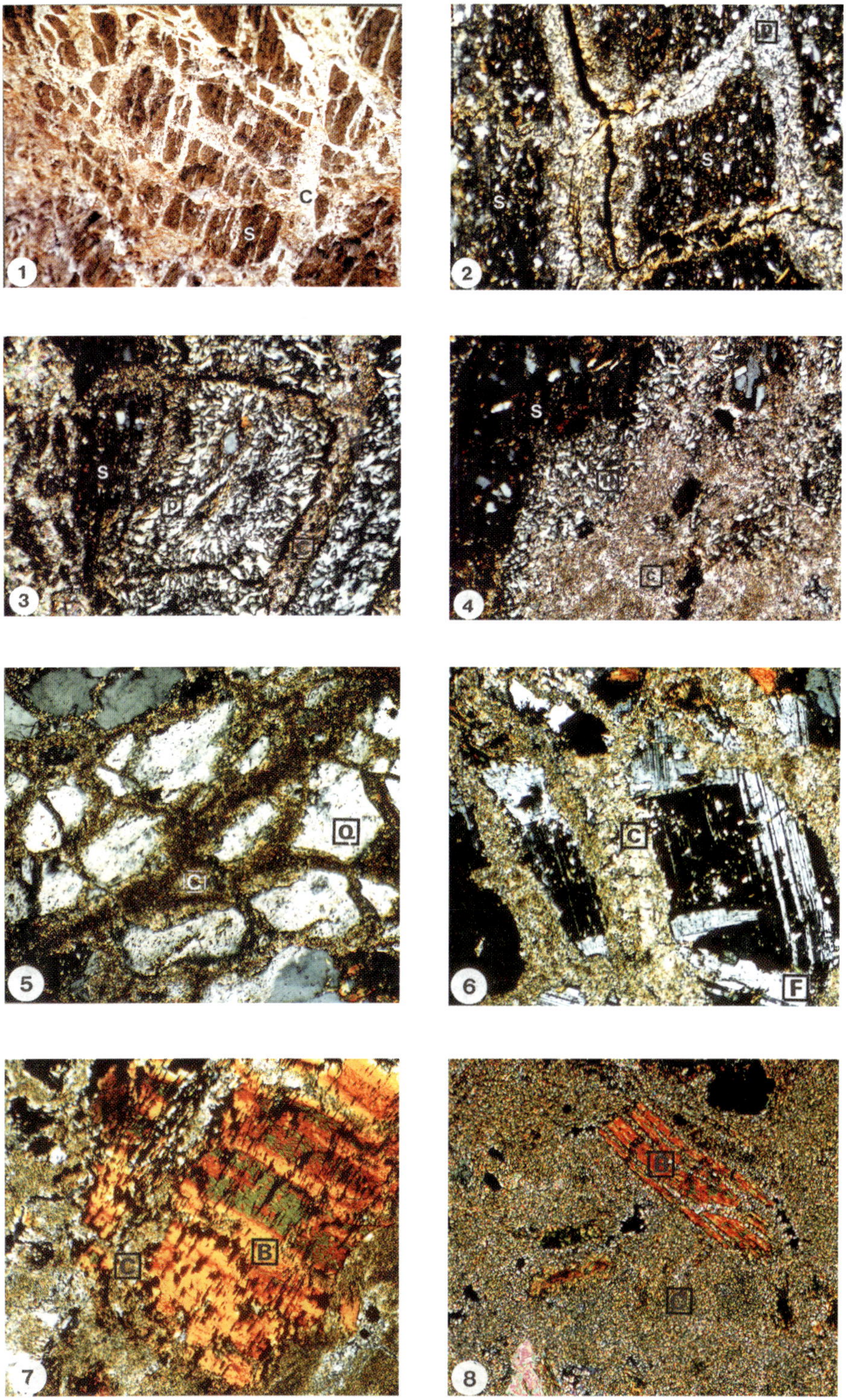
1
S
C
2
S
S
D
3
S
D
C
4
S
D
C
5
Q
C
6
C
F
7
C
B
8
B
C

3.1.1 Field Observations

At the base of the section, slightly weathered to non-weathered greenschists are observed. These are cut with respect to cleavage by a system of parallel and transverse diaclases. Within the weathered greenschists, calcareous accumulation progressively develops in the form of a grid above which a calcareous crust stands out. The first impression is that the horizontal and vertical bars of the calcareous grid represent filling of cracks. But careful observation shows that, in fact the calcareous grid does not disturb the structural organization of the parent rock. Moreover, numerous relict fragments of schists are preserved without displacement and they remain in their original orientation within the calcareous units, which become thicker from the bottom to the top of the section (Plate I, 1).

These observations do not allow reference to a system of fissures filled with precipitated calcite. In fact, each schist polyhedron, which is delimited by diaclases, is progressively replaced by calcite from the outer to the inner zone: the replacement process seems to start from fissures, joints and cleavages which represent privileged pathways for soil solutions to carry the dissolved Ca.

3.1.2 Microscope Examination

The greenschist is cut into polyhedra by fissures (Plate I, 1) which develop clay rinds in the initial stage of weathering (in the zone of incipient schist calcitization; Plate I, 2). These clay rinds consist of randomly oriented bundles of fibres of palygorskite and will be considered later. With further development, palygorskite entirely pervades the schist polyhedra, although schist relicts can still be observed. Concurrently, areas of palygorskite replacement in the polyhedra are seen to be replaced by calcite (Plate I, 3). It should be noted that during the replacement of the original silicate material of the schist by palygorskite and then by calcite, the volume of the polyhedra remains constant. This is the evidence for concluding that the progressive replacement is isovolumetric (epigenetic replacement or calcareous «épigénie»; Millot et al. 1977). In the final stage, at the top of the calcrete, volumes are no longer preserved. Calcite forms a continuous matrix enclosing relicts of schist and palygorskite (Plate I, 4), which attests for the progressive process.

Examples exhibiting mineral replacement by calcite were also found in the Ifni granite of the Anti-Atlas region, Morocco, for instance. First case: calcite replaces a quartz crystal whose separated fragments retained the same optical orientation (same extinction; Plate I, 5). Second case: it replaces a feldspar with two orthogonal macle systems without any visible displacement (Plate I, 6). Third case: calcite replaces a biotite with lines of ferruginous segregations perpendicular to the cleavages which are well preserved in the calcitized zones (Plate I, 7). Fourth case: biotite is almost completely replaced by calcite (Plate I, 8). These are a few of many examples testifying to the epigenetic replacement of original minerals of a granite by calcite.

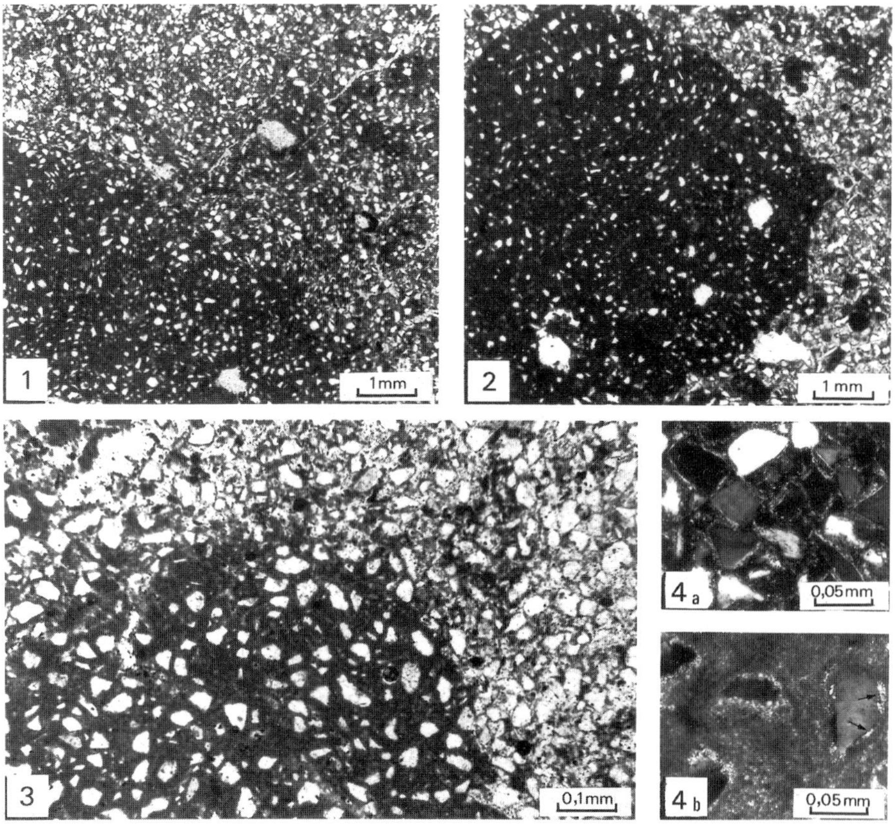

Plate II, 1–4. Epigenetic replacements by calcite in calcretes developed from sediment. (After Truc et al. 1985). **1** scattered spots of calcitic micrite in a clay–quartz loam; **2** indurated micritic nodule: quartz content decreased to 30% of the total; **3** detail of the calcitization front; **4a** clay coatings (cutanes) around quartz grains; **4b** corrosion of quartz grains and epigenetic replacement of quartz by calcite (relict cutanes are shown by *arrows*). (After Truc et al, 1985)

3.2 Isovolumetric Replacement in Non-carbonate Materials

Further demonstration of calcareous isovolumetric replacement came from studies of nodules or calcretes which developed from loose materials without easily recognizable geometric fabric. Under the microscope, the distribution of quartz grains is identical in both the sediment and the enclosed nodules (Truc et al. 1985). The following example from Chabet Smala section in Tunisia shows the progressive transition from a loam to a compact calcareous slab (Regaya 1984).

In the clay–quartz loam, calcareous accumulation first appears in the form of scattered spots without clear outlines (Plate II, 1) and then in the form of nodules (Plate II, 2, 3). In the calcitized zones, the abundance and size of the quartz grains decrease as a result of corrosion (30 to 50%), and the amount of clay minerals composed of palygorskite is only 1–2% of the total rock. The quartz skeleton is never displaced or removed by calcitization as suggested by Gile et al. (1966), Allen (1985) and Machette (1985): quartz is clearly replaced by calcite (Plate II, 4a,b). Towards the top of the section, nodules coalesce and form first a continuous calcareous layer and then a slab in which quartz grains represent only 5% of the volume of the rock and from which palygorskite is absent.

Systematic comparison of several series of thin sections from calcretes of various loose sediments shows the corrosion of quartz grains, as well as the permanence of their distribution until replacement. Volumes are preserved or, in other words, the appearance of a calcitic volume corresponds to the disappearance of an equal volume of clay–quartz material; such correspondence is by definition a calcareous isovolumetric replacement.

3.3 Characteristics of Calcareous Isovolumetric Replacement

The characteristics of calcareous replacement were pointed out by Georges Millot after many studies in collaboration with different teams of geologists, soil scientists, geomorphologists and geochemists. They are as follows:

1. In semiarid climates, alternating humid and dry seasons induce dissolution of silicates during humid periods and crystallization of carbonates during dry ones, with an intermediate stage of palygorskite, especially when the parent rocks are rich in Al-Mg silicates (Paquet 1983). This proposal is based on modelling (Millot et al. 1977, 1978; Al Droubi et al. 1978) and experimentation (Halitim et al. 1983). The latter authors clearly demonstrated that during wet periods and under the influence of Ca–Mg solutions, the dissolution of primary silicates is accompanied by the formation of a fibrous Mg silicate, usually palygorskite, in natural environments.
2. Calcareous isovolumetric replacement may affect all types of rocks, e.g. plutonic, volcanic, or sedimentary rocks, and results in a monometallic (Ca) and monomineralic (calcite) state in calcretes and in a bimetallic (Ca, Mg) and monomineral (dolomite) state in dolocretes (Millot and Paquet 1987), the latter not being discussed in this chapter.
3. Calcareous isovolumetric replacement is a function of weathering related to the planation of landforms (Ruellan et al. 1979; Millot 1980). In fact, because this replacement changes the nature of the parent material at the base of the weathering cover, although it does not modify either its volume or its fabric, there is no apparent effect on landforms. However, when regulated by water table level, calcareous replacement induces the «hidden» and «hypodermic» (as defined by Millot (1980), i.e. under the surface) levelling of the bed rock.

These characteristics, as well as examples of the geochemistry of landscapes and paleoclimatic and paleogeographic reconstructions, are successively examined below.

4 Genesis, Stability and Instability of Palygorskite

A discussion of epigenetic replacement requires the examination of the role of palygorskite which represents the predominant clay mineral in calcretes of semiarid areas (references in Jones and Galan 1988). This fibrous mineral may represent up to 95% of the clay fraction.

4.1 Chemical Composition of Palygorskites

Chemical analysis of individual palygorskite particles collected from calcretes in Precambrian greenschist gave the information to be considered below. There are very slight chemical variations from one particle to another (Paquet 1983; Table 1), however, palygorskite from the inner part of the calcrete corresponding to maximal calcitization, may have the highest Ca and Mg contents. From the tentative distribution of ions between tetrahedra and octahedra on the basis of 21 oxygens per half unit cell the structural formulae of samples (4), (5) and (8) are as follows:

$$(4)\ (Si_{7.88}Al_{0,12})\ (Al_{1.92}Mg_{1.64}Fe_{0.28})\ O_{21}\ Ca_{0.06}\ K_{0.08}$$
$$(5)\ (Si_{7.80}Al_{0.20})\ (Al_{1.68}Mg_{1.86}Fe_{0.30})\ O_{21}\ Ca_{0.20}\ K_{0.14}$$
$$(8)\ (Si_{7.86}Al_{0.14})\ (Al_{1.86}Mg_{1.86}Fe_{0.22})\ O_{21}\ Ca_{0.10}\ K_{0.10}$$

In these formulae, the Fe contents are higher than those usually reported (Weaver and Pollard 1975; Newman 1987), which seems to be specific for palygorskite from weathered materials or soils (Paquet 1983; Paquet et al. 1987). Such high Fe contents have also been recently reported by Singer et al. (1995) for soil sepiolites.

Octahedral compositions of palygorskite and smectite particles from calcretes were compared on the basis of $(Al + Fe^{3+})^{VI}$ vs. Mg^{VI} values for six positive charges per half unit cell (Paquet et al. 1987). The small chemical variability from particle to particle, as mentioned above, was found for both the smectite

	Basal part of calcrete				Middle part of calcrete			Surficial part of calcrete			
	1	2	3	4	5	6	7	8	9	10	11
SiO_2	69,65	68,62	69,67	70,34	68,68	69,26	68,52	70,00	70,95	70,42	70,65
Al_2O_3	15,66	15,61	15,43	15,45	14,14	14,28	15,30	15,15	14,93	15,82	15,43
Fe_2O_3	3,49	3,75	3,53	3,44	3,61	3,28	3,24	2,57	2,03	2,52	2,02
MgO	9,63	9,76	10,65	9,72	10,93	11,06	10,87	10,77	10,65	10,23	10,03
CaO	1,12	1,03	0,35	0,48	1,57	1,52	1,57	0,78	0,46	0,28	0,72
K_2O	0,45	0,64	0,37	0,57	0,97	0,60	0,99	0,72	0,97	0,82	0,69

Table 1. Chemical analysis of palygorskite fibers extracted from a calcrete developed on Precambrian greenschists

and the palygorskite. A large overlap of the chemical compositions of palygorskite and smectite was visible, with preferential uptake of Mg by palygorskite.

4.2 Main Occurrences of Palygorskite in Calcretes

Palygorskite may be found in several microscopic facies of calcretes, and occurs in the following ways: (1) as coatings or cutans surrounding soil micro-aggregates or primary minerals (Beattie and Baldane 1958; Van den Heuvel 1966; Beattie 1970; Singer and Norrish 1974) and called «palygorskans» by Brewer (1964); (2) as micro-aggregates dispersed in the calcareous matrix (Van den Heuvel 1966; Turchenek and Oades 1974); (3) as oriented or unoriented bundles of fibres in pores and voids (Eswaran and Barzanji 1974; Yaalon and Wieder 1976); and (4) as weathering rinds surrounding primary minerals or rock fragments and, in turn, surrounded by calcitization zones (Millot et al. 1977; Paquet 1983). In some calcretes, palygorskite may be accompanied or replaced by sepiolite, a Mg-rich fibrous mineral (Van den Heuvel 1966; Gardner 1972; Bachman and Machette 1977; Hay and Wiggins 1980; Jones 1983; Paquet et al. 1987).

4.3 Conditions of Palygorskite Genesis in Calcretes

The genesis of palygorskite in calcretes can be related to three main processes, according to various authors: (1) weathering of more or less Mg-rich primary minerals, such as amphiboles and pyroxenes (Elgabaly 1962; Eswaran and Barzanji 1974), or biotite, chlorite and phengite (Millot et al. 1977); (2) weathering of secondary minerals, such as smectites (Frye et al. 1974; Yaalon and Wieder 1976; Watts 1976, 1980; El Prince et al. 1979; Viani et al. 1983; Jones 1986); swelling minerals (Bachman and Machette 1977) and mixtures of chlorite, kaolinite, smectite and mica (Millot et al. 1969, 1977; Jones and Galan 1988); and (3) direct authigenesis from soil solutions in evaporative conditions (Van den Heuvel 1966; Gardner 1972; Singer and Norrish 1974; McGrath 1984; Thellier et al. 1988). The two first processes have been illustrated at the rock scale (Millot et al. 1977) and at the crystal scale (Paquet 1983). In the calcrete formed on Precambrian greenschists, taken as an example in the above dicussion, microprobe analyses were carried out on thin sections along a transect from schist matrix to a microfissure (Plate I, Fig. 4). The following three domains could be observed on the spectrum (Fig. 2): a domain in which Si, Al, K and Mg are abundant and correspond to the phyllosilicates of the parent rock; a domain in which Mg clearly increases and Si and Al decrease, indicating the presence of palygorskite; and a domain in which Si and Al contents are low, Mg decreases noticeably and Ca appears in abundance, corresponding to the calcitized zone.

Microscope and microprobe investigations lead to the following interpretations: (1) palygorskite is a mineral formed by direct weathering of primary minerals of parent rocks (Plate I, Fig. 2), or as the result of transformation of, followed by recrystallization from clay minerals of the parent materials of calcretes; (2) the formation of palygorskite takes place before and during calci-

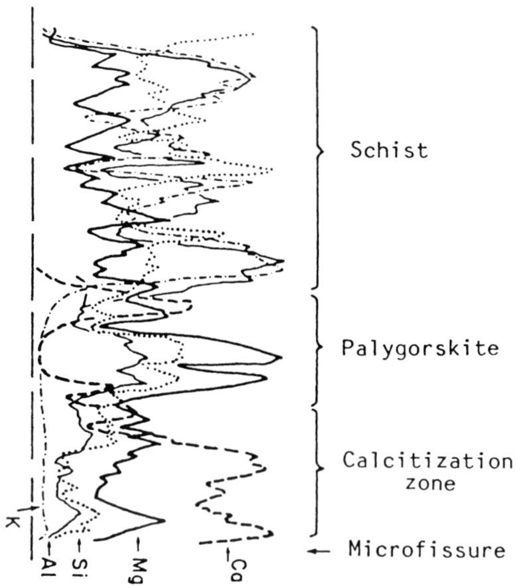

Fig. 2. Microprobe analysis along a section from the schist to a microfissure. (After Paquet 1983)

tization (Fig. 2; Plate I, Fig. 4); and (3) at the end of calcitization, the amount of palygorskite tends to decrease, being replaced by calcite, like the silicates. This is contrary to the assertion of McGrath (1984) according to whom the amount of palygorskite increases as calcitic induration develops. In fact, as calcitization becomes intense, especially in calcareous crusts and slabs, palygorskite is no longer stable; just as is observed in loose horizons overlying calcretes, where seasonal variations are more perceptible and soil solutions more dilute. In such horizons, palygorskite transforms into smectite which is more stable in surficial environments, whereas calcite is progressively dissolved (Paquet and Millot 1972).

The conditions governing the genesis and stability of palygorskite have been clarified by means of bidimensional (Velde 1985; Jones and Galan 1988) or tridimensional stability diagrams (La Iglesia 1977; Weaver and Beck 1977). These diagrams have been calculated from different types of data: free formation energy for palygorskite (Tardy and Garrels 1974); rate of aqueous dissolution for pedogenic palygorskite (Singer 1977); and solute-ion activity (Jones 1986). The diagrams show that genesis and stability of palygorskite are dependent on three principal variables – high pH, and high Si and Mg activities in soil solutions. These three factors are most beneficial during dry periods of semiarid climates when evaporation concentrates solutions. In contrast, when the medium is less alkaline or more dilute, the neoformed clay minerals are smectites (Paquet 1983).

Some authors have reported the occurrence and even the predominance of sepiolite in calcretes (Van den Heuvel 1966; Hay and Wiggins 1980; Jones 1983; Milnes 1992). The presence of sepiolite is attributed (1) to an abundance of dissolved Mg and Si in conjunction with increased concentration of soil solutions

(Jones and Galan 1988), or (2) to the absence or immobilization of solutes for colloidal aluminium (Jones 1983; Mc Grath 1984) as the result of a reduction in permeability and, at least for some loci, or (3) to «closure» of the system by calcite precipitation (Velde 1985). According to Bachman and Machette (1977), sepiolite has in any case to be considered as a late-stage product of pedogenesis, whose content increases with age within calcretized geomorphic surfaces. These authors have also asserted that both sepiolite and palygorskite are absent in the younger surficial horizons overlying calcretes.

5 Occurrences of Calcite and Calcrete Formation

As evidenced by the denomination «calcrete», calcite is the predominant mineral in calcretized horizons. Calcite whose precipitation necessitates high pH values (often higher than 9, even than 10; Ruellan 1970; Mohr et al. 1972), exhibits both macroscopic and microscopic forms of accumulation.

5.1 Calcite Habits

According to Monger et al. (1991), the major macroscopic habits consist of filaments, coatings on relict grains, fillings of tubes, nodules disposed vertically or diagonally, internodular fillings and petrocalcic horizons. The microscopic habits are multiple, such as filaments representing calcified root hairs, micritic calcite-impregnated pore walls («pore hypocoatings» or «neocalcitanes»; Seghal and Stoops 1972), coatings of relict grains in the form of micritic, or sparry calcitanes in nodules, and dense micritic calcite surrounding skeleton grains. Internodular fillings of calcitanes and non-coalescent micritic calcite were also observed, as well as laminae of pure micritic calcite, sparry calcite in voids and structures of microcodium in laminar zones.

5.2 Modes of Induration

Nahon (1991) demonstrated that induration proceeds by progressive glaebulization, e.g. by multiplication, association and coalescence of calcitic plasmic concentrations into a continuous indurated horizon. Calcitization develops from replacement of primary minerals of parent rocks by precipitation of calcite from soil solutions that diffuse from inter-aggregate fissures until they reach the small pores of the plasma of the host rock. In the beginning, calcitic plasmic concentrations, e.g. spots and nodules ranging in size from a few microns to 2 cm are formed, and their roughly spherical shape testifies to centripetal and isotropic calcitic concentration. In argillaceous material, the size is controlled by the size of the aggregates.

For Nahon (1991), the progressive glaebulization, as considered from a spatial or temporal point of view, is accompanied by an evolutionary mineral sequence in which calcite represents the predominant and stable mineralogical phase. As characterized by IR spectrometry in micro-sequences of temperate climate (Nahon et al. 1980; Dupuis et al. 1984; Ducloux et al. 1984), this mineral sequence shows the following stages: amorphous Ca carbonate, which is in a precocious, very hydrated and metastable form, disordered calcite and/or aragonite, and true calcite. Goudie (1973) described aragonite in calcretes from South Africa and Milnes (1992) reported small quantities of the same carbonate in a calcrete from South Australia. However, aragonite is usually considered as ephemeral (Nahon et al. 1980), like fibrous calcite (Bocquier 1973; Nahon 1976; Durand 1979; Pouget and Rambaud 1980; Verges et al. 1982). Calcrete formation thus results from progressive glaebulization, but it may also form from calcite accumulation in the macro- and microporal systems of the soil profile, which evokes the neocalcitanes, calcitanes and fillings mentioned above.

In fact, Milnes (1992) showed that the calcite concentration of an indurated horizon is related to the accumulation of micritic calcite in pores, which progressively stops water infiltration. Calcite is considered to be at once illuviated from overlying pedologic horizons and precipitated from soil solutions. Chadwick et al. (1987) indicated that large voids are preferential sites for calcite precipitation, since they dry more rapidly than small ones and because they are in more direct contact with the atmosphere, which is depleted in CO_2. However, Wieder and Yaalon (1974) suggested that dispersed clay minerals constituting a microporal system may also serve as nucleation points for micritic calcite formation.

Precipitation of calcite and thus development of calcitization of indurated layers requires a three-stage process (Salomons et al. 1978): Ca enrichment of soil solutions, Ca transfer by soil solutions, and precipitation of calcite. Evaporation is usually considered to be the main factor in calcite precipitation although other mechanisms may also intervene, especially the decrease of CO_2 partial pressure resulting in a pH increase. This «triple process» governs the formation of calcrete, whether the latter should be considered as resulting from epigenetic replacement by calcite of minerals from parental materials (Millot et al. 1977; Salomons et al. 1978; Reheis 1988; Nahon 1991; Reheis et al. 1992), or from impregnation by calcite of a more/less porous layer (Gile et al. 1966; Allen 1985; Machette 1985; Milnes 1992).

Recently, Wang et al. (1993, 1994) proposed a model combining «dynamic, mass-continuity, hydric, geochemical, mineralogical and textural factors». Because of this combination it is possible to «predict the features of a calcrete as a function of any desired climatic changes in temperature, wet-dry relative durations, PCO_2, water-table position and slope». First restricted to calcretes developed at the expenses of quartzite, this dynamic model was then extended to calcretes replacing silicate rocks in semiarid regions (Wang et al. 1994) and could successfully produce the mineral and textural zonations identified in natural calcretes.

5.3 Role of Microorganisms and Biologic Interventions in Calcrete Units

Laminar crusts («croûtes zonaires» or ribbon pellicles; Ruellan 1970) in the uppermost part of calcretes are of doubtful origin. According to Rabenhorst et al. (1984, 1991), their origin has been traditionally ascribed to inorganic processes of carbonate precipitation. However, more recently some authors have referred to a biogenic origin. According to Gile et al. (1966) and Ruellan (1967), laminar crusts are composed of alternating light and dark thin laminae of micritic calcite; the latter would precipitate from thin layers of free water perched above an almost impermeable calcareous horizon. Each lamina might represent a physical (calcite dissolution–reprecipitation) episode in the uppermost part of the calcrete.

A biogenic origin could involve any of several types of microorganisms forming laminar structures, for example: cyanobacterial mats with stromatolite features (Campbell 1979; Krumbein and Giele 1979), filamentous Cyanophyceae (Wright 1989; Verrechia 1990; Verrechia et al. 1995) and microcodium (Estaban 1974; Monger et al. 1991). All these microorganisms could directly or indirectly induce precipitation of calcite (Folk 1993) and their development might be favoured in micro-environments, such as pores where humid conditions are preserved even during the dry season (Milnes 1992). Moreover, many cases of calcrete exhibit proof of biological activity affecting either the laminar part, as seen above, or the inner part of the calcrete; such cases are calcified roots (Klappa 1980; Wright et al. 1988, 1995; Rossinsky et al. 1992), insect burrows (Milnes 1992), fungal hyphae (Callot et al. 1985), microfossils (Milnes and Ludbrook 1986) and Bulimes macro-fauna (Truc 1989).

5.4 Origin of Calcium

If calcrete develops from dolomitic marly rocks, its Ca obviously originates partly or totally from these rocks. Ca was also demonstrated by Bourgeon (1992) to be autochthonous in the case of calcretes on Ca-rich silicate rocks. The author studied calcrete-bearing red soils which developed from amphibolic gneisses and amphibolites on pediments of the southern part of the Mysore Plateau (India) in semiarid conditions. Carbonate accumulation was shown to be the result of integration of three processes: upward movement of slightly weathered Ca-rich materials through organic activity, slow transfers of these materials by rainwash on pediments, and weathering of these materials resulting in enrichment of soil solutions in soluble elements, especially Ca.

In contrast, if no relation can be observed between the thickness and the calcite content of a calcrete and the Ca amount of the corresponding parent rock or of the silt–clay horizon at the top of the profile, Ca must have entered the weathering profile – as in the case of the calcrete developed from silicate parent rocks described above. If surrounding landforms are carbonate-rich, they provide the calcium necessary for calcrete formation either in dissolved form, related to short- or long-distance transfer by soil solutions (Nahon 1991), or in suspended form, as in the case of slope debris, colluvium and alluvium (Millot et al. 1977). If no carbonate-rich surrounding landform can be observed, the only other possibility is aeolian influx.

Many authors have reported the role of calcareous dust in the formation of calcretes in semiarid or desert areas, especially in the United States (Brown 1956; Gile et al. 1966; Reeves 1970; Gardner 1972; Gile and Hawley 1972; Machette 1985; Schlesinger 1985; Liu et al. 1994), but also in Australia (Jessup 1961; Jessup and Wright 1971; Milnes 1992), North Africa (Coque 1962; Millot et al. 1977) and in many other parts of the world. Chadwick and Davis (1990) and McFadden et al. (1991) even reported on a relationship between the rate of calcareous dustfall and the development of calcrete. For other authors, Ca could be provided by sea spray (Morelli et al. 1971; Jame 1972; Oliva 1974).

Rabenhorst et al. (1984) showed that the combination of micromorphological and isotopic analyses is helpful in distinguishing between pedogenic and lithogenic carbonates. The origin of Ca, if external to the weathering profile in which calcrete develops, may be identified through isotopic analysis. In fact, the $^{87}Sr/^{86}Sr$ ratio is typical for each rock type and calculation of that ratio in the calcareous glaebules of calcrete gives useful information on the Ca source (Nahon 1991), especially if the ratio is compared with the Nd isotopic composition (Grousset et al. 1992). This might be a promising investigation method for tracing the origin of Ca in the calcretes.

6 Calcite–Silicate Mineral Assemblages in Calcretes

It is now well-established that calcretes are calcareous layers occurring within weathering profiles developed in semiarid to arid climates with marked dry seasons inducing strong evaporation and high pHs. Calcite predominates in the mineralogical composition of calcretes, but it is accompanied by clay minerals which are essentially fibrous in shape (very frequently palygorskite), to the extent that it is tempting to consider the latter as a diagnostic mineral for calcretes. Sepiolite and occasional smectite are scarcer. Relict primary minerals such as quartz, feldspars and micas are usually observed as traces.

6.1 Isotopic Composition

The conditions under which this mineral assemblage could form can be estimated from C and O isotopic compositions of the calcite (Gauthier-Lafaye et al. 1993). The studies including isotopic analyses of C and O of calcite emphasized the importance of evaporation in the formation of the carbonates (Salomons et al. 1978; Talma and Netterberg 1983; Schlesinger 1985; Mc Fadden et al. 1991). Lawrence and Taylor (1971) also showed that in clay minerals formed in weathering profiles, C and O isotopic compositions are similar to those of local meteoric waters, especially when the clay minerals, such as kaolinite, halloysite or montmorillonite, formed in an open system.

Samples	Clay content (% of total)	$\delta^{18}O$ (SMOW)	$\delta^{13}C$ (PDB)
Calcrete A			
Calcite (1)	100%	+22,0	−5,8
Calcite (2)	75%	+26,8	−11,3
Palygorskite (2)	20% (+ Sm. 5%)	+28,0	
Palygorskite (3)	90% (+ Sm. 10%)	+34,2	
Calcrete B			
Calcite	100%	+28,0	−9,83
Smectite	100%	+27,0	

Table 2. Mineralogical and isotopic compositions of the Portuguese samples[1]. (After Gauthier-Lafaye et al. 1993)

However, no isotopic data on clay minerals from calcretes are presently available. The calcretes are from environments in which the isotopic composition of meteoric water has been modified by evaporation, which induces enrichments in the heavy isotopes, as shown by Schlesinger (1985) for calcites of calcretes. If clay minerals of calcretes are in isotopic equilibrium with soil solutions, then they should be more/or less enriched in ^{18}O and thus reflect the rate of evaporation in solutions from which they have formed.

6.2 Results

Stable isotope analysis has been applied to the study of two calcretes developed from basaltic rocks in Portugal and containing the following mineral assemblages: calcite–palygorskite in calcrete A and calcite–smectite in calcrete B. The

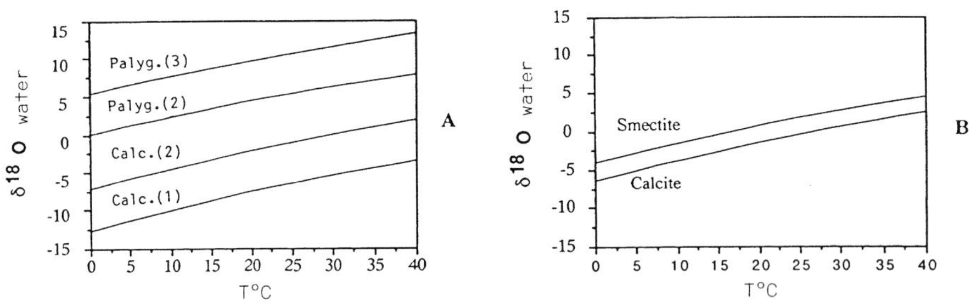

Fig. 3. Oxygen isotopic variation of water in isotopic equilibrium with the analysed minerals vs. temperature, in calcretes A and B. *Palyg.* Palygorskite; *Calc.* calcite. (After Gauthier-Lafaye et al. 1993)

1 Isotopic compositions are measured per thousand as compared to standard mean ocean water (SMOW) for oxygen, and Peedee Formation *Belemnitella americana* (PDB) for carbon. For oxygen, precision is ±0.2 per thousand in the case of carbonates and ±0.5 per thousand in the case of silicates; for carbon, precision is 0.3 per thousand.

results are presented in Table 2 and Fig. 3 (Gauthier-Lafaye et al. 1993). In calcrete A, oxygen isotopic equilibrium is roughly achieved between the calcite (1) and the local meteoric water ($\delta^{18}O$=0.4%, after Yurtsever and Gat 1981) or a slight enrichment in ^{18}O is noticed. The enrichment in ^{18}O is more significant in calcite (2) for which $\delta^{18}O$=26,8%. For palygorskites, the enrichment in ^{18}O is pronounced, especially in the case of palygorskite (3) in which $\delta^{18}O$=37.2%. Such an enrichment is ascribed to the meteoric water which is enriched in ^{18}O. For such an isotopic composition, evaporation is estimated to be about 80% by the Rayleigh evaporation model. In calcrete B, calcite and smectite yield isotopic compositions indicating rather similar formation conditions. The oxygen isotopic compositions correspond to meteoric water whose evaporation is estimated at about 40–50% of the total water. In the case of calcrete A, the enrichment in ^{18}O is more important for the clay mineral than for the calcite.

6.3 Conclusions

The ^{18}O content of clay minerals and associated calcites in calcretes attests to the fact that these two types of minerals are not in isotopic equilibrium with the local meteoric waters: they display marked enrichments in ^{18}O. Such enrichments can be related to evaporative episodes before mineral formation.

The formation of palygorskite and smectite requires more marked evaporative conditions than does calcite precipitation, as the ^{18}O content of calcite is more similar to that of meteoric waters. This would substantiate the hypothesis of McFadden et al. (1991) that pedogenic calcite dissolves and precipitates in open systems. It is further suggested that palygorskite formation takes place in a more confined environment than that of smectite.

7 Calcrete and Geochemistry of Landscapes

During his frequent field-trips to tropical Africa and the Sahara, Georges Millot was often intrigued by the large flat surfaces with low slopes and frequent induration, which characterize semiarid and arid regions. These large surfaces which are connected to hillsides by a system of concave slopes and downslopes to the terraces of the fluvial drains, develop on hard parent rocks and are called "pediments". During the 1970s, debates took place on the origin of these vast indurated surfaces: were these surfaces indurated because they were levelled, or have they been affected by planation because they were «reinforced» by calcrete formation? The process of calcareous epigenetic replacement recognized by Georges Millot brought some evidence into the debate (Ruellan et al. 1979; Millot 1980).

7.1 Rectification of the Basement

Calcareous epigenetic replacement works at the bottom of weathering covers. As shown above, the nature of the parental material is changed but not its structure or volume. Thus, there is no direct or perceptible action on landforms other than a preparatory one. In fact, epigenetic calcareous replacement, in which the weathering front is regulated by the underground water table, reduces the surface roughness of bed rock; it induces a buried and «hypodermic» rectification of the basement.

7.2 Pedological Action and Ablation

At the top of calcretes, pedological processes operate; these include biological reworking, structural modification and above all, dissolution of calcite during rain periods. Thus, the thickness and volume of calcrete decreases; calcrete is degraded from the top, leading to the development of a loose, sandy or silt-clay horizon which represents the original minerals which escaped dissolution or calcareous replacement. That loose horizon may contain rock fragments which, together with scattered vegetation, play a fundamental role in preventing rainwater from concentrating in rills. Thus, in semiarid areas, instead of rill wash, a rain wash takes place and progressively induces the ablation of the loose superficial horizon. The result is progressive rectification of the landforms, e.g. planation.

7.3 Integration of the Processes (Millot 1980, 1983)

In summary, one can assert that rain wash lowers the surface of the loose horizon by ablation, that the basal part of this loose horizon is reconstructed by dissolution of the surficial layer of the calcrete, due to hydrologic and biological action, and that the calcrete is expanded downward by calcareous epigenetic replacement of the basement. Thus, three regular and roughly parallel surfaces develop progressively, i.e. a soil surface, a calcrete surface and the surface corresponding to the basal zone of epigenetic replacement. The ultimate combination of these results is planation of landforms.

In the course of time, calcrete evolution progresses due to two effects. In the uppermost part of the calcrete, a dissolution front liberates a surficial, loose and mobile horizon which takes part in the rectification of slopes when affected by erosion. In the basal part of the calcrete, calcitization leads to the replacement of parental materials and thus to the smoothing of the basement. Consequently, calcrete formation is an important factor in landform planation.

8 Calcretes and Paleoenvironments

8.1 Pedogenesis and Water Table Diagenesis

Calcrete formation by replacement of parental material is a geochemical process working during weathering (pedogenesis), as well as during «hypodermic» alteration (water table diagenesis; Mann and Horwitz 1979; Arakel 1985; Truc et al. 1985; Rossinsky et al. 1992). In fact, calcrete formation depends on two types of process. First, during pedogenesis in hard parent rocks, soil solutions circulate along different pathways, such as along schistosity or fracturation joints, microfissures and diaclases. Calcareous epigenetic replacement develops in the form of calcareous grids whose growth and thickening result in a continuous indurated layer. In loose and porous rocks, soil solutions circulate in the macro- and microporal systems and the resulting calcareous forms are spherical (mottles, soft and hard nodules). They coalesce progressively, resulting also in a continuous calcareous layer. The second set of processes operates through the water table, whose dynamics fluctuate with permeability and with the seasonal supply from rainfall. During dry periods, the mineral load of water induces – as with soil solutions – dissolution of silicates and precipitation of carbonates such as calcite or dolomite. Through repeated alternations of dry and wet periods, massive calcretes develop progressively.

When studying a calcrete, it is not always easy to identify what results from pedogenesis and what results from water table diagenesis. For this reason, Klappa (1983) proposed a unique term, «pedo-diagenesis». However, even if the geochemical mechanisms concerned are similar in both cases, some differences in structural sequencing may be helpful. In fact, in calcretes related to pedogenesis, pedo-biological sequences can be observed. These are increasingly numerous towards the top of the profiles, before carbonate dissolution in the surficial loose horizons. In calcretes related to water table diagenesis, such pedo-biological sequences are lacking and only mineral replacements are observed (Truc et al. 1985).

Otherwise, the combination of petrographic and isotopic analyses (Prikryl et al. 1988) is thought to be very useful to differentiate the pedogenetic from the diagenetic origin of calcretes. Recent studies which integrate stable and neodymium isotopic compositions also seem to be quite promising.

8.2 Paleocalcretes: Jebel Chambi (Tunisia)

The dual origin of calcretes has been observed in many geological formations, a classic example being seen in the section of the Jebel Chambi in Tunisia (Sassi et al. 1984). The Jebel Chambi is located in the area called «Ile de Kasserine» in central Tunisia. No continental or marine deposit has been identified up to now, either in the Jebel Chambi or in the other emerged zones («ensemble de la Jeffara»), for the period between Late Cretaceous and Late Eocene/Oligocene. And it was traditionally recognized that a red and white thick detrital forma-

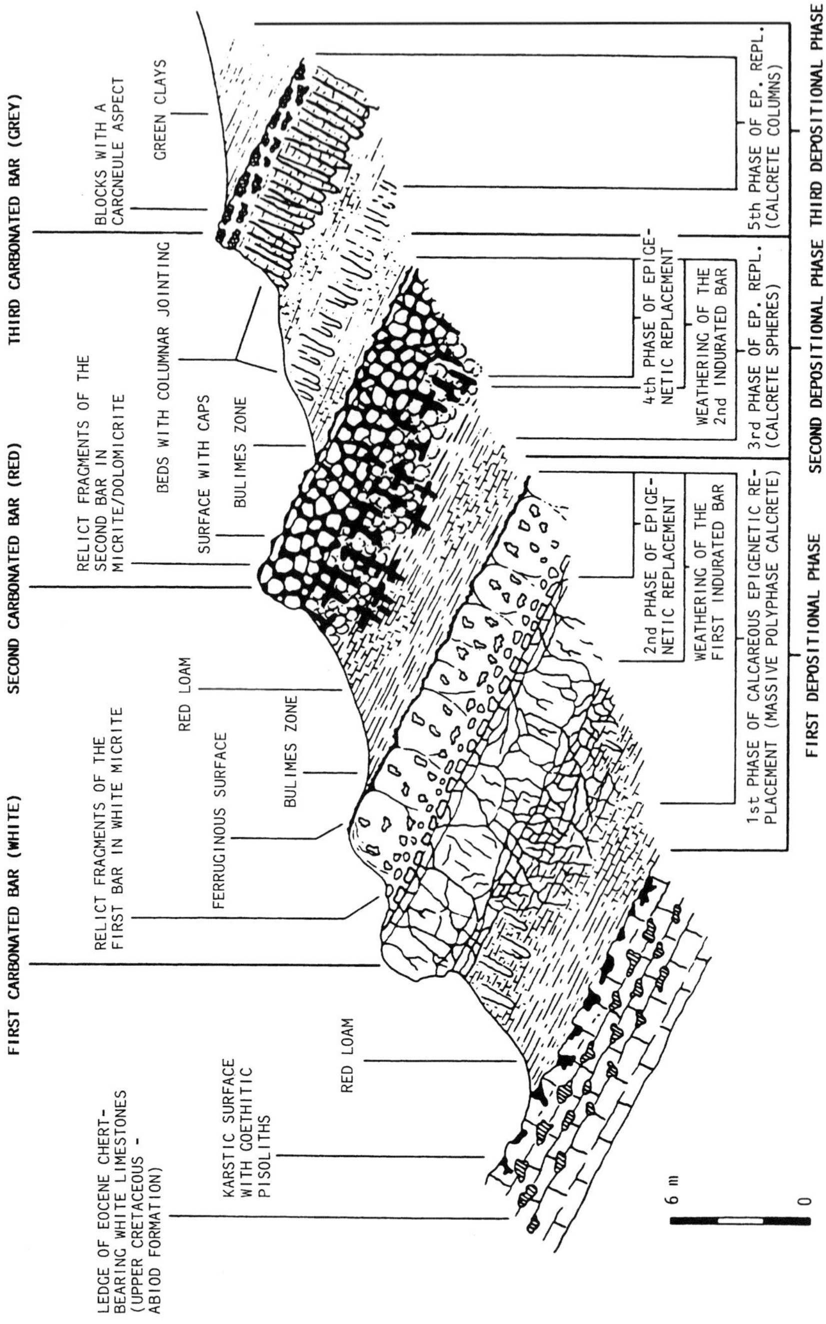

Fig. 4. «Sedimentation-calcrete formation» cycles. Example from the South Chambi section. (After Sassi et al. 1984)

tion, Miocene in age, rests on the Abiod Limestone which represents the youngest Mesozoic unit.

The detailed study of the Jebel Chambi and of neighbouring jebels led to the conclusion that the thick detrital formation involved three cycles of «sedimentation-calcrete formation» (Fig. 4). In each of the cycles, there was an alternation of accumulation of red detrital sediments which were progressively calcretized and of degraded calcretes. Outcrops, rock samples and thin sections all showed evidence of replacement by calcite or dolomite either by weathering or by water table diagenesis. Water circulation was favoured because of the nature of the sediment (Truc et al. 1985). Periods of degradation of the calcretes were illustrated by ablation, dissolution and incipient karstification. These alternations may be connected either to tectonic movements which affected the Jebel Chambi during the Tertiary or to climate variations. Palygorskite is the only clay mineral in the calcrete; its neogenesis is evident since the surrounding landscape and the loamy parent materials are depleted in palygorskite.

A terrestrial gastropod fauna including *Bulimes* which definitively belongs to the Middle Eocene, is found in the loamy beds as well as in the calcrete-bearing ones (Truc et al. 1985). Therefore, a formation which was classically ascribed to marine or lacustrine Miocene is actually a calcrete-bearing, continental, Eocene formation. The case of the Jebel Chambi is not unique; in fact, such calcrete-bearing, continental, Eocene formations were later identified in the Chaîne Nord des Chotts, Tunisia (Abdeljaouad 1989), in the High Plains of Oranie (Bensalah et al. 1987), and in the Hammada du Guir, Morocco (Elyoussi et al. 1990). Evidence for epigenetic replacement, abundance of palygorskite, and biostratigraphic data thus characterize a detrital Eocene unit with calcretes and Bulimes fauna in the three North African countries.

8.3 Paleoenvironmental Consequences

The identification of calcareous epigenetic replacement in an indurated continental formation, whatever its age may be, leads to following conclusions. From a paleoclimatic point of view, the development of calcareous epigenetic replacement of parent material indicates alternating climatic conditions and long dry periods. From a paleogeographic point of view, calcareous bars or slabs cutting or replacing detrital sediments cannot be automatically considered to be of lacustrine or marine origin (example of the Jebel Chambi). From a stratigraphic or cartographic point of view, the geometric relations between the calcretized sediments and the other sedimentary formations in the same basin should henceforth be recorded on geological maps (example of North Africa as a whole).

9 Conclusions

Due to the studies of Georges Millot, a new insight into the genesis and evolution of calcretes, which are usually considered as the result of impregnation of more or less porous media by calcite, was obtained. The principal lessons which may be drawn from the work of Georges Millot are the following:

1. Calcretes represent a weathering system which is specific to semiarid climatic conditions.
2. Calcretes develop in many cases by isovolumetric replacement of the minerals of parent rocks and of weathering products by calcite (calcareous «épigénie»).
3. Regulated by the underground water table, this epigenetic replacement reduces roughness of basement rocks below the calcretes. At the top, after biological and pedogenetic reworking, calcrete degrades, resulting in a loose and mobile silty-clay horizon which is subject to ablation. Calcrete formation thus facilitates the regularization of landforms and is an active factor in planation.

Due to the identification of the phenomenon of epigenetic replacement in calcretes, it can be stated that many other supergene deposits may be reinterpreted in terms of epigenetic replacement (iron crusts, nickeliferous and manganiferous ore deposits, silcretes, etc.) and that numerous calcareous formations considered to be lacustrine or palustrine may actually be characterized as palaeocalcretes. Thus, the latter could provide important information on palaeoclimates and palaeoenvironments. Even the palaeoclimatic stages of calcrete formation are being dated now (Liu et al. 1994; Mack et al. 1994).

Acknowledgements. The authors are deeply thankful to Prof. Larry Frakes for his critical review of the paper and his valuable help with the English translation.

References

Abdeljaouad S (1989) Dolomitisation et calcitisation successives dans l'Eocène détritique continental de la Tunisie méridionale. Alternances d'épigénies par diagenèse de nappe et d'altérations météoriques. Bull Soc Géol Fr 8 (V,4):837–847

Al Droubi A, Grondin JL, Fritz B, Tardy Y (1978) Calcul des équilibres dans le système $CaCO_3$-H_2O-CO_2. Rappel des conditions de dissolution et de précipitation de la calcite. Sci Géol Bull (Strasb) 31:195–202

Allen BL (1985) Micromorphology of aridisols. In: Douglas LA, Thompson ML (eds) Soil micromorphology and soil classification. SSSA Spec Publ (Madison) 15:197–216

Arakel AV (1985) Vadose diagenesis and multiple calcrete soil profile development in Hutt Lagoon area, Western Australia. Rev Géol Dyn Géogr Phys 26:243–254

Bachman GO, Machette NM (1977) Calcic soils and calcretes in the southwestern United States. US Geol Surv Open-File Rep 77–794, 163 pp

Beattie JA (1970) Perculiar features of soil development in Parna deposits in the eastern Riverina, New South Wales. Aust J Soil Res 8:145–156

Beattie JA, Haldane AD (1958) The occurrence of palygorskite and barite in certain Parna soils of the Murrumbidgee region, New South Wales. Aust J Sci 20:274–275

Bech J, Nahon D, Paquet H, Ruellan A, Millot G (1980) Sur l'extension géographique et climatique des phénomènes d'épigénie par la calcite dans les encroûtements calcaires. Exemple de la Catalogne. CR Acad Sci Paris 291 (D):371–376

Bensalah M, Benest M, Gaouar A, Truc G, Morel JL (1987) Découverte de l'Eocène continental à Bulimes dans les Hautes Plaines oranaises (Algérie): conséquences paléogéographiques et structurales. C R Acad Sci Paris 304 (II):35–38

Bocquier G (1973) Genèse et évolution de deux toposéquences de sols tropicaux du Tchad Interprétation biogéodynamique. Mém ORSTOM, Paris, 62, 350 p

Boulet R (1974) Toposéquences de sols tropicaux en Haute-Volta: équilibre et déséquilibre pédobioclimatique. Mém ORSTOM, Paris, 85, 272 pp

Bourgeon G (1992) Les sols rouges de l'Inde péninsulaire méridionale Pédogenèse fersiallitique sur socle cristallin en milieu tropical. Publ Dépt d'Ecologie, Institut français de Pondichéry, 31, 271 pp

Brewer R (1964) Fabric and mineral analysis of soils. Wiley, New York, 470 pp

Brown CN (1956) The origin on caliche on the northeastern Llano Estacado, Texas. J Geol 64:1–15

Callot G, Guyon A, Mousain D (1985) Inter-relations entre aiguilles de calcite et hyphes mycéliens. Agronomie 5:209–216

Campbell SE (1979) Soil stabilization by a prokaryotic desert crust: Implications for Precambrian land biota. Orig Life 9:335–348

Chadwick OA, Davis JO (1990) Soil-forming intervals caused by eolian sediment pulses in the Lahontan basin, northwestern Nevada. Geology 18:243–246

Chadwick OA, Hendricks DM, Nettleton WD (1987) Silica in duric soils: I. A depositional model. Soil Sci Soc Am J 51:975–982

Coque R (1962) La Tunisie présaharienne (étude géomorphologique). Thèse Lettres, Armand Colin, Paris, 476 pp

Ducloux J, Dupuis T, Butel P, Nahon D (1984) Carbonates de calcium amorphe et cristallisés dans les encroûtements calcaires des milieux tempérés Comparaison des séquences minérales naturelles et expérimentales. CR Acad Sci Paris 298 (II):147–149

Dupuis T, Ducloux J, Butel P, Nahon D (1984) Etude par spectrographie infra-rouge d'un encroûtement calcaire sous galet. Mise en évidence et modélisation expérimentale d'une suite minérale évolutive à partir de carbonate de calcium amorphe. Clay Min 19:605–614

Durand R (1979) La pédogenèse en pays calcaire dans le nord-est de la France. Sci Géol Mém (Strasb) 55:198

Elgabaly MM (1962) The presence of attapulgite in some soils of the western desert of Egypt. Soil Sci 93:387–390

El Prince AM, Mashhady AS, Aba-Husayn MM (1979) The occurrence of pedogenic palygorskite (attapulgite) in Saudi Arabia. Soil Sci 128:211–218

Elyoussi M, Truc G, Paquet H, Millot G, Triat JM (1990) Un piémont détritique à encroûtements carbonatés. La Hammada du Guir au Maroc. C R 3e Forum nat Géomorphologie, Méditerranée, h sér, pp 34–35

Estaban M (1974) Caliche textures and Microcodium. Soc Geol Ital Bull 92:105–125

Eswaran H, Barzani AF (1974) Evidence for the neoformation of attapulgite in some soils of Iraq. Trans Xth Int Congr Soil Sci, Moscow 7:154–161

Folk RL (1993) SEM imaging of bacteria and nanno-bacteria in carbonate sediments and rocks. J Sediment Petrol 63:990–999

Frye JC, Glass HD, Leonard AB, Coleman DD (1974) Caliche development and clay mineral zonation of the Ogallala Formation in central-eastern New-Mexico. New-Mexico Bureau Mines and Minerals Resources, Circ 144, 16 pp

Gardner LR (1972) Origin of the Mormon Mesa Caliche, Clark County, Nevada. Geol Soc Am Bull 83:143–155

Gauthier-Lafaye F, Taieb R, Paquet H, Chahi A, Prudencio I, Sequeira Braga MA (1993) Composition isotopique de l'oxygène de palygorskites associées à des calcrètes: conditions de formation. CR Acad Sci Paris 316 (II):1239–1245

Gile LH (1967) Soils of an ancient basin floor near Las Cruces, New Mexico. Soil Sci 103:265–276

Gile LH, Hawley JW (1972) The prediction of soil occurrence in certain desert regions of the southwestern United States. Soil Sci Soc Am Proc 36:119–124

Gile LH, Peterson FF, Grossman RB (1966) Morphological and genetic sequence of carbonate accumulation in desert soils. Soil Sci 101:347–360

Goudie A (1973) Duricrusts in tropical and subtropical landscapes. Clarendon Press, Oxford, 174 pp

Grousset FE, Rognon P, Coudé-Gaussen G, Pedermay P (1992) Origin of peri-Saharan dust deposits traced by their Nd and Sr isotopic composition. Palaeogeogr Palaeoclimatol Palaeoecol 93:203–212

Halitim A, Robert M, Pédro G (1983) Etude expérimentale de l'épigénie calcaire des silicates en milieu confiné Caractérisation des conditions de son développement et des modalités de sa mise en jeu. Sci Géol Mém (Strasb) 71:63–73

Hay RL, Wiggins B (1980) Pellets, ooids, sepiolite and silica in three calcretes of the Southwestern United States. Sedimentology 27:559–576

James NP (1972) Holocene and Pleistocene calcareous crust (caliche) profiles: criteria for subaerial exposure. J Sedim Petrol 42:817–836

Jessup RW (1961) Evolution of the two youngest (Quaternary) soil layers in the south eastern portion of the Australian arid zone. I. The Parakylia layer. J Soil Sci 12:52–63

Jessup RW, Wright MJ (1971) Cenozoic sediments, soils and climates at Whyalla, South Australia. Geoderma 6:27–308

Jones BF (1983) Occurrence of clay minerals in surficial deposits of southwestern Nevada. Sci Géol Mém (Strasb) 72:81–92

Jones BF (1986) Clay minerals diagenesis in lacustrine sediments. US Geol Surv Bull 1578:291–300

Jones BF, Galan E (1988) Palygorskite-sepiolite. In: Bailey SW (ed) Hydrous phyllosilicates exclusive of micas. Geol Soc Am Rev Mineral 19:631–674

Klappa CF (1979) Calcified filaments in Quaternary calcretes: organo-mineral interactions in the subaerial vadose environment. J Sediment Petrol 49:965–968

Klappa CF (1980) Rhizoliths in terrestrial carbonates: classification, recognition, genesis and significance. Sedimentology 27:613–629

Klappa CF (1983) A process response model for the formation of pedogenetic calcretes In: Wilson RCL (ed) Residual deposits: Surface related weathering processes and material. Geol Soc Spec Publ II, Blackwell, London, pp 211–220

Krumbein WE, Giele C (1979) Calcification in a coccoid cyanobacterium associated with the formation of desert stromatolites. Sedimentology 26:593–604

Lawrence JR, Taylor HP (1971) Deuterium and oxygen-18 correlation: clay minerals and hydroxides in Quaternary soil compared to meteoric water. Geochim Cosmochim Acta 35:993–100

La Iglesia A (1977) Precipitacion por disolucion homogenea de silicatos de aluminio y magnesio a temperature ambiente sintesis de la palygorskita. Estudios Geol 33:535–544

Liu B, Philips FM, Elmore D, Sharma P (1994) Depth dependence of soil carbonate accumulation based on cosmogenic ^{36}Cl dating. Geology 22:1071–1074

Machette MN (1985) Calcic soils of the American southwest soils and Quaternary geology of the southwestern United States. Spec Pap Geol Am 203:1–22

Mack GH, Cole DR, James WC, Giordano TH, Salyards SL (1994) Stable oxygen and carbon isotopes of pedogenic carbonate as indicators of Plio-Pleistocene paleoclimate in the southern Rio Grande rift, south central New Mexico. Am J Sci 294:621–640

Mahdoudi ML, Lang J, Pascal A (1989) Pétrographie et signification des encroûtements carbonatés dans les séries rouges mésozoïques du Haut-Atlas central (secteur de Telouet- Ighrem, Maroc). Sci Géol Mém (Strasb), pp 143–156

Mann AW, Horwitz RC (1979) Groundwater calcretes deposits in Australia: some observations from Western Australia. J Geol Soc Aust 26:293–303

McFadden LD, Amundson RG, Chadwick OA (1991) Numerical modeling, chemical and isotopic studies of carbonate accumulation in soils of arid regions In: Nettleton WD (ed) Occurrence, characteristics and genesis of carbonate, gypsum and silica accumulations in soils. SSSA Spec Publ (Madison) 26:17–35

McGrath DA (1984) Morphological and mineralogical characteristics of indurated caliches of the Llano Estacado MS Thesis, Texas Tech Univ, 123 pp (in Jones and Galan 1988)

Millot G (1980) Les grands aplanissements des socles continentaux dans les pays subtropicaux et désertiques. Mém h. sér Soc Géol Fr 10:295–305

Millot G (1983) Planation of continents by intertropical weathering and pedogenetic processes. IInd international seminar on lateritisation processes, São Paulo, 1982. In: Melfi AJ, Carvalho A (eds) Lateritisation processes, pp 53–63

Millot G, Paquet H (1987) Le remplacement à volume constant ou épigénie dans les altérations météoriques et les gîtes minéraux supergènes. Yerbilimleri 14:1–11

Millot G, Paquet H, Ruellan A (1969) Néoformation de l'attapulgite dans les sols à carapaces calcaires de la Basse Moulouya (Maroc oriental). CR Acad Sci Paris 268 (D):2771–2774

Millot G, Nahon D, Paquet H, Ruellan A, Tardy Y (1977) L'épigénie calcaire des roches silicatées dans les encroûtements carbonatés en pays subaride, anti-atlas, Maroc. Sci Géol Bull (Strasb) 30:129–152

Millot G, Nahon D, Paquet H, Ruellan A, Tardy Y (1978) Geochemistry of calcareous epigenesis in calcretes: argillization, silicate hydrolysis, calcitization. 10th Int Congr Sedimentology, Jerusalem, Abstr, p 439

Millot G, Bocquier G, Boulet R, Chauvel A, Leprun JC, Nahon D, Paquet H, Pédro G, Ruellan A, Tardy Y (1979) Géochimie de la surface, pédogenèse, aplanissements et formes du relief dans les pays méditerranéens et tropicaux. Sci Géol Mém (Strasb) 53:39–43

Milnes AR (1992) Calcrete. In: Martini IP, Chesworth W (eds) Weathering, soils and paleosols, developments in earth surface processes, vol 2. Elsevier, Amsterdam, 618 pp

Milnes AR, Ludbrook NH (1986) Provenance of microfossils in aeolian calcarenites and calcretes in southern South Australia. Aust J Earth Sci 33:145–159

Mohr ECJ, Van Baren FA, Van Schuylenborgh J (1972) Tropical soils A comprehensive study of their genesis, 3rd edn. Mouton-Ichtiar Baru, Van Hoeve, The Hague, 481 pp

Monger HC, Daugerthy LA, Gile LH (1991) A microscopic examination of pedogenetic calcite in an aridisol of southern New Mexico. In: Nettleton WD (ed) Occurrence, characteristics and genesis of carbonate, gypsum and silica accumulations in soils. SSSA Spec Publ (Madison) 26:37–60

Morelli J, Buat Menard P, Chesselet R (1971) Mise en évidence de l'atmosphère marine d'aérosols enrichis en potassium et calcium, ayant la surface de la mer pour origine. CR Acad Sci Paris 272 (B):812–815

Nahon D (1976) Cuirasses ferrugineuses et encroûtements calcaires au Sénégal occidental et en Mauritanie Systèmes évolutifs: géochimie, structures, relais et coexistence. Sci Géol Mém (Strasb) 44:232

Nahon DB (1991) Introduction to the petrology of soils and chemical weathering. Wiley, New York, 313 pp

Nahon D, Paquet H, Ruellan A, Millot G (1975) Encroûtements calcaires dans les altérations des marnes éocènes de la falaise de Thiès (Sénégal): organisation et morphologie. Sci Géol Bull (Strasb) 28:29–46

Nahon D, Ducloux J, Butel P, Augas G, Paquet H (1980) Néoformation d'aragonite, première étape d'une suite minéralogique évolutive dans les encroûtements calcaires C R Acad Sci Paris 291 (II):725–727

Netterberg F (1980) Geology of southern African calcretes: I. Terminology, description, macrofeatures and classification. Trans Geol Soc S Afr 83:255–283

Newman ACD (1987) Chemistry of clay minerals. Miner Soc Monogr 6. Longman Sci Techn, 480 pp

Oliva P (1974) La distribution des croûtes et encroûtements dans l'Anti-Atlas occidental Une approche géomorphologique du problème des croûtes sur roches non calciques. CNRS Trav RCP 249:47–77

Paquet H (1983) Stability, instability and significance of attapulgite in the calcretes of mediterranean and tropical areas with marked dry seasons. Sci Géol Mém (Strasb) 72:131–140

Paquet H, Millot G (1972) Geochemical evolution of clay minerals in the weathered products and soils of mediterranean climates. Proc IVth Int Clay Conf Madrid, pp 271–284

Paquet H, Duplay J, Valleron-Blanc MM, Millot G (1987) Octahedral compositions of individual particles in smectite-palygorskite and smectite-sepiolite assemblages In: Schultz LG, Van Olphen H, Mumpton FA (eds) Proc VIIIth Int Clay Conf Denver 1985. Clay Minerals Society, Bloomington, pp 73–77

Pascal A, Mahdoudi ML, Lang J, Paquet H, Millot G (1989) Palygorskites continentales épigéniques dans le Jurassique moyen du Haut-Atlas central (Maroc) Double signification des palygorskites dans les séries géologiques. C R Acad Sci Paris 309 (II):899–906

Pouget M, Rambaud D (1980) Quelques types de cristallisation de calcite dans les sols à croûte calcaire (steppes algériennes). Apport de la microscopie électronique. C R Réunion Groupe d'Etudes sur les Carbonates, Univ Bordeaux, pp 371–379

Prikryl JD, Posey HH, Kyle JR (1988) A petrographic and geochemical model for the origin of calcite cap rock at Damon Mound Salt Dome, Texas, USA. Chem Geol 74:67–97

Rabenhorst MC, Wilding LP, West LT (1984) Identification of pedogenic carbonates using stable carbon isotope and microfabric analyses. Soil Sci Soc Am 50:693–699

Rabenhorst MC, West LT, Wilding LP (1991) Genesis of calcic and petrocalcic horizons in soils over carbonate rocks. In: Occurrence, characteristics and genesis of carbonate, gypsum and silica accumulations in soils. Soil Sci Soc Am, Spec Publ 26:61–74

Reeves CC Jr (1970) Origin, classification and geologic history of caliche on the southern High Plains, Texas and eastern New Mexico. J Geol 78:352–362

Regaya K (1984) Les accumulations calcaires dans les limons de Matmata de la région de Gabès en Tunisie. Sci Géol Bull (Strasb) 37:387–398

Reheis MC (1988) Pedogenic replacement of aluminosilicate grains by $CaCO_3$ in Ustollic Haplargids, south central Montana, USA. Geoderma 41:243–261

Reheis MC, Sowers JM, Taylor EM, Mc Fadden LD, Harden JW (1992) Morphology and genesis of carbonate soils on the Kyle Canyon fan, Nevada, USA. Geoderma 52:303–342

Rossinsky V Jr, Wanless HR, Swart PK (1992) Penetrative calcretes and their stratigraphic implications. Geology 20:331–334

Ruellan A (1967) Individualisation et accumulation du calcaire dans les sols et les dépôts quaternaires du Maroc. Cah ORSTOM, Sér Pédol 5:421–462

Ruellan A (1970) Contribution à la connaissance des sols de régions méditerranéennes: les sols à profil calcaire différencié des plaines de la Basse-Moulouya (Maroc oriental). Mém ORSTOM 54:302 pp

Ruellan A, Nahon D, Paquet H, Millot G (1977) Géochimie de la surface et formes du relief. VI. Rôle des encroûtements et épigénies calcaires dans le façonnement du modelé en pays aride. Sci Géol Bull (Strasb) 30 (4):283–288

Ruellan A, Beaudet G, Nahon D, Paquet H, Rognon P, Millot G (1979) Rôle des encroûtements calcaires dans le façonnement des glacis d'ablation des régions arides et semi- arides du Maroc. C R Acad Sci Paris 289 (D):619–622

Salomons W, Goodie A, Mook WG (1978) Isotopic composition of calcrete deposits from Europe, Africa and India. Earth Surf Proc 3:43–57

Sassi S, Triat JM, Truc G, Millot G (1984) Découverte de l'Eocène continental en Tunisie centrale: la formation du Jebel Chambi et ses encroûtements carbonatés. C R Acad Sci Paris 299 (II):357–364

Schlesinger WH (1985) The formation of caliche in soils of the Mojave Desert, California. Geochim Cosmochim Acta 49:57–66

Seghal JL, Stoops G (1972) Pedogenetic calcite accumulation in arid and semiarid regions of the Indo-Gangetic alluvial plain of erstwhile Punjab India: their morphology and origin. Geoderma 8:59–72

Singer A (1977) Dissolution of two Australian palygorskites in dilute acid. Clays Clay Min 25:126–130

Singer A, Norrish K (1974) Pedogenic palygorskite occurrence in Australia. Am Miner 59:508–517

Singer A, Kirsten W, Bühmann C (1995) Fibrous clay minerals in the soils of Namaqualand, South Africa: characteristics and formation. Geoderma 66:43–70

Talma AS, Netterberg F (1983) Stable isotope abundances in calcretes In: Wilson RCL (ed) Residual deposits: surface related weathering processes and material. Geol Soc, London, Spec Publ 11. Blackwell, London, 208 pp

Tardy Y, Garrels RM (1974) A method of estimating the Gibbs energies of formation of layer silicates. Geochim Cosmochim Acta 38:1101–1106

Thellier C, Fritz B, Paquet H, Gac JY, Clauer N (1988) Chemical and mineralogical effects of saline water movement through a soil during evaporation. Soil Sci 146:22–29

Truc G (1989) Malacofaunes à Bulimes du Maghreb Révisions des données paléontologiques, stratigraphiques et paléoclimatiques relatives à ces mollusques terestres de l'Eocène moyen et supérieur. GSA. «89», 8e Conf Soc Géol Afrique, Rabat, résumé, 2 pp

Truc G, Triat JM, Sassi S, Paquet H, Millot G (1985) Caractères généraux de l'épigénie carbonatée de surface par altération météorique liée à la pédogenèse et par altération sous couverture liée à la diagenèse. CR Acad Sci Paris 300 (II):283–290

Turchenek LW, Oades JM (1974) Occurrence of palygorskite in a ground-water rendzina in south east South Australia. Trans Xth Int Congr Soil Sci, Moscow 5 (III):229–230
Van den Heuvel RC (1966) The occurrence of sepiolite and attapulgite in the calcareous zone of a soil near Las Cruces, New Mexico. Clays Clay Min (13th Natl Conf 1964), pp 193–207
Velde B (1985) Clay minerals: a physico-chemical explanation of their occurrence. Developments in sedimentology, vol 40. Elsevier, Amsterdam, 427 pp
Verges V, Madon M, Bruand A, Bocquier G (1982) Micromorphologie et cristallogenèse de microcristaux supergènes de calcite en aiguille. Bull Minér 105:351–356
Verrechia E (1990) Incidence de l'activité fungique sur l'induration des profils carbonatés de type calcrète. L'exemple du cycle oxalate-carbonate de calcium dans les encroûtements calcaires de Galilée (Israël). C R Acad Sci Paris 311 (II):1367–1374
Verrechia E, Ribier J, Freytet P, Patillon M, Rolko K (1991) Une origine biologique pour certaines croûtes zonaires: contribution des cyanophycées à leur édification. Conséquences sur l'évolution des profils de type calcrète. CR Acad Sci Paris 313 (II):1619–1625
Verrechia EP, Freytet P, Verrechia KE, Dumont A (1995) Spherulites in calcrete laminar crusts: biogenic $CaCO_3$ precipitation as a major contributor to crust formation. J Sediment Res A 465:690–700
Viani BE, Al-Masshady AS, Dixon JB (1983) Mineralogy of Saudi Arabian soils: central alluvial basins. Soil Sci Soc Am J 47:149–157
Wang Y, Nahon D, Merino E (1993) Geochemistry and dynamics of calcrete genesis in semi- arid regions. Chem Geol 107:349–351
Wang Y, Nahon D, Merino E (1994) Dynamic model of the genesis of calcretes replacing silicate rocks in semi-arid regions. Geochim Cosmochim Acta 23:5131–5145
Watts NL (1976) Paleopedogenic palygorskite from the basal Permo-Triassic of northwest Scotland. Am Mineral 61:299–302
Watts NL (1980) Quaternary pedogenic calcretes from the Kalahari (southern Africa): mineralogy, genesis and diagenesis. Sedimentology 27:661–686
Weaver CE, Beck KC (1977) Miocene of the SE United States: a model for chemical sedimentation in a peri-marine environment. Developments in sedimentology, vol 22. Elsevier, Amsterdam, 234 pp
Weaver CE, Pollard LD (1975) The chemistry of clay minerals. Developments in sedimentology, vol 15. Elsevier, Amsterdam, 213 pp
Wieder M, Yaalon DH (1974) Effect of a matrix composition on carbonate nodule crystallization. Geoderma 11:95–121
Wright VP (1982) Calcrete palaeosols from the lower Carboniferous Llanelly Formation, South Wales. Sedim Geol 33:1–33
Wright VP (1989) Terrestrial stromatolites and laminar calcretes: a review. Sedim Geol 65:1–13
Wright VP, Platt NH, Wimbledon WA (1988) Biogenic laminar calcretes evidence of calcified root-mat horizons in paleosols. Sedimentology 35:603–620
Wright VP, Platt NH, Marriott SB, Beck VH (1995) A classification of rhizogenic (root-formed) calcretes with examples from the Upper Jurassic–Lower Cretaceous of Spain and Upper Cretaceous of southern France. Sediment Geol 100:143–158
Yaalon DH and Wieder M (1976) Pedogenic palygorskite in some brown (calciorthid) soils of Israël. Clay Min 11:73–79
Yurtsever Y, Gat JR (1981) Atmospheric waters. In: Stable isotope hydrology, Technical Reports Series 210, IAEA, Vienna, pp 103–142

3 Laterites and Bauxites

Bruno Boulangé, Jean-Paul Ambrosi and Daniel Nahon

1 Introduction

The term «bauxite» was introduced by Berthier (1821) and refers to rock samples belonging to the laterite type. Lateritic bauxites have been studied by many, mainly by prospection and exploration geologists, and by geologists, soil scientists and geomorphologists interested in the mechanisms of bauxite formation and evolution. Many overviews have been published on the topic during this century (Lacroix 1913; Harrasowitz 1926; Harrison 1934; Patterson 1967; Valeton 1972; Bardossy and Aleva 1990). It is now well known that bauxites and laterites result from pedogenic processes, i.e. from weathering of a large variety of sedimentary, metamorphic and igneous rocks, under tropical and humid climates (Bocquier et al. 1984).

Lateritic bauxites, which are very often indurated, consist of Al oxi-hydroxides (gibbsite $Al(OH)_3$, boehmite AlOOH) and Fe oxi-hydroxides (goethite FeOOH, haematite Fe_2O_3) resulting from weathering of the parent rocks. The degree of weathering varies according to their chemical and mineralogical compositions, and to the climatic and drainage conditions. The alkalis and alkali-earths are completely removed, while Si is either completely or partly leached. In the former case, gibbsite forms from the remaining Al, and in the latter case, the residue of Si combines with Al to form kaolinite. The weathering of the parent rock leads to the formation of «secondary» or neoformed minerals which can be associated with «relict» minerals inherited from parent rocks. Several weathering units (horizons, facies) can be defined from the spatial (textural and structural) repartitioning of the mineralogical phases. The formation of these different weathering units reflects the allitization and monosiallitization mechanisms which have been defined by Pédro (1964), the term bauxite strictly referring to a horizon of Al accumulation that has an economic value. Georges Millot (1964) first distinguished two types of bauxitization of parent rocks i.e. the «direct» and the «indirect» one. These two processes lead to the formation of aluminous horizons called «original bauxites» (Fig. 1) which can undergo structural, mineralogical and chemical transformations resulting in the formation of «degraded» bauxites (pseudobrechic, nodular, pisolitic bauxites of Boulangé 1984; Boulangé and Millot 1988). During these transformations,

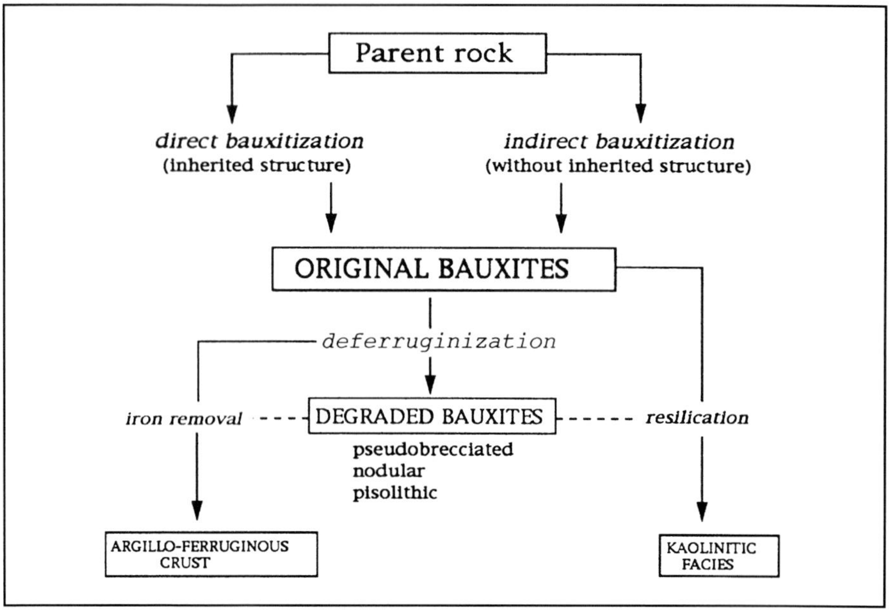

Fig. 1. The bauxitic facies and their genetic relationships

new accumulation units form according to the type of chemical transfer. For example, Fe migration into the subjacent argillaceous saprolites leads to the formation of the argillo-ferruginous crust. Resilications can create new kaolinite facies. These various facies form according to the physico-chemical conditions of the weathering environment which depend on the climatic (temperature, humidity, vegetation) and topographic parameters (slope, drainage).

2 Formation of the Original Bauxites

2.1 Weathering of the Parent Rock

During direct bauxitization, a pervasive hydrolysis of the parent minerals leads to the formation of gibbsite and goethite which are the only neoformed minerals. All major elements of the parent rocks, including Si but not Al and Fe, are removed. Al and Fe accumulate within the boundaries of the framework grains and in in-filling fissures and cleavages, and create a network of septa which encloses dissolution voids (Delvigne 1965; Nahon 1991). These elements can also be transported over short distances outside the boundaries of the original crystals. These transfers result in a random distribution of gibbsite in dissolution voids within the Fe–Mg minerals, and in an occurrence of gibbsite and goethite in dissolution cracks of quartz grains.

During indirect bauxitization, removal of the elements released by hydrolysis of the parent minerals is incomplete. Fe, Al and part of the Si accumulate in situ and form goethite and kaolinite. This process does not alter the texture of the parent rocks: kaolinite exists as pseudomorphs of the parent minerals, and goethite as septa which coat the inter-mineral fissures. The Fe contained in these goethite networks is then re-distributed and forms haematite concretions in the micropores (Tardy and Nahon 1985). This kaolinite alterite is a precursor for the bauxite sensu stricto which eventually forms as a result of the dissolution of the kaolinite and the neoformation of gibbsite. During this bauxitization stage, the texture of the parent rocks is rapidly lost, but the original lithologic structure is preserved (veins, fractures, folds).

To summarize, direct or indirect bauxitization involves the following processes: complete or partial dissolution of the parent minerals, loss of cohesion of the parent rocks associated with an increase of porosity, and conservation of broken but unaltered relicts of parent minerals. The fragments of single grains present single extinctions under XPL, indicating that they did not move and still retain their initial volumes (Nahon 1986a,b). However, variations of volume up to 20 to 30% related to the degree of porosity, are sometimes noticed (Colin et al. 1992). As a result, gibbsite forms in horizons where the structure and occasionally the texture of the parent rocks are preserved. Those horizons which are not yet in the final stage of bauxitization, are called lithomarge with conserved structure (Tardy 1969), coarse saprolite (Trescases 1975), or isalterite (Chatelin 1974; Boulangé 1984). This type of weathering under humid climates leads to the formation of a continuous and simple profile from parent rocks to the top soil with plant cover. Biological activity is responsible for the disaggregation of alterites, redistribution of the various components (gibbsite, kaolinite, goethite), and for downward transfers of elements into an accumulation horizon. These relative and absolute accumulations (D'Hoore 1954) contribute to an increase of the Al content up to economic concentrations.

2.2 Transfers and Accumulation in Isalterites

Dissolution of parent minerals creates an important intra- and inter-mineral porosity, adding to the fissural porosity of the parent rocks. The porosity is subsequently lined or filled with material precipitated from dissolved species or from particles of percolating fluids. From bottom to the top of isalterites, i.e. from the fissural porosity to the alveolar porosity, different types of accumulation have been observed (Bocquier et al. 1984). These transfers generate different sub-units inside the isalterites.

At the bottom of the isalterites, these transfers are characterized by an accumulation of silico-aluminous and amorphous material and of halloysite in the trans-mineral fissures of the parent rocks, while weathering proceeds. These accumulations increase the Al content of the parent rocks. Boulangé (1984) reported an increase of 2 to 5% of the Al content relative to the initial granite. Transfers and redistributions of elements occur through porosity created by weathering of the parent minerals from the top to the bottom of the isalterites. Accumulation resulting from these chemical and particle transfers forms yellow to red sepic (isotropic) matrices which consist either of alumino-ferruginous amorphous products, or of argillo-ferruginous material. In these accumulations, kaolinite, which is poorly crystallized, is associated with goethite and forms small disk-shaped crystals (Ambrosi and Nahon 1986). Desiccation and re-hydration of the kaolinitic and goethitic products lead to a zonation of the rocks which results from Fe migration and concentration as haematite in internal red kaolinite zone characterized by microporosity. The external yellowish zone, which is Fe depleted, displays a higher porosity, and is composed of an aluminous plasma where gibbsite replaces kaolinite (Boulangé et al. 1975). These segregations within the cutans are mainly related to variations of water activity and pore size (Tardy and Nahon 1985; Trolard and Tardy 1987). Solutions circulating in the cutans induce dissolution of goethite and kaolinite. Fe migrates over short distances and precipitates as haematite in the microporosity of the internal zone. Kaolinite displays a lower dissolution towards the internal zone and is replaced by gibbsite which precipitates as prismatic crystals (Fig. 2). In each void, successive accumulations form complex cutans which display similar organizations, i.e. with Fe and kaolinite in the internal zone and gibbsite in the external zone, reflecting the succession of dry and wet conditions of the environment. These accumulations in the poral spaces increase the cohesion of the isalterite. They also increase the Al content, adding to the bauxitic character of the isalterite.

In the bauxite facies, the Fe and Al contents are related to the chemistry of the parent rocks: a parent rock with a high Al content will generate a bauxite with a high Al content. Isovolume mass balances on the basis of the method

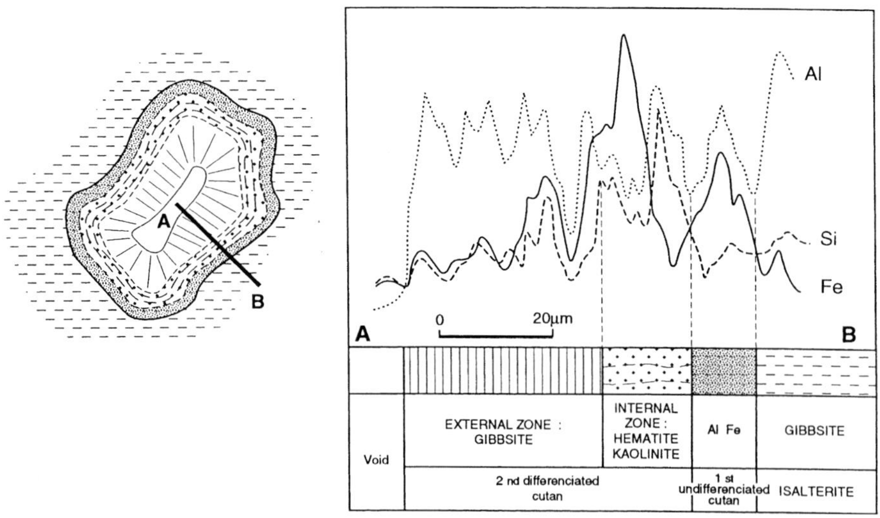

Fig. 2. Microprobe analysis in a ferrigibbsian cross section: relative distribution of Fe, Si and Al

Layers	Thickness (m)	Al_2O_3 (%)	Fe_2O_3 (%)
Fragmentary bauxite	1	130	290
Upper massive bauxite	0.5-1	140	80
Lower massive bauxite	10-15	120	200
Ferruginous layer	0.2-0.5	60	250
Upper isalterite	3-5	60	250
Lower isalterite	>2	20	80
Granite		44 g in 100 cm^3	4.5 g in 100 cm^3

Table 1. Isovolume estimation of Al_2O_3 and Fe_2O_3 gains in the isalterites and the original bauxite of the Mt. Tato (Ivory Coast; after Boulangé 1984)

developed by Millot and Bonifas (1955) have shown that more than 50% of the Fe and Al contained in a bauxite developed on a granite, come from transfers which took place after weathering of the parent rock. Boulangé (1984) noticed that accumulation of Fe oxides is always greater than accumulation of Al (Table 1). Iso-elemental mass balances (Zr, Ti) give better results when the variations of the rock volume are important (Colin et al. 1993).

Under humid tropical climates with alternating rainy and dry seasons, the original bauxites are progressively transformed through a succession of processes: absolute accumulation of kaolinite (halloysite) in the fissures of the parent rocks, dissolution of the framework minerals and neoformation of gibbsite, secondary cutan accumulations of kaolinite and goethite, and transformation and alteration of these argillo-ferruginous accumulations into haematite and gibbsite (Fig. 3). A simple weathering profile with a continuous sequence including the parent rock, the isalterite and the soil is, therefore, replaced by another profile with an accumulation horizon (original bauxite) on top. If the same climatic conditions persist, the bauxite profile becomes progressively thicker at the expenses of the parent rock, as the alteration proceeds down the profile (Nahon and Millot 1977). This type of profile develops on hilly landscapes with convex slopes, which are characteristic of humid tropical zones.

3 Evolution of the Original Bauxites: Formation of Degraded Bauxites

Original bauxites formed under humid climates with alternating wet and dry seasons, show various evolution patterns depending on whether the climate becomes wetter or drier.

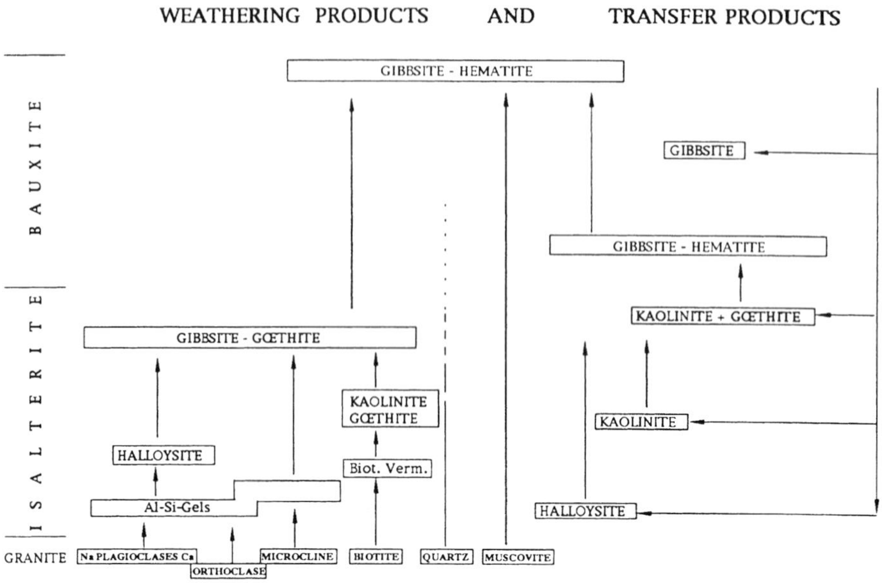

Fig. 3. Sketch of a bauxite profile developed from a granite (Mount Tato, Ivory Coast): layers and chemistry (means for each layer and facies)

3.1 Evolution of the Original Bauxites Under Humid Climates

Under humid climates, original bauxites undergo localized Fe leaching which determines the development of alumino-ferruginous pseudobrechic volumes. Short-distance transfer of Fe takes place towards the centre of the pseudobrechic volumes which are progressively indurated, to form nodules located in an Al matrix which is lighter in colour (Parron et al. 1983). The nodules develop concentric (Nahon 1976) and pisolitic structures which are mainly composed of goethite with up to 25% mol of AlOOH. As Fe leaching proceeds, the nodular and pisolitic structures tend to disappear and the Al matrix becomes predominant.

Above the bauxite levels, vegetation (forest) protects the clay-sandy cover which formed by relative accumulation of residual Si which was observed within the original bauxite. The biochemical cycle of Si through the vegetation feeds kaolinite formation in the cover (Lucas et al. 1993). This mechanism also affects the underlying bauxite unit, and the degree of kaolinization of the Al matrix is related to the level of residual silica. Under permanent humid conditions, as the parent rocks are progressively bauxitized, the degradation of the top of the bauxite horizon leads to the development of a thicker soft clay-sandy cover. Evolution of this cover towards podzolic soils, which involves clay migration, accumulation, aggregation and destruction of the aggregates, can also be observed (Pédro and Chauvel 1978; Ambrosi and Nahon 1986; Lucas and Chauvel

1992). At the bottom of the bauxite profile, changes in the conditions of weathering of the parent rocks due to impeded vertical drainage can also result in the resilication of gibbsite with its transformation into kaolinite.

Changes in humid conditions erase the characteristic organization of the original bauxite, which is transformed into a pseudobrechic or nodular degraded bauxite, depending on the degree of Fe leaching. Kaolinite formation at the top and/or in the lower part of the profile induces the loss of the bauxite character. These processes which are more or less intense, can lead to an extreme evolution, such as a fersiallitic soil generated at the expenses of the totally eroded bauxite.

3.2 Evolution of Bauxites under Tropical Climates with Alternating Seasons

When a tropical climate with alternating dry and humid seasons persists, the original bauxites are exposed to climatic conditions similar to those having prevailed during their formation. As the formation of original bauxite and the accumulation of material in the saprolite proceed, bauxites undergo important mineralogical and structural transformations caused by the alternation of dry and wet periods, and by geomorphological variations.

The transformation of gibbsite into boehmite is the first change which can be observed. Two opposite views concerning the stability of gibbsite and boehmite exist according to thermodynamic data: gibbsite is thought to be more stable than boehmite (Garrels and Christ 1965; Sarazin 1979), or gibbsite is less stable than boehmite (Rossini et al. 1952; Nahon 1976). Chesworth (1972) and Trolard and Tardy (1987) claimed that water state and water saturation have to be considered in the calculations. Gibbsite is stable in a saturated environment at 25° C and 1 atm, and boehmite becomes more stable than gibbsite in an under-saturated environment.

Environments which are water under-saturated can predominate in the upper part of bauxite levels and on hillsides that are bauxitized. Boehmite stability which is not favoured by an alternation of dry and wet seasons, is however increased by an intense ferruginization which leads to a boehmite–haematite assemblage (Tardy et al. 1988b). Part of the released Fe from profiles located higher in the landscape migrates laterally and is mainly retained by the original bauxites located in the slopes, which become characterized by a boehmite–haematite association. When the clay-sandy top horizon is not too thick, the alternation of dry and rainy seasons favours the evolution of the original gibbsite–haematite association towards a boehmite–haematite association characterized by high Fe amounts of up to 50%.

Reorganization of the boehmite–haematite material leads to the formation of Al pisoliths (Boulangé and Bocquier 1983). The Al–Fe crusts are cut into pseudobrechic volumes and nodules. The centripetal degradation of these fragments and nodules feeds a new inter-nodular matrix. The alternating wet and dry conditions result in the development of new spherical nodular volumes by

Fe migration from a saturated central zone to a more dehydrated external zone. Consequently, progressive Fe leaching leads to reorganization of the matrix into pisoliths with a cortex becoming more and more complex. A complete Fe leaching either on the scale of the microsystem or the horizon, induces formation of a boehmitic matrix with pisolitic structures.

The original bauxites progressively lose their structure and are replaced by degraded bauxites which display intermediate evolutionary stages including pseudobrechic, nodular and Fe or strictly Al pisolitic facies. The last stage of bauxitization is represented by a boehmite-rich bauxite with pisolitic structures. This bauxite is eventually affected by dissolution processes; the Al which has been released, may migrate over short distances and precipitate as gibbsite in the pore spaces, or may migrate downwards even to the lower saprolite in which it forms kaolinite with Si.

The formation of pisoliths represents a stage in the degradation of the laterite bauxites. Fe and Al are progressively released and migrate downwards and downslope. In these zones, accumulation of Fe and combination of the Al with Si results in the formation of argillo-ferruginous crusts representing the last stage of these transformations.

3.3 Evolution of the Bauxites Under Semi-Arid Climates

Under semi-arid climates characterized by predominantly dry periods, the scarce vegetation does not have any efficient geochemical and protective role. Relicts of original bauxites or indurated degraded bauxites are often exposed by mechanical erosion. Gibbsite is not stable in the uppermost part of the profile due to low water activity, Al migrates downwards and residual Fe accumulates. These processes result in the formation of a Fe-rich crust at the top of the bauxitic profiles (Boulangé 1984). Exposure of degraded bauxites which are strictly Al and contain boehmite and gibbsite, improves the stability of boehmite (Ambrosi 1990).

According to the degree of induration or degradation of the original and degraded bauxites, mechanical erosion carries away the residual materials (nodules, pisoliths, clays, etc.) from bauxites and alterites and the parent rocks can be exposed.

4 Mass Balances of the Different Evolutions

The chemical trends which are observed in the sequence: parent rock, alterite, original bauxite, degraded bauxite and argillo-ferruginous lithology, are illustrated by two examples: (1) bauxites developed on granite in the Ivory Coast (Lakota); and (2) bauxites developed on clay-sandy sediments in Brazilian Amazonia (Porto Trombetas).

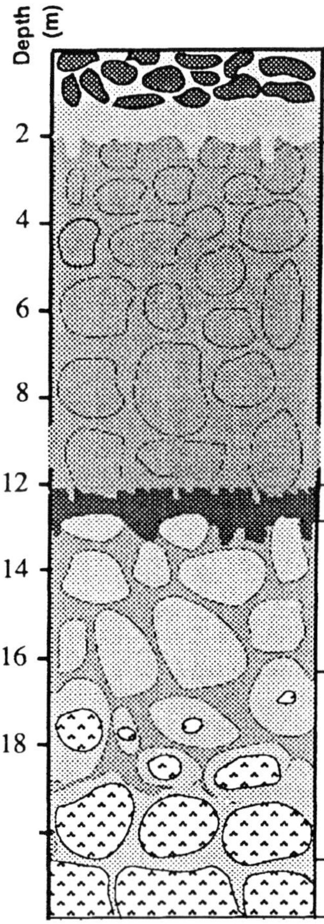

	SiO2 l	SiO2 c	Al2O3	Fe2O3	TiO2
fragmentary bauxite	0.60	3.00	55.70	9.61	0.90
upper massive bauxite	0.10	2.40	61.10	4.60	0.56
lower massive bauxite	1.50	3.10	57.30	7.93	0.89
ferruginous layer	8.80	5.20	43.40	17.05	1.07
upper isalterite	9.00	14.30	43.20	9.70	0.92
lower isalterite	20.40	16.50	36.50	7.00	0.75
	21.00	25.10	35.60	2.40	0.25
	25.50	30.00	30.60	1.70	0.15
	33.00	37.40	16.90	1.00	0.10
Granite	29.30	39.80	16.60	2.78	0.31

Fig. 4. Sketch of the mineralogical evolution during the original bauxite development from a granite weathering (Mount Tato, Ivory Coast)

4.1 The Lakota Bauxites, Ivory Coast

The Lakota deposit (700 000 tons) is located at the top of a 380-m-high hill (Mont Tato). The profile displays an isalterite about 5 m thick which is covered by a massive original bauxite about 15 m thick (Fig. 4). In the Lakota area, pisolitic bauxite horizons or argillo-ferruginous crusts occur at the top of many other hills whose summits are lower than Mont Tato.

In Mount Tato, some of the parent minerals have been transformed into gibbsite in an indirect way (albite and orthoclase feldspars, biotite), and others have been transformed directly into gibbsite (microcline, muscovite; Fig. 4). This alteration is generally pseudomorphic, and the texture and structure of the parent rock are preserved. A network of septa enclosing residual voids

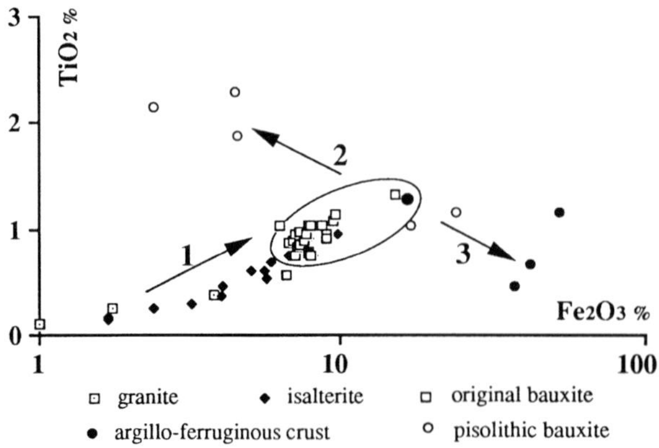

Fig. 5. Fe_2O_3-TiO_2 relationship in the original and degraded bauxites of the Lakota area (Ivory Coast). *1* Original bauxite development; *2* pisolitic degraded bauxite development; , ferruginous crust development

forms. These voids are rapidly filled by gibbsite and haematite cutans (ferri-gibbsitans) which strengthen the weathered rock. As the structure is preserved, isovolumetric calculations can be used to evaluate the gains and losses of Fe and Al in each horizon. These transfers are very large and can represent up to 50% of the original bauxites. The calculations show that the formation of an original bauxite of about 15 m requires 45 m of granite on the top of the hill where lateral accumulations are not expected (Boulangé 1984).

The degraded pisolitic bauxites have been formed at the expense of the original bauxites. During this transformation, some of the elements which are associated with unweathered minerals do not move. During the early stage of weathering, Ti of primary minerals such as biotite and ilmenite, is retained as very resistant anatase. The Fe_2O_3/TiO_2 ratios show a relative accumulation of Fe and Ti, and show that their contents increase in similar proportions (Fig. 5). In the original bauxite, Fe_2O_3 contents range from 7 to 10% and the TiO_2 contents remain close to 1%. The Fe and Al enrichment (ferrigibbsitans) therefore takes place without any extra supply of Ti. On the contrary, the pisolitic lithologies show high TiO_2 concentrations which are related to the residual behaviour of anatase during the Fe leaching occurring during pisolith formation. The formation of the Al–Fe crusts is, on the contrary, associated with an absolute accumulation of Fe without Ti (Fig. 5). The multiplication of the TiO_2 contents by two or three in the pisolitic bauxites indicates a decrease of volume ranging from 50 to 70% during the transformation of the original bauxites.

4.2 The Porto Trombetas Bauxites, Brazil

Located in Amazonia, the Porto Trombetas deposit (1000 million tons) developed on clayey and Cretaceous clay-sandy sediments. The profile displays from the top to the bottom (Fig. 6): a clay-sandy horizon with kaolinite (8–10 m), a nodular bauxite (1–3 m), an horizon with ferruginous nodules (1 m), a bauxite (6 m) and a kaolinite horizon (1 m) on the sediments. The studies of this deposit (Boulangé and Carvalho 1989), and of the very similar Juruti deposit (Lucas 1989), show that the succession of horizons cannot result from a sedimentary

LAYERS		SiO_2 %	Al_2O_3 %	Fe_2O_3 %	TiO_2 %	Zr ppm
CLAY SANDY KAOLINITIC HORIZON		36.62	36.61	8.25	2.77	1518
NODULAR BAUXITE	matrix	26.20	43.35	8.30	2.25	1400
	Al nod.	4.30	61.20	2.20	1.10	640
FERRUGINOUS NODULES	matrix	22.80	29.00	31.10	2.20	1240
	Fe nod.	12.85	22.70	47.95	0.89	710
BAUXITE		1.37	53.80	15.25	0.80	430
		3.89	57.90	6.60	1.36	553
KAOLINITIC HORIZON		28.63	40.37	10.23	2.57	1033
SEDIMENT		39.87	36.65	7.92	0.70	215

Fig. 6. Sketch of a bauxite profile on sediments (Porto Trombetas, Amazonia, Brazil): layers and chemistry (means for each layer and facies)

process, but from a long geochemical process which took place under equatorial or humid tropical climates. The sandy-clayey sediments are intensively weathered and are transformed into an original bauxite. The vegetation plays an important role in the dynamic of this bauxite, mainly in the Si behaviour. The root system of the plants takes a part of the residual Si which is always present (4–6%) in the original bauxites (residual quartz and kaolinite which has not been transformed into gibbsite). Si is returned to the soil through the litter and contributes to the resilication of gibbsite. This original bauxite is, therefore, successively affected by Fe leaching and resilication processes: (1) Fe accumulates in an underlaying sedimentary layer rich in quartz which corresponds to the horizon with Fe nodules observed today; and (2) resilication of the thick top clay-sandy horizon takes place. Large quantities of Al are concentrated at the base of this clay-sandy horizon in the nodular bauxite.

Crust, ferruginous nodules and upper kaolinite represent facies derived from degradation of an original bauxite, and they are characterized by variations of the Fe and Ti (Fig. 7a). The TiO_2 and Zr contents which are high in the

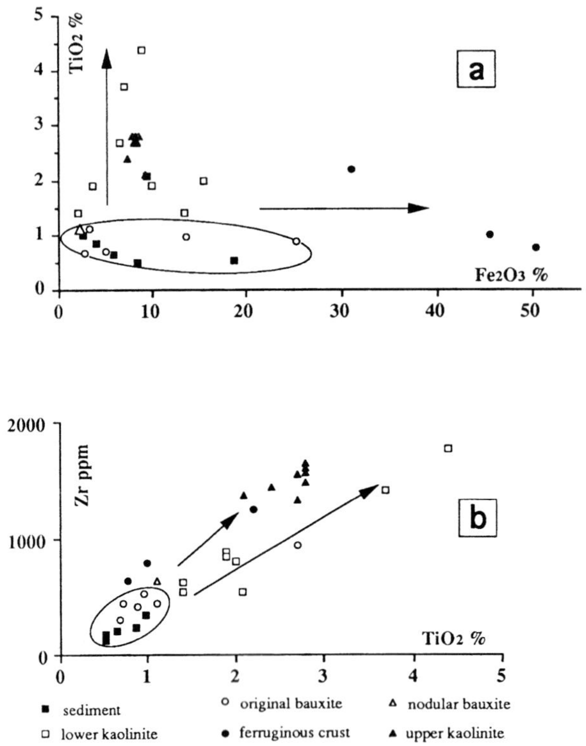

Fig. 7a Fe_2O_3-TiO_2; **b** TiO_2-Zr relationships in the original and degraded bauxites of the Porto Trombetas mine (Brazil) derived from weathering of sediments

upper kaolinite and low in the Fe nodules, corroborate this filiation (Fig. 7b). In the lower kaolinite, disappearance of the structure and decrease of the rock volume, which are related to the resilication process, lead to a relative accumulation of residual minerals (anatase and zircon) shown by higher TiO_2 and Zr contents in this layer.

Iso-elemental calculations (Zr) show that the volume of the clay-sandy horizon is reduced by a factor of nine when compared to the sediments. Therefore, formation of a top clay-sandy horizon about 10 m thick requires 90 m of sediments (Boulangé and Carvalho 1989). In such a context, Si loss is very high and represents 90% of the total mass of the sediments. The quantity of residual Si in the original bauxites is high enough to form the top layer. It is, therefore, not necessary to consider transport of sedimentary and aeolian material as has been done by Kotschoubey and Truckenbrodt (1981) in Paragominas, and Lucas (1989) at Juruti.

5 Alteration Rate and Age of Bauxites

Estimates of the rate of progression of the weathering fronts vary, according to the calculation parameters used. Fritz and Tardy (1973) obtained a bauxitization rate of about 3 mm/1000 years, using a thermodynamic model. Leneuf

(1959) gave a value of 5–50 mm/1000 years for the weathering of granite into kaolinite in the Lakota area of the Ivory Coast. For Gac (1980), the average rate of rock weathering is about 13.5 mm/1000 years in the Chari basin in Tchad. Isovolume calculations gave a rate for formation of the original bauxites in Mount Tato of about 14 mm/1000 years (Boulangé 1984). The formation of 15 m of original bauxite requires 3 to 5 million years. Data on the rate of kaolinite weathering are scarce: for instance a value of 1 mm/1000 years was obtained using experimental methods (Pédro 1964) and 3.5 mm/1000 years using a thermodynamic model (Fritz and Tardy 1973). The formation of the Porto Trombetas profile discussed here required about 100 m of original sediments, and lasted 30 to 100 Ma according to these values. Lucas (1989) calculated that 30 Ma were necessary to weather 75 m of sediments.

The two examples presented above are probably limit cases. The Porto Trombetas deposit shows that bauxitization is an old process which lasted during a long period under humid tropical climatic conditions. The Mount Tato deposit shows that original bauxites can form even today under a tropical climate with alternating dry and rainy seasons, and also that this bauxitization strongly depends on the geomorphological conditions. However, presence of degraded lithologies in the Lakota area shows that bauxitization which has generated these rocks, is also very old in Africa. Various authors agree that laterite bauxite deposits have been formed during the Cretaceous. During this time span, the climatic conditions were varyied (Tardy et al. 1988a; Kobilsek 1990), but the bauxite profiles continued to evolve either by thickness increase or progressive degradation.

6 Evolution of the Bauxite Landscapes

In Amazonia, original bauxites and degraded rock lithologies exhibit a vertical succession under a humid tropical climate. The thickness of the formations indicates that conditions did not change over a long period. A continuous lowering of the weathering sequence results in the geochemical planation of the bauxite landscapes. This planation is oriented and increased by the horizontal structures of the sedimentary parent rocks.

In Africa, the degraded rocks (pisolitic bauxite and argillo-ferruginous crust) generally show that bauxitization is also an old process under a tropical climate with alternating dry and humid seasons. At the landscape scale, the spatial organization of the lithologies is related to altitude; from top to bottom of the slope, original bauxites, degraded, nodular or pisolitic bauxites and argillo-ferruginous crust can be successively observed. This distribution results from the evolution of the relief of the original bauxites. As the topographic surfaces are levelled by continuous lowering of the weathering sequence of the original bauxites, degradation domains are observed on the slopes. The loss of material, mainly of Fe and Al, leads to an important volume decrease of the original rocks. The progressive reduction of volume related to this evolution first creates a surface covered by a pisolitic bauxite and, then, after a complete dissolution of these pisoliths, creates a second surface covered by an argillo-ferruginous crust. The degradation processes which develop from the bottom to the top of the hill, contribute to the levelling of the landscape (Boulangé and Millot 1988). In Africa, remnants of pisolitic and argillo-ferruginous crusts indicate a long geochemical and climatic process.

7 Conclusion

Equatorial and tropical climatic conditions lead to the formation of original bauxites with economical value, to the degradation both of these bauxites and of the degraded lithologies (pisolitic bauxite, ferruginous crust, kaolinite) which result from simple changes in the geochemical systems related to lowering of the weathering profiles.

The bauxite landforms are progressively erased by chemical erosion when the hydrographic system does not change. On the profile or hillside scale, the vertical or lateral distribution of the original and degraded bauxites corresponds to geochemical processes which are mainly controlled by climatic factors.

An original bauxite profile can be preserved and become thicker only if the conditions of vertical drainage remain optimal, i.e. if a lowering of the hydrographic system occurs. On the region or continent (craton) scale, the development and conservation of a bauxite deposit requires a tectonic control.

References

Ambrosi JP (1990) Modélisation thermodynamique de l'altération latéritique dans le système Fe_2O_3-Al_2O_3-SiO_2-H_2O. Thèse Univ Poitiers, 377 pp

Ambrosi JP, Nahon D (1986) Petrological and geochemical differentiation of lateritic iron crust profiles. Chem Geol 57:371–393

Bardossy G, Aleva GJJ (1990) Lateritic bauxites. Development in economic geology, vol 27. Elsevier, Amsterdam, 624 pp

Berthier P (1821) Analyse de l'alumine hydratée. Ann Mines 6:531–534

Bocquier G, Muller JP, Boulangé B (1984) Les latérites. Connaissances et perspectives actuelles sur les mécanismes de leur différenciation. Livre Jubilaire du Cinquentenaire, Assoc Franç Etude du Sol, pp 123–138

Boulangé B (1984) Les formations bauxitiques latéritiques de Côte d'Ivoire. ORSTOM, Paris. Trav Doc 175:341pp

Boulangé B, Bocquier G (1983) Le rôle du fer dans la formation des pisolites alumineux au sein des cuirasses bauxitiques latéritiques. In: Coll Int CNRS Pétrologie des altérations et des sols. Sci Géol Bull (Strasb) 72:29–36 pp

Boulangé B, Carvalho A (1989) The genesis and evolution of the Porto Trombetas bauxite deposits in the Amazon Basin, Para, Brazil. Trav ICSOBA 19 (22):71–79

Boulangé B, Millot G (1988) La distribution des bauxites sur le craton ouest-africain. Sci Géol Bull (Strasb) 41:113–123

Boulangé B, Paquet H, Bocquier G (1975) Le rôle de l'argile dans la migration et l'accumulation de l'alumine de certaines bauxites tropicales. CR Acad Sci Paris 280 (D):2183–2186

Chatelin Y (1974) Les sols ferrallitiques. Tome III: l'altération. ORSTOM, Paris. Init Docum Techn 24:144 pp

Chesworth W (1972) The stability of gibbsite and bœhmite at the surface of the earth. Clays Clay Min 20:369–374

Colin F, Brimhall GH, Nahon D, Lewis CJ, Baronnet A, Danty K (1992) Equatorial rainforest lateritic soils: a geomembrane filter. Geology 20:523–526

Colin F, Vieillard P, Ambrosi JP (1993) Quantitative approach to physical and chemical gold mobility in equatorial rainforest lateritic environment. Earth Planet Sci Lett 114:269–285

Delvigne J (1965) Pédogenèse en zone tropicale. La formation des minéraux secondaires en milieu ferrallitique. Dunod, Paris. Mém ORSTOM 13:177 pp

D'Hoore J (1954) L'accumulation des sesquioxydes libres dans les sols tropicaux. Public INEAC Sér Sci 62:132 pp

Fritz B, Tardy Y (1973) Etude thermodynamique du système gibbsite, quartz, kaolinite, gaz carbonique. Application à la genèse des podzols et des bauxites. Sci Géol Bull (Strasb) 26 (4):339–367

Gac Y (1980) Géochimie du bassin du Lac Tchad. Bilan de l'altération, de l'érosion et de la sédimentation. ORSTOM, Paris. Trav Doc 123:251 pp

Garrels RM, Christ CL (1965) Solutions, minerals and equilibria. Harper and Row, New York, 254 pp

Harrison JB (1934) The katamorphism of igneous rocks under humid tropical conditions. Imp Bur Soil Sci, Harpenden, 79 pp

Harrasowitz H (1926) Laterit. Fortschr Geol Paläontol 4:253–566

Kobilsek B (1990) Géochimie et pétrographie des bauxites latéritiques d'Amazonie brésilienne. Comparaison avec l'Afrique, l'Inde et l'Australie. Thèse, Univ Strasbourg, 201 pp

Kotschoubey B, Truckenbrodt W (1981) Evolução poligenetica das bauxitas do distrito de Paragominas, Acailandia (estados de Para e Maranhao). Rev Bras Geocienc 11:193–202

Lacroix A (1913) Les latérites de la Guinée et les produits d'altération qui leurs sont associés. Nouv Arch Mus 5:255–356

Leneuf N (1959) L'altération des granites calco-alcalins et des grano-diorites en Côte d'Ivoire forestière et les sols qui en sont dérivés. ORSTOM, Paris, 210 pp

Lucas Y (1989) Systèmes pédologiques en Amazonie brésilienne. Equilibres, déséquilibres et transformations. Thèse, Univ Poitiers, 157 pp

Lucas Y, Chauvel A (1992) Soil formation in tropically weathered terranes. In: Butt CRM, Zeegers H (eds) Handbook of exploration geochemistry, vol 4. Regolith exploration geochemistry in tropical and subtropical terranes. Elsevier, Amsterdam, pp 57–77

Lucas Y, Luizão FJ, Chauvel A, Rouiller J, Nahon D (1993) The relation between biological activity of the rain forest and mineral composition of soils. Science 260:521–523

Millot G (1964) Géologie des argiles. Masson, Paris, 499 pp

Millot G, Bonifas M (1955) Transformations isovolumétriques dans les phénomènes de latéritisation et de bauxitisation. Bull Serv Carte Géol Als Lorr (Strasb) 8:3–10

Nahon D (1976) Cuirasses ferrugineuses et encroûtements calcaires au Sénégal Oriental et en Mauritanie. Systèmes évolutifs: géochimie, structures, relais et coexistence. Sci Géol Mém (Strasb) 44:232

Nahon D (1986a) Evolution of iron crusts in tropical landscapes. In: Coleman SM, Dethier DP (eds) Rates of chemical weathering of rocks and minerals. Academic Press, New York, pp 169–191

Nahon D (1986b) Microgeochemical environments in lateritic weathering. International meeting on geochemistry of the earth surface and processes of mineral formation, Granada, Spain, pp 141–156

Nahon D (1991) Introduction to the petrology of soils and chemical weathering. Wiley, London, 313 pp

Nahon D, Millot G (1977) Géochimie de la surface et forme du relief V Enfoncement géochimique des cuirasses ferrugineuses par épigénie du manteau d'altération des roches mères gréseuses. Influence sur le paysage. Sci Géol Bull (Strasb) 30 (4):275–282

Parron C, Guendon JL, Boulangé B, Bocquier G (1983) Evolutions minérales et microstructurales dans les bauxites du Midi de la France. Mécanismes de la bauxitisation sur substrat carbonaté. Trav Lab Sci Terre Fac St Jérôme, Marseille, X:54, 50 pp

Patterson SH (1967) Bauxite reserves and potential aluminium resources of the world. Geol Surv (Wash) 1228:176 pp

Pédro G (1964) La genèse des hydroxydes d'aluminium par altération expérimentale des roches cristallines et le problème des latérites. Int Geol Congr New Dehli 22 (14):1–13

Pédro G, Chauvel A (1978) Genèse de sols beiges (ferrugineux tropicaux lessivés) par transformations de sols rouges ferrallitiques de Casamance (Sénégal) Modalités de leur propagation. Cah ORSTOM Sér Pédol 3:231–249

Rossini FD, Wagman DD, Evans WH, Levine S, Jaffe I (1952) Selected values of chemical thermodynamic properties National Bureau of Standards. Circ 500:1268

Sarazin G (1979) Géochimie de l'aluminium au cours de l'altération des granites et des basaltes sous climat tempéré. Thèse, Univ Paris, 169 pp

Tardy Y (1969) Géochimie des altérations. Etude des arènes et des eaux de quelques massifs cristallins d'Europe et d'Afrique. Mém Serv Carte Géol Als Lorr (Strasb) 31:199 pp

Tardy Y, Nahon D (1985) Stability of Al-gœthite, Al-hématite, Fe^{3+}-kaolinite in bauxites, ferricretes and laterites. An approach of the mechanism of the concretion formation. Am J Sci 285:865–903

Tardy Y, Melfi AJ, Valeton I (1988a) Climats et paléoclimats tropicaux. Rôle des facteurs climatiques et thermodynamiques (température et activité de l'eau) sur la composition minéralogique des bauxites et des cuirasses ferrugineuses, au Brésil et en Afrique. CR Acad Sci Paris 306 (II):289–295

Tardy Y, Bardossy G, Nahon D (1988b) Fluctuations de l'activité de l'eau et successions de minéraux hydratés et déshydratés au sein des profils latéritiques ferrugineux et bauxitiques. CR Acad Sci Paris 307 (II):753–759

Trescases JJ (1975) L'évolution géochimique supergène des roches ultrabasiques en zone tropicale. Formation des gisements nickélifères de Nouvelle Calédonie. ORSTOM, Paris, Mém 78:259

Trolard F, Tardy Y (1987) The stabilities of gibbsite, bœhmite, aluminous gœthites and aluminous hematite in bauxites, ferricretes and laterites as function of water activity, temperature and particle size. Geochim Cosmochim Acta 51:945–957

Valeton I (1972) Bauxites. Developments in soil sciences, vol 1. Elsevier, Amsterdam, 226 pp

4 Geochemical Processes in Tropical Landscapes: Role of the Soil Covers

RENÉ BOULET, YVES LUCAS, EMMANUEL FRITSCH AND HÉLÈNE PAQUET

1 Introduction

The relationship between morphogenesis and pedogenesis has concerned geomorphologists and pedologists for a long time. As suggested for example by Tricart and Michel (1965), soil science relates to geomorphology as geomorphology does to structural geology. Geomorphological evolution constitutes a framework to soil formation and evolution. Soils rarely develop directly in situ at the expense of a parent rock. Much more frequent is the formation of soils within regoliths, alterites, slope debris, alluviums, colluviums or aeolian accumulations, etc.

In the frame of a wide scientific programme entitled «Géochimie de la surface et formes du relief», Georges Millot stimulated investigations of a multidisciplinary group composed of geologists, soil scientists, geographers and geochemists working in tropical Africa on humid forest and on desert areas (Millot et al. 1976; Bocquier et al. 1977; Boulet et al. 1977; Nahon and Millot 1977; Millot 1979; Millot et al. 1979; Millot 1980a,b). These investigations produced abundant information about weathering processes and the geochemical cycle of elements, and also about the role of weathering and pedogenesis in the development of landforms. For Millot (1979), the main point was that weathering and pedogenesis do not only generate dissolution or disaggregation processes before surficial ablation takes place, they also operate in another way, namely as hidden «subterranean» or «hypodermic» laboratories which transform the nature and structure of the weathering cover before superficial erosional factors alter landforms.

Since then, these investigations have been complemented by numerous studies in humid tropical areas. The combination of all the results shows that well-known weathering covers include two great groups of soil profiles whose roles in the evolution of landforms are distinct. The first group represents associations between soil covers and landforms which preserve a permanent pedological differentiation, during the process of deepening, i.e. they preserve the same vertical sequences of horizons. However, these soil cover–landform associations evolve continuously, which means that each horizon develops in its lower part at the expense of the underlying horizon and is transformed in its upper part by the overlying one. For that reason, these soil cover–landform associations are called soil covers in dynamic equilibrium (Fig. 1A): «dynamic» because they evolve con-

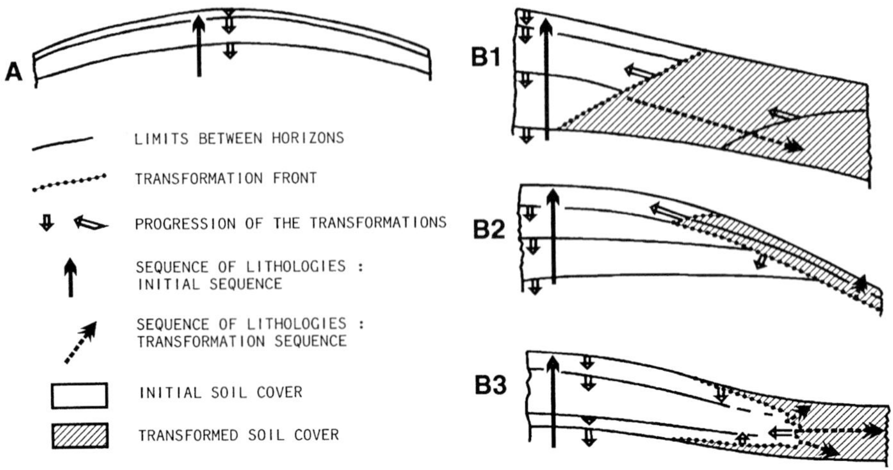

Fig. 1. Equilibrium and disequilibrium of the soil cover. **A** Soil cover in equilibrium: each horizon progresses vertically at the expense of the underlying horizon. **B1–3** Transformation system: the initial soil cover is in disequilibrium and is replaced by a new one. **B1** Transformation starts from the base of the initial soil cover. System from Burkina Faso, see text. **B2** Transformation starts from the top of the initial soil cover. System from French Guyana, see text. **B3** Transformation starts from the base and the top of the initial soil cover. System from the Manaus region, see text

tinuously and «in equilibrium» because the sequence of horizons always remains unchanged. This sequence of horizons does not display any unconformity, being parallel to the soil surface. Lateral variations are progressive. They can be described, with a good approximation, by a vertical sequence of horizons (e.g. soil profile) or by a vertical sequence of units of organization derived from each other.

In the second group, morpho-pedogenetic evolution results in progressive replacement of an original soil cover–landform association by another often very different one. For that reason, they are called transformation systems and, for the most part, soil covers display two (or more) sequences of horizons which are discordant to each other (Fig. 1B). The first sequence is vertical and is characterized, like soil covers in dynamic equilibrium, by horizons roughly parallel to the soil surface. The second one is crosscutting and develops laterally either from the top, from the base, or from both, by truncating the first sequence which is no longer developing. The characterization of the system thus necessitates a continuous section from the top of the interfluve to the drainage line.

2 Soil Covers in Dynamic Equilibrium

Soil covers in dynamic equilibrium develop in pedoclimatic conditions which are stable enough to keep the sequence of transformations from parent rock to soil surface constant, and which result in an unchanged sequence of horizons, although the latter are in constant transformation. The pedoclimatic conditions are determined by following factors: climate, vegetation, edafauna, soil porosity, local water level, the four last being related. Water level depends on the evolution of landforms, and on the variation in tectonics and eustatic levels. Climatic features and both tectonic and geomorphologic evolution are related to pedological processes in the long term, but variable from a geographical point of view. Soil covers in dynamic equilibrium are supposed to be found in two circumstances: in areas where climatic and tectonic changes are not strong enough to create disequilibrium and in profiles where the materials are young enough not to have been affected by climatic or tectonic variation responsible for modifications of the equilibrium conditions. Examples relating from the driest to the most humid present-day climates will be presented.

2.1 "Eutrophic" Brown Soils and Vertisols on Migmatites of the Sudanian Tropical Zone of West Africa

In south-central Burkina Faso, ancient kaolinite covers have disappeared or are only preserved as scarce relicts in the Garango I toposequence of Boulet (1974). In these areas, under a mean annual rainfall of 900 mm distributed between May and October, calc-alkaline biotite- and amphibole-rich migmatites alter into clay min-

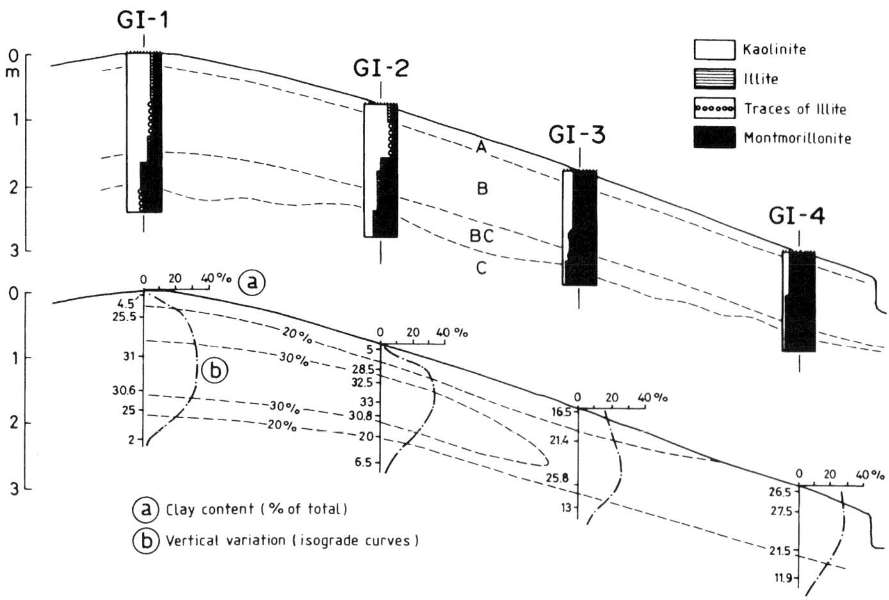

Fig. 2a–b. Toposequence of Garango I (Burkina Faso, Africa). **a** Horizon organization and clay mineralogy. **b** Clay isogrades and vertical variations in the four major profiles. (Boulet 1974; reproduced with permission of ORSTOM)

erals, among which montmorillonite is predominant together with traces of kaolinite and illite (Fig. 2a). Landforms consist of short interfluves with convex upper parts and gentle slopes of 1.5–2%. The thin soil cover (about 2 m) displays vertic features increasing downslope with associated vertic "eutrophe" soils upslope and vertisols with calcareous nodules downslope. The horizons are parallel to the soil surface and show progressive lateral variations from up- to downslope. These transformations are mainly colour changes from bright brown to olive brown, calcareous nodules being observed only in vertisols with smectite contents ranging from 50% in the "eutrophe" brown soils to about 100% in the vertisols.

Under the microscope, the formation of a clay plasma can be observed in the lower part of the soil which develops by weathering of primary minerals in a horizon whose petrographic fabric is preserved. In the horizon B, pedoturbation homogenizes the material by alternating wetting and desiccation; the material becoming brown with a prismatic fabric. Upslope, the clay content of the surficial part of the soil decreases to less than 5%, which results in a sandy, poorly cohesive horizon vulnerable to mechanical erosion (Fig. 2b). In the Garango I toposequence, field tests of humidity have shown that the soil cover remains moist down to the base. The soil cover deepens at the expense of weathering material and the upper part is eroded, principally upslope, which induces a decrease in the relief.

Such smectite soil covers develop on granitic or magmatitic rocks in the edapho-climatic conditions of the Sudanian zone of West Africa, which is characterized by high temperature (mean annual temperature=28 °C), mean annual rainfall between 600 and 1000 mm, a rainy season of 5 months and natural vegetation of savanna type with widely spaced trees. Such soil covers display a vertical dynamic; they are in equilibrium relative to present-day pedoclimatic conditions, as corroborated by the study of the hydric regime and by the permanence of the sequences of elementary structures. They are quite recent and, in any case, they postdate the thick kaolinite mantles which still cover a large part of the same area. It will be shown that these soil covers record a lack of equilibrium and are subject to transformation to the North, where the mean annual rainfall decreases.

2.2 Soil Covers Consisting of Tropical Ferruginous Soils

Soil covers composed of tropical ferruginous soils occur in northwestern Central Africa, in a granitic or granito-gneissic area where the ancient kaolinite soil cover has been eroded (Lucas 1980). They are located near the main drainage lines, e.g. on the most evolved hillsides of that rejuvenated zone. The present-day climate is characterized by a mean annual rainfall of 1200–1300 mm and by a dry season of 5–6 months; landforms display slopes of less than 4% and scattered domes of outcropping rock. Upslope near the domes, lithosols and little evolved soils can be observed; most of the slope is occupied by more or less indurated tropical ferruginous soils. On the lower part of the slopes, small areas of hydromorphic soils are found. The surficial horizons of the tropical ferruginous soils are beige or yellowish-brown and sandy in the upper 20–40 cm; the clay content progressively increases with depth (reaching about 30% of the total soil material). Loose horizons which develop to a depth of 80–120 cm overlie a light beige and red mottled «carapace» in which the water table is observed during rainy periods. The volume of preserved petrographic structures, already great in the «carapace», increases rapidly with depth (110–120 cm) where a kaolinite weathering horizon occurs. The «carapace» is likely to represent the first stages of formation of iron crusts by Fe segregation under temporarily hydromorphic conditions. The formation of these iron crusts will be described in the next section.

2.3 Soil Covers with Iron Crusts

Iron crusts are almost omnipresent in Africa, south of the Sahara. However, in the Sahelian zone (Leprun 1979), as well as in the humid equatorial zone (Bitom 1988; Bilong et al. 1992), iron crusts alter more readily than they form. Recent studies coordinated by Tardy (1990, 1994) and carried out in the frame of the PIRAT Programme (Interdisciplinary Programme of Investigation on Periatlantic Intertropical Biogeodynamics) have localized the areas where Fe-crust formation presents a positive balance. The laterite Fe-crust-bearing profiles develop under a tropical climate with contrasted seasons, a mean annual rainfall higher than 1100 mm (1600 mm in some cases), a mean annual temperature of 28 °C, and a dry season of 6 months/year (Tardy 1993). Presently, the most important area of iron crust formation is Central Africa. Other areas were identified in northwestern Ivory Coast, with a mean annual rainfall of 1600 mm (Eschenbrenner 1987) and in southern Mali, with a mean annual rainfall of 1100 mm (Freyssinet 1990).

The theoretical Fe-crust-bearing laterite profile displays the following horizons from the parent rock to the top (Tardy 1993).

1. A coarse saprolite with disjoined minerals in quartz-rich rocks; quartz excepted, the primary minerals are more or less weathered, but texture remains sandy («arène»). In rocks devoid of quartz, rock fragments in the process of weathering are observed in altering material with fine textures.
2. A «lithomarge» or fine saprolite in which primary minerals, but not quartz, are altered into kaolinite, goethite or haematite.
3. A mottled horizon in which the formation and accumulation of Fe nodules begins and in which different domains are juxtaposed: discoloured domains eluviated in Fe and clay, microporous domains where Fe accumulates in the form of haematite, and voids more or less filled by illuviated clay enriched in Fe as goethite. Haematite accumulates in microporous kaolinite plasmas and is associated with kaolinite up to the top of the iron crust.
4. A poorly indurated «carapace» more Fe-rich than the mottled clay, but preserving its structure. The «carapace» represents a transitional stage between mottled clay and iron crust.
5. An iron crust which is highly indurated and Fe-rich (haematite) and which may display various lithologies (massive, nodular, vermiform).
6. A horizon of dismantlement and degradation of the iron crust, principally through goethitic cortification.

Even if developed in pedoclimatic conditions favourable to their formation, as for instance in Central Africa, the iron crust-bearing soil covers display zones of weakness along which they degrade and dismantle. These zones correspond to thalwegs, as well as to the borders and central parts of plateaus (Beauvais 1991; Tardy 1993). In addition, the soil cover under consideration is no longer in equilibrium. In the other zones, iron crust with its suite of horizons develops at the expense of fine saprolite («lithomarge») and is degraded in its upper part; iron crust is thus lowered in the landscape and operates like a soil cover in dynamic equilibrium. According to Tardy and many other authors (references in Nahon 1991), the rate of lowering is about 1 m per 100000 years. Iron crust-bearing soil covers are obviously more resistant to erosion than loose ones, as corroborated by the existence of very numerous residual hills in tropical areas. These residual hills are no longer in equilibrium, but they are far from being indestructible. In fact, a decrease or an increase in rainfall results in their disappearance either by geochemical dissolution, or by transformation to loose material, as described by Beauvais and Tardy (1991).

2.4 Soil Covers Consisting of Red Ferrallitic Soils

Fauck (1972) studied red ferrallitic soils developed from sandstones, which are frequent in West Africa (Casamance, Burkina Faso, Dahomey, Nigeria, etc.). These soils develop under mean annual rainfall ranging from 1200–1800 mm and are distributed in one rainy season in Casamance and Burkina Faso, and in two rainy seasons in Dahomey and Togo, the mean annual temperature being 28 °C. Such soil profiles are homogeneous and cover landforms composed of plateaus with low slopes (several percent). Chauvel (1977; Fig. 10, section C) reported the occurrence of such plateaus with thick and homogeneous red ferrallitic soil covers for a mean annual rainfall of 1200 mm in Casamance. Iron crusts which may even be observed above the parent rocks, are abundant in Casamance, rare in Dahomey and lacking in Burkina Faso. However, they do not seem to be connected to the pedogenesis of red ferrallitic soils and they are probably more ancient.

The red ferrallitic soils usually display the following horizons from bottom to top:

1. A red, ochrous and white variegated horizon which is transitional to the parent rock and often contains residual fragments of more or less weathered sandstone.
2. A rubified, homogeneous, clay-sandy or clayey horizon, whose thickness is about 2 m, but which can reach 5 m, and which is rich in micro-aggregates. These are resistant to dispersion, and their fabric provides a large porosity. They were also called pseudo-sands or micro-nodules (Chauvel 1977).
3. A 20–60 cm very progressive transition to the upper lying horizon.
4. A humic horizon, which is 10–20 cm thick, more or less dark grey, sandy or clay-sandy, and porous.

Fauck emphasized that the texture of rubified horizons is little dependent on that of the parent rocks. In fact, he reported examples of such horizons in Burkina

Faso, for which the clay content is about 44%, whereas it is 4% in the corresponding parental sandstones. The clay enrichment is principally explained by relative accumulation caused by the strong dissolution of quartz, by formation of kaolinite from Si and Al provided by hydrolysis of kaolinite in the upper horizon, and by clay translocations, although this last point has scarcely been considered.

Dense forest is the type of vegetation which is considered to favour the formation of such soil covers and to keep their equilibrium. Under other types of vegetation such as tree-savanna, bush-savanna or cultivation, these soil covers are surficially degraded. Chauvel pointed out that in Casamance, under cultivation, the incipient stages of transformation are related to palaeoclimatic variations. These transformations take place at a mean annual rainfall lower than 1000 mm and will be described in a further section.

Red ferrallitic soils also develop at the expense of basic rocks, mainly basalts. They are especially widespread in Brazil under humid subtropical and tropical climates, where mean annual rainfall and temperature range between 2000 and 1500 mm and between 25 and 16 °C, respectively. Compared to soils developed on sandstones, they display a very clayey texture, a frequent occurrence of gibbsite, higher Fe contents and stronger micro-aggregation. These soils are called «ground-coffee soils» (Melfi 1968; Carvalho 1970; Pédro et al. 1976). They are developed above a thick gibbsite–kaolinite weathering zone of the basalt. Between that basaltic weathering zone and the soil, an iron crust in process of dismantlement is frequently observed, which then constitutes the parent rock of the red soil. The formation of red latosols at the expense of iron crusts was also reported by Nahon et al. (1989) in South Amazonia, at the forest–savanna boundary, for profiles developed on gneiss. The frequent occurrence of lines or levels of Fe nodules at the bottom of latosols (Volkoff 1984/1985) leads to the supposition that iron crust-bearing soil covers were more extensive in Brazil in the past. Their present-day scarce occurrence was ascribed by Tardy et al. (1988) to temperatures lower than those observed in iron crust-bearing soil covers of Africa and to the dissimilar evolution of palaeoclimates, e.g. from arid to humid in South America and from humid to arid in West Africa and Central Africa.

2.5 Soil Covers Consisting of Yellow Ferrallitic Soils

Thick yellow ferrallitic soils covering more or less undulating plateaus have been identified in numerous humid tropical areas. The following list is not exhaustive: plateaus of middle Amazonia located on both sides of the Amazon river and developed at the expense of quartz–kaolinite sedimentary formations, undulating plateaus developed on the basement of northwestern French Guyana, plateaus developed on the basement in the forest area of South Cameroon, and plateaus developed at the expense of the calc–schist formations of the Bouenza country in Congo, etc.

The soil cover of the Amazonian plateaus developed under humid tropical climates (mean annual rainfall 2500 mm) will be taken as an example (Lucas 1989; Lucas et al. 1996). The vertical profile displays three main units (Fig. 3): (1) the lower unit is composed of a white alterite which is a quartz- and kaolinite-bearing sediment already affected by pedological transformations; (2) the middle unit is characterized from bottom to top by extensive dissolution of quartz grains, by authigenesis of kaolinite in the form of a compact clay plasma and by the progressive development of ferruginous and gibbsite nodules; and (3) the upper unit is

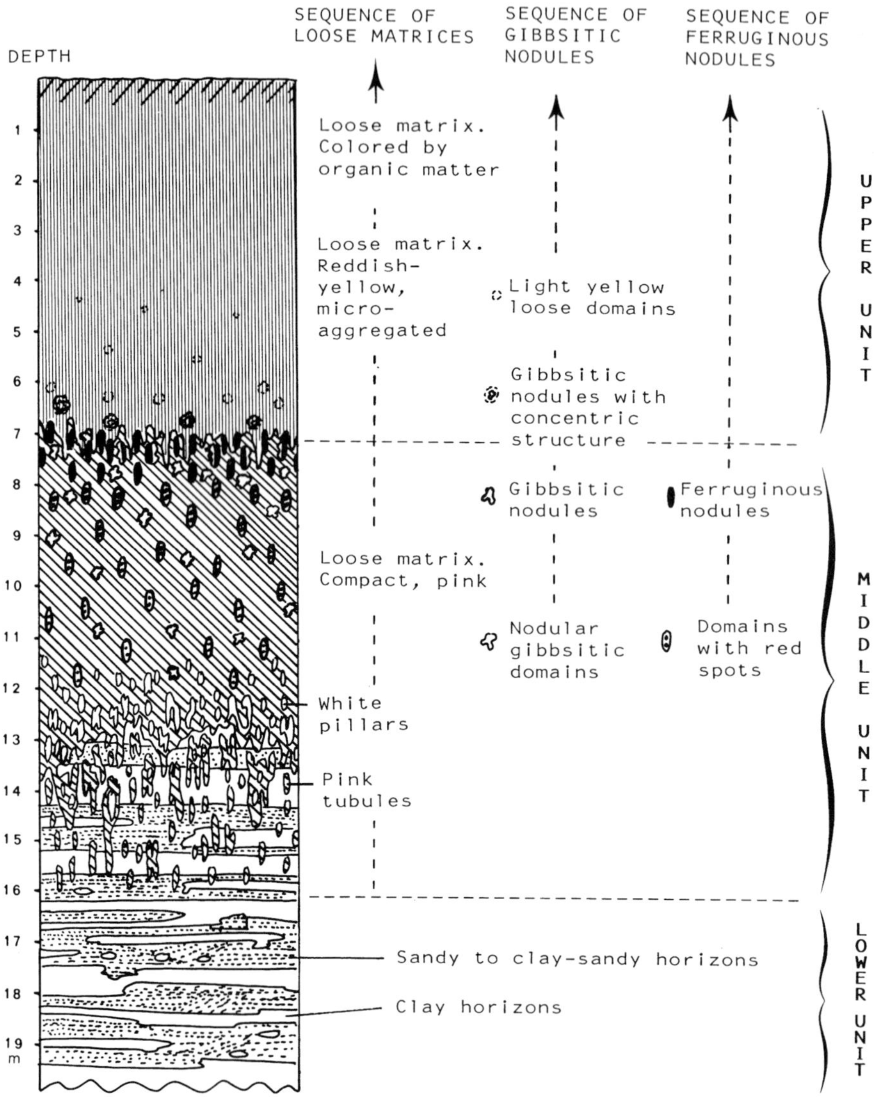

Fig. 3. Sketch of the vertical organization of soils on plateaus, North-Manaus area, Amazonia. (After Lucas 1989)

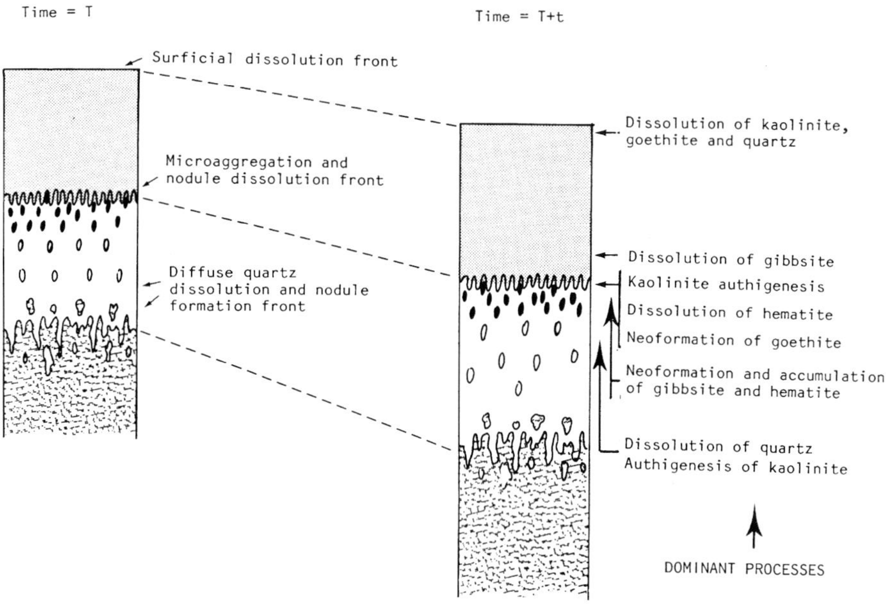

Fig. 4. Evolution of soils by geochemical lowering process within a vertical profile (soils on plateaus, North Manaus area, Amazonia; Lucas 1989)

yellow-red, clay-rich, discontinuously micro-aggregated and shows dissolution of ferruginous and gibbsite nodules at its base. The lower part of the two upper units corresponds to transformation fronts, each of them progressively moving downward within the underlying unit due to several processes (Fig. 4). Thus, the evolution of the soil cover proceeds together with a lowering of its position in the parent rock and with a lowering of the topographic surface, in accord with losses of material which is characteristic of the transformation. The importance of the rate of lowering can be estimated by calculations based on the export of dissolved elements out of the catchment area (Lucas et al. 1989). Lowering of the topographic surface corresponding to the formation of the present-day profile extends between 21 and 222 m. According to the hypothesis based on the quartz content of the sediment, the most probable value is around 40 m; under present-day conditions, the rate of lowering is about 20 cm per 100000 years. Thus, pedogenesis represents an active factor in the planation of landforms, although few important modifications, except thickening, take place inside soil covers. That is why the latter are called «soil covers in dynamic equilibrium».

In these soil covers, most of the genetic sequences of organization (in the meaning of Brewer 1964) are continuous from the base to the top of the profile (in the example described here, there is progressive differentiation of plasmas, progressive development of nodules and transformation to loose plasma, etc.). However, one can observe some evidence of organization sequences that disappear within the profile; they represent relict features of earlier processes different from present-day ones. For example, black haematite nodules or fragments of iron crust observed in the upper part of the middle unit of the soil covers of the Manaus area, cannot be related to any precursor. A detailed study suggests that they represent relicts of a now almost destroyed Fe crust. Other examples of such transformations were described in intertropical Africa (Bitom 1988; Beauvais and Tardy 1991; Bilong et al. 1992).

3 Soil Covers in Chemical Disequilibrium (Transformation Systems)

When the factors governing pedoclimates change sufficiently for soil cover to remain no longer in dynamic equilibrium, the soil cover is transformed to another soil cover which tends towards a dynamic equilibrium with the new climatic conditions. These transformations begin where pedoclimatic changes are maximum, e.g. mainly downslope, but also occasionally in the upper part of interfluves. Transformations then develop laterally. The «transformation system» corresponds here to both the original and the transformed soil covers.

This type of soil cover has been identified and described in both the humid and the dry tropical zones with contrasted seasons. In the latter, lateral transformations play an important role in the planation of landforms, as emphasized often by Millot (1980a,b, 1982).

3.1 Transformation Systems in Africa Between Sahara and the Humid Tropical Zone

3.1.1 Surficial Transformation Systems: Eluviation–Erosion Shift

The transformation of smectite soil covers has been studied in the dry Sahelian zone (Boulet 1974). In Burkina Faso, smectite soil covers which have developed at the expense of granite, are in dynamic equilibrium for mean annual rainfall higher than 600 mm. Where there is less rainfall, these soil covers do not hold moisture over their full thickness and the upper parts undergo transformation. The transformation system is characterized by two soil types in the case of landforms with low slopes (about 0.3%): solonetz whose upper horizon is sandy and overlies a columnar clayey lower horizon, and subarid brown soils whose upper horizon is clay-rich and displays a cubic structure (Fig. 5). Subarid brown soils result from erosion of the surficial sandy horizon of the solonetz, by suffosion caused by water circulation at the sand–clay contact. Moisture–desiccation alternations induce transformation of the structure from columnar to cubic, after the small columnar units are brought to outcrop due to the wetting–desiccation alternations. The eroded sand accumulates up to only 50 m further downslope because of low competency of rainwash, and because the small columnar units are rebuilt due to water circulation at the sand–clay contact. Circulation also induces separation of the plasma from the skel-

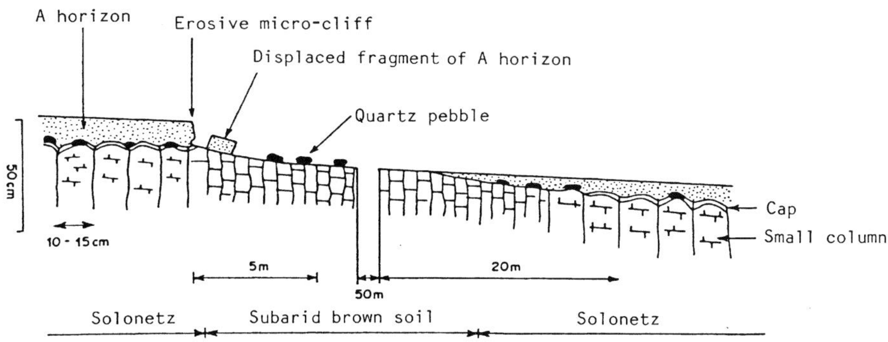

Fig. 5. An eluviation–erosion shift system. Toposequence of Tassamakat in Burkina Faso, Africa. (Boulet 1974; reproduced with permission of ORSTOM)

eton in the clay material, thus feeding the sandy upper horizon. Mechanical erosion follows pedogenesis which prepares the sandy materials to be released by hypodermic or surficial water circulation. This results in progressive thinning of soils and in formation of long pediments with very low slopes which announce desert plains.

3.1.2 Internal Transformation Systems

3.1.2.1 Lateral Transfer Systems: Eluvial–Illuvial Systems

Most of the transformation systems identified in Africa between the Sahel and the tropical zone are composed of initial, loose or Fe crust-bearing kaolinite covers which are transformed either by material transfer within the hillslope or by reorganization in situ or over a short distance.

Soil covers displaying discordant horizons were first identified by Boulet (1974) and re-examined by Nahon (1991) in Burkina Faso. Boulet described the transformation of a kaolinite-rich, loose initial soil cover by release of material upslope and accumulation downslope. Such a soil cover is located at Garango, at the foot of a granitic inselberg developed under a Sudanian–Sahelian climate (mean annual rainfall of 900 mm). It is composed of a ferrallitic upslope domain which is characterized by a vertical genetic sequence of elementary organizations (Fig. 6, IV). In the upper fourth of the hillside, the upslope domain transforms laterally and abruptly to an eluviated sandy material which wedges out at depth. This transition is a transformation front along which are observed plasma–skeleton disjunction, export of plasma through the sandy material downslope and relative accumulation of skeleton. This transformation front develops from base to top and from the up- to downslope zone within the horizons of the ferrallitic soil. The exported plasma migrates in two ways. The coarser-grained particles decant to an accumulation horizon in the lower part of the eluviated sandy material. The finer ones, together with dissolved elements, migrate laterally to the downslope zone of the sandy material where they accumulate in form of tongues; the latter extend upslope and display

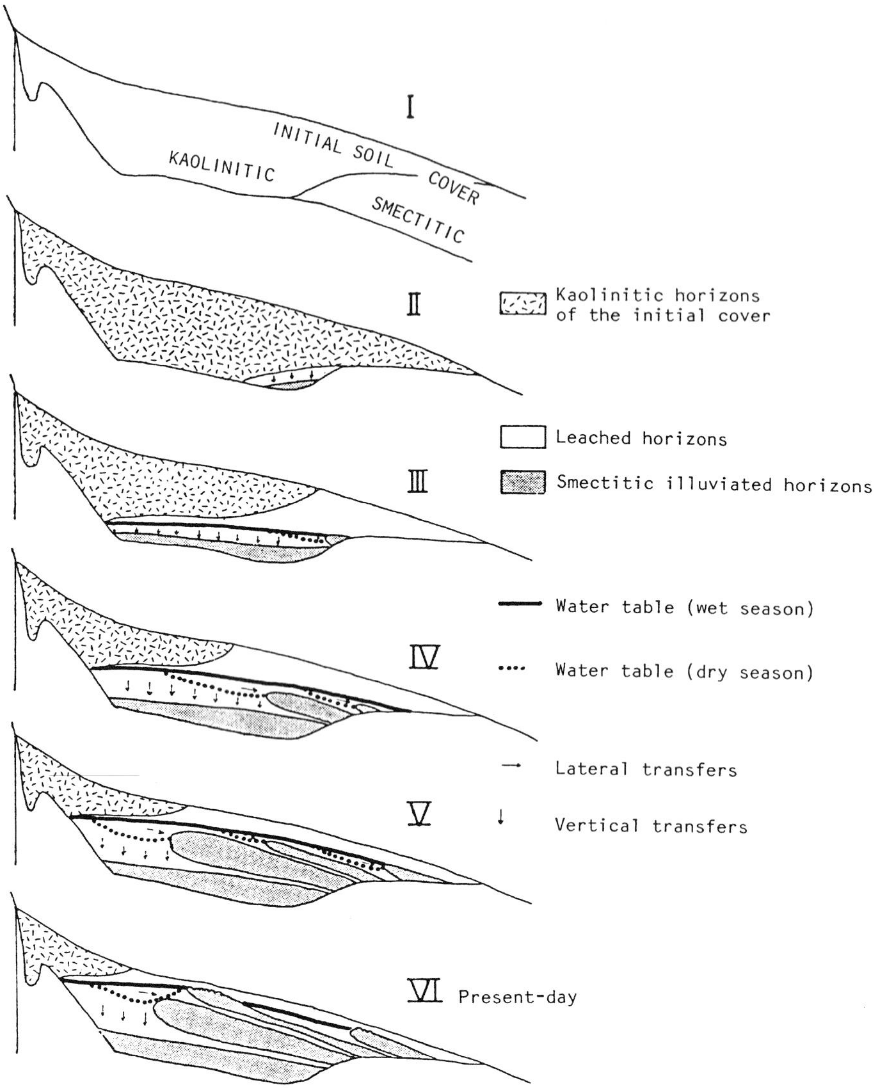

Fig. 6. Evolution of the soil cover (I–VI) at Garango II (Burkina Faso, Africa) reconstructed from present-day organization and dynamics. (Boulet 1974; reproduced with permission of ORSTOM)

important authigenesis of smectites. Loss of material connected to eluviation of the upslope domain results in the collapse of landforms which develop from down- to upslope and provide sandy material. Similar observations in different latitudinal locations prompted recognition of the occurrence of such transformation systems ascribed to a disequilibrium of the initial ferrallitic soil covers as a consequence of evolution to a dry climate. These transformation systems are widespread over

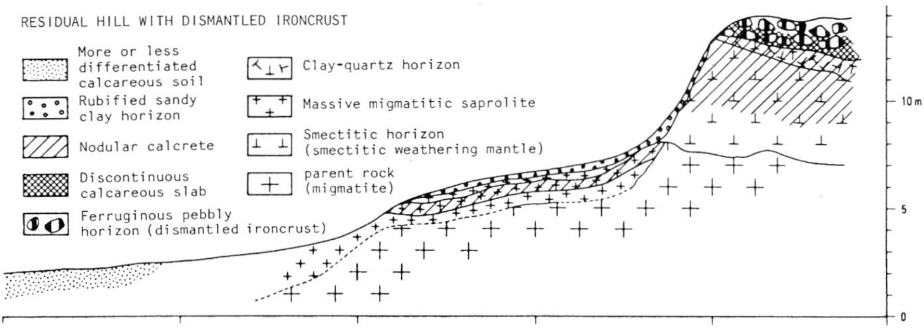

Fig. 7. Toposequence of Inkebdene. (Mauritania, Africa; Nahon 1976; reproduced with permission of ORSTOM)

sandstones in Burkina Faso and northern Togo. They were also studied in northern Cameroon (Brabant and Gavaud 1976; Brabant 1990) and recognized in northwestern Central Africa. In northeastern Brazil, similar systems were identified by Soubiès and Chauvel (1984). In Chad, the soil catenas studied by Bocquier (1971) ought to be considered as transformation systems of the same type, but where complete transformation of the kaolinite upslope part was achieved.

Nahon (1976) studied the distribution of iron crusts and calcretes in the desert zone of Senegal and Mauritania, where they develop separately for a mean annual rainfall between 600 and 700 mm, according to whether the parent rocks are Fe- or Ca-rich. Under a desert climate, with a mean annual rainfall between 20 and 40 mm, this author showed the conjunction of both types of indurated formation over the same migmatitic parent rock. At Inkebdene (Fig. 7), a residual hill is capped by a dismantled Fe crust whose clay fraction contains both kaolinite and smectite. The smectite content increases towards the base where calcareous nodules and relict fragments of calcareous slab encompass Fe blocks. There is also a transition from this blocky level to a smectitic horizon, in which quartzitic sands and pebbles as well as calcareous nodules are abundant. Beneath, a 2–3 m thick smectite material including calcareous nodules overlies the slightly weathered parent rock. Compared to the southern iron crusts, this profile is peculiar because of the disappearance of the kaolinitic level which is thinned and invaded by smectite. According to Leprun (1977, 1979), the quartzitic pebble horizon could be considered as relict material after «geochemical melting» of iron crust, as discussed in the Burkina Faso example. To the north, where the climate is more arid, iron crusts progressively disappear and are no longer observed, except as spherical fragments encompassed in calcretes.

Iron crusts and calcretes were often considered as indurated formations protecting landforms from erosion. In contrast, Nahon (1976) demonstrated that iron crusts and calcretes are factors of corrosion rather than protection of landscapes. Leprun (1977, 1979) studied iron crusts developed at the expense of the crystalline basement in the dry Sudanian–Sahelian zone of West Africa (mean annual rainfall

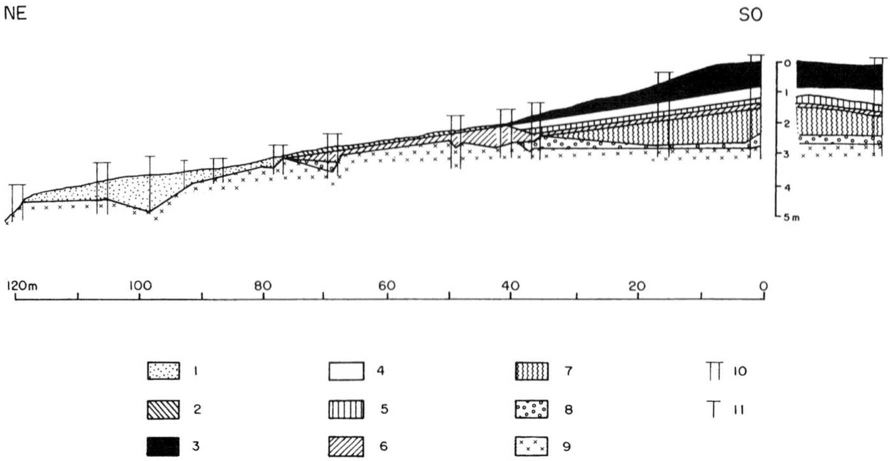

Fig. 8. Toposequence of Thion (Burkina Faso, Africa; Leprun 1979; reproduced with permission of Sciences Géologiques). *1* sandy horizon; *2* A1 horizon; *3* iron crust; *4* eluvial horizon; *5* illuvial horizon; *6* polyhedral horizon; *7* vertic weathering horizon with calcareous nodules; *8* saprolite; *9* migmatite; *10* studied section; *11* borehole

between 600 and 900 mm). He demonstrated that they are affected by internal «geochemical melting» either from their base (Fig. 8) or from their intermediate zone. This destruction proceeds by Fe removal under the influence of percolating waters, resulting in fragmentation of the iron crust into smaller and smaller units, and in dissolution of these fragments. The only primary minerals, mainly quartz in the form of whitened sands and pebbles, are preserved in situ. The released clay fraction accumulates at the base of the illuvial horizon or is exported (Fig. 8). The terminal stage consists of the disappearance of iron crust except for a few fragments. Mechanical surficial degradation has obviously played a major role in the adjustment of the top of the iron crust (Leprun 1972), but internal geochemical degradation is the main factor in the destruction and disappearance of iron crusts.

Degradation does not similarly affect all parts of the landforms in the Sudanian–Sahelian zone. For example at Thion in Burkina Faso, an upward disappearance of the Fe crust can be observed. In the case of iron crusts developed on smooth slopes, degradation operates in the intermediate zone of the slope, whilst the upslope part evolves into a residual hill and the downslope iron crust sinks. If the iron crust landform is convex–concave, which is rare, the upslope part transforms into a Fe fine gravel ridge whilst the downslope part of the iron crust is preserved. But as in the desert, destruction of the iron crust extends until it has been completely destroyed, leaving only some relics of its previous presence in the form of ferruginous blocks or nodules scattered on the surface of desert regs, or in the form of residual hills. In northwestern Ivory Coast, a multidisciplinary programme, including hydrologists and soil scientists, provided a thorough study of a 1.35 km^3 watershed located on a gneissic basement under mean annual rainfall of 1350 mm. The pedological study showed that soil cover is divided into two domains including several systems (Fritsch et al. 1990a; Fig. 9): (1) an upslope ferrallitic domain composed of a micro-aggregated red soil system and an indurated system representing the relicts of a plateau framed by a Fe crust and bordered by an escarpment; and (2) a downslope ferruginous and hydromorphic domain characterized by a surficial degradation system, an iron crust system on the slope, a hydromorphic system and an alluvial system. At the border of the plateau, the upslope ferrallitic domain displays an outcropping iron crust which is separated from underlying alterite by a red-clay horizon; in the mid part of the plateau, there is a transition from the iron crust to a fine-gravel horizon with red-clay matrix. Owing to the presence of the red-clay horizon devoid of ferruginous differentiations and to the lateral transition from the iron crust to the fine-gravel horizon, the iron crust is supposed to be transformed by the red-clay material.

In summary, the surficial degradation system is unconformably related to the upslope ferrallitic domain and extends over the remainder of the slope. It displays the following variations from base to top, and from up- to downslope: variation in colour from ochrous to yellow and to white, decrease in content of fine-grained components (kaolinite and Fe oxi-hydroxides) and modification in structure from fragmentary to massive. The Fe-crust system on the slope of dihedral form, crops out or is close to the surface in the mid slope. Upslope, it develops at the expense of red horizons and of the upper part of the weathering zone of the ferrallitic domain and is discordant relative to the boundary of the soil-weathering zone. It displays a ferruginous differentiation resulting in the formation of red and white mottling, of

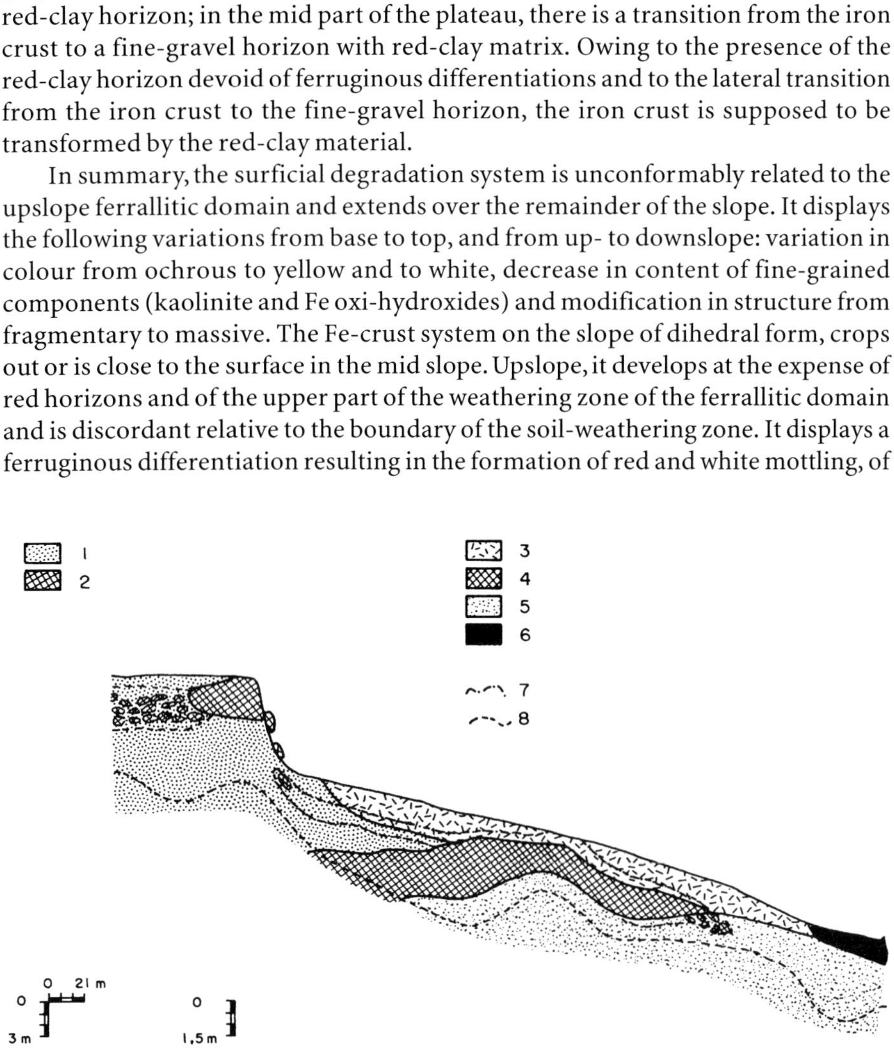

Fig. 9. Hyperbav (Ivory Coast, Africa; Fritsch et al. 1990). Reproduced by permission of ORSTOM. Ferrallitic domain: *1* red soil system. *2* iron crust system on plateau. Ferruginous domain: *3* surficial degradation system. *4* iron crust system on slope. *5* hydromorphic system. *6* alluvial system. *7* fine gravel unit. *8* top of alterite.

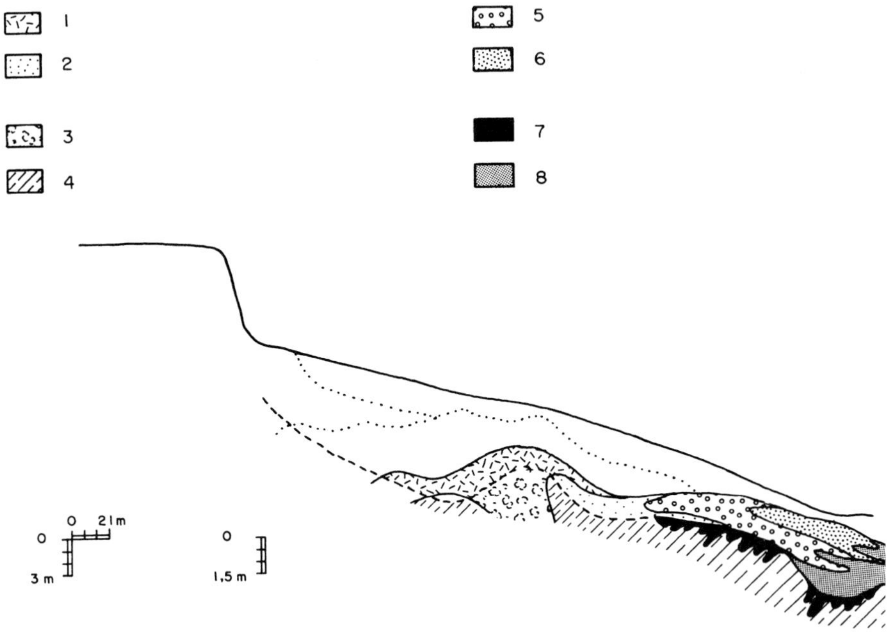

Fig. 10. Hyperbav (Ivory Coast, Africa). Transformation and differentiation in the hydromorphic system (Fritsch et al. 1990; reproduced with permission of ORSTOM). Pseudogley and gley subsystem: *1* variegated to reticulated slightly indurated horizon with white patches; *2* yellow, ochreous and sometimes red horizon with white to grey matrix (A to AS); *3* mottled isalterite: red reticles with yellow rims and white matrix; *4* hydromorphic isalterite: yellow, vitreous white, greenish grey or bluish subvertical layers. Eluvial–illuvial subsystem: *5* white to grey horizon with ochreous, yellow and light yellow mottlings; *6* white to grey rolling horizon (S) with or without light yellow mottlings; *7* grey subvertical veins with vertic features within the alterite; *8* grey horizon (SA to AS) with illuviation cutans.

a «carapace» and of an iron crust. The iron crust can be observed in two places: on the dihedral top, where it is dark-red to black and is near the surface or crops out; the second location being downslope and at depth where the iron crust is ochrous yellow and is dismantled when invaded by the hydromorphic system. The hydromorphic system (Fig. 10) appears in the form of a tongue which crops out downslope and occurs at depth upslope. It develops at the expense of the two preceding systems and of the weathering zone, crosscutting the soil–weathering transition zone upslope and the iron crust system downslope. From up- to downslope are observed: (1) a succession of transformations with channeled boundaries, which result in bleaching and increasingly sandy texture; and (2) well downslope, an individualization of kaolinitic and smectitic accumulation horizons in the form of superposed tongues similar to those found at Garango. Two major types of process operate in such a complex soil cover. One corresponds to absolute Fe accumulation in the iron crust system; this accumulation is also litho-dependent since it is located over a Fe-rich compartment of the parent rock. The other corresponds to loss of fine-grained components (Fe oxides and clay minerals) which are largely exported out of the basin. These exportations, which are especially important in the lower part of the slope, result in topographic prominance of the iron crust as a consequence of collapse of the topographic surface, and in formation of secondary thalwegs. The study of hydrodynamics on the slope scale (Planchon and Janeau 1990) shows that the development of secondary thalwegs corresponds to an internal determinism depending on water circulation along privileged pathways within the soil profile. Thus, formation of drainage axes is induced by the internal dynamics of the soil cover. The latter is a hinge element in the latitudinal sequence of transformation systems. In fact, such a soil cover displays eluviated–illuviated systems which are more extensive northward, where they invade most of the landscape in some cases.

3.1.2.2 In Situ Reorganized Systems

In southwestern Casamance, with a mean annual rainfall of 1200 mm, the landscape is composed of plateaus covered with micro-aggregated ferrallitic soils, whose horizons are concordant and which are developed above clay-rich sandstones (Chauvel 1977). To the northeast, climatic conditions become drier and more seasonally contrasted: beige soils appear in the slightly depressed mid part of the plateaus and progressively occupy the major part of the latter, whilst red soils are restricted to peripheral patches. This is a transformation system initiated by climatic evolution towards the dry type and in which beige soils develop laterally at the expense of red ones. Transformation results from collapse of the micro-aggregated structure due to hydric constraints during the dry season, from removal of Fe occupying exchange sites in kaolinite and from the consecutive dispersion of the clay fraction, which is redistributed in the profile. At this stage a beige soil is formed whose low permeability favours rainwash and erosion and induces evolution of landforms, as shown in Fig. 11.

In tropical areas with contrasting seasons, transformation systems operate in two ways. One is direct and results from internal processes within the soil cover, i.e. degradations, transfer of material in the form of dissolved elements or microparticles, and transformation in situ together with volume loss. The action of these mechanisms is not very important for the evolution of landscapes. In contrast, the second way is indirect and obviously more important. In fact, the internal processes supply surficial mechanical erosion and especially sheet erosion with an easily mobilizable material in which the Fe-clay, plasma-skeleton or Fe-skeleton bonds have already been destroyed; they also induce a decrease in permeability in the upper horizons, thus favouring rainwash.

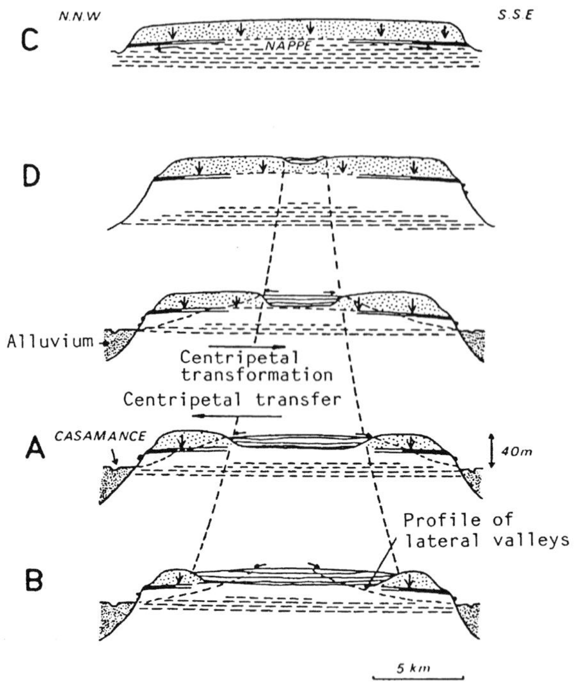

Fig. 11A–D. The supposed successive evolutionary stages of a plateau in Middle Casamance, Africa. (Chauvel 1979; reproduced with permission of ORSTOM)

3.2 Transformation Systems in the Humid Tropical Zone

3.2.1 Ferrallitic Soil–Podzol Transformation Systems

3.2.1.1 In the Manaus Area, Brazil

With mean annual rainfall of 2500 mm, the plateaus of the Manaus area display a soil cover which is in dynamic equilibrium and is composed of concordant horizons, as mentioned above (Lucas 1989). However, the 25–30 m high valley sides are characterized by a transformation system which progressively replaces clay ferrallitic soils by gigantic podzols and plateaus by low hills.

Incipient transformation corresponds to formation of steep slip-off slopes (Fig. 12a), which are affected by progressive removal of clay (Fig. 13) and represent an eluvial system favouring lateral export of soil solutions. At this stage, water circulates only in the form of inferoflux on the flattened bottoms of valleys, in the basal part of rather thin sandy horizons and without surficial flowing. Convex morphology together with flowings on valley floors indicate an essentially geochemical evolution of landforms. This interpretation is coherent with estimations of rainwash, as drawn from the data of neutron humidimetry obtained along the slopes of the same area (Pimentel da Silva et al. 1992); there is no evidence of water redistribution by rainwash under forest. As transformation develops, the impoverished soils of the lower parts of slopes extend progressively upslope and typical bleaching

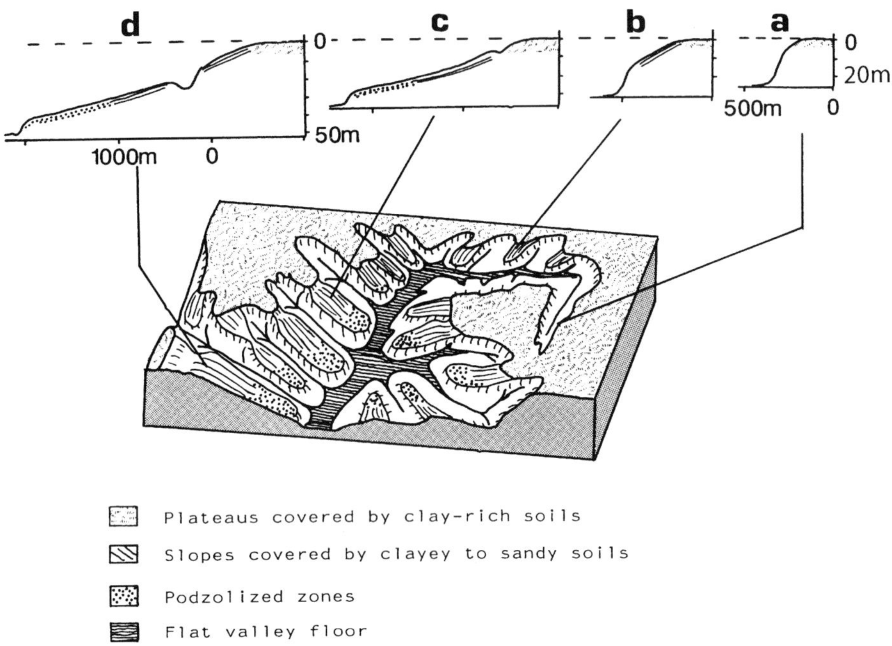

Fig. 12a–d. Schematic pattern of the different types of slopes in the North Manaus area (Amazonia). (Lucas 1989)

of podzols is identified from a given grain-size distribution (2–4% of the <2 μm fraction). The significant removal of material from a thick part of the soil, as related to this evolution, results in progressive formation of long and slightly concave

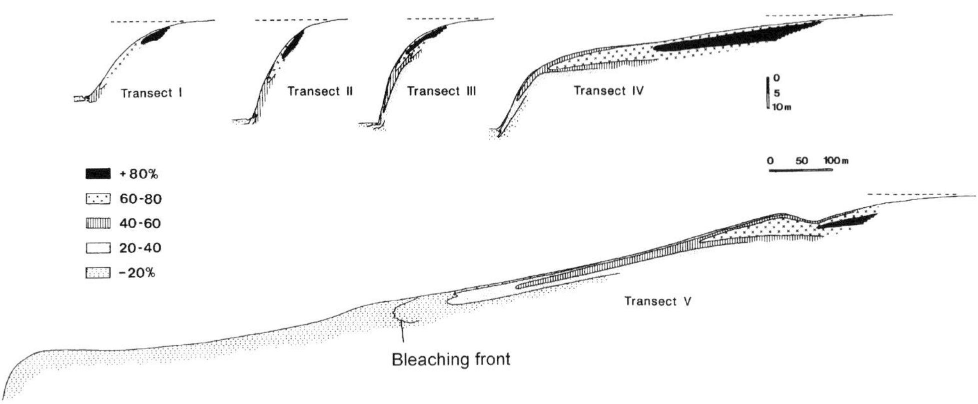

Fig. 13. Distribution of the <2 μm fraction in the slope soils, North-Manaus, Amazonia (same scale for all the transects; Lucas 1989)

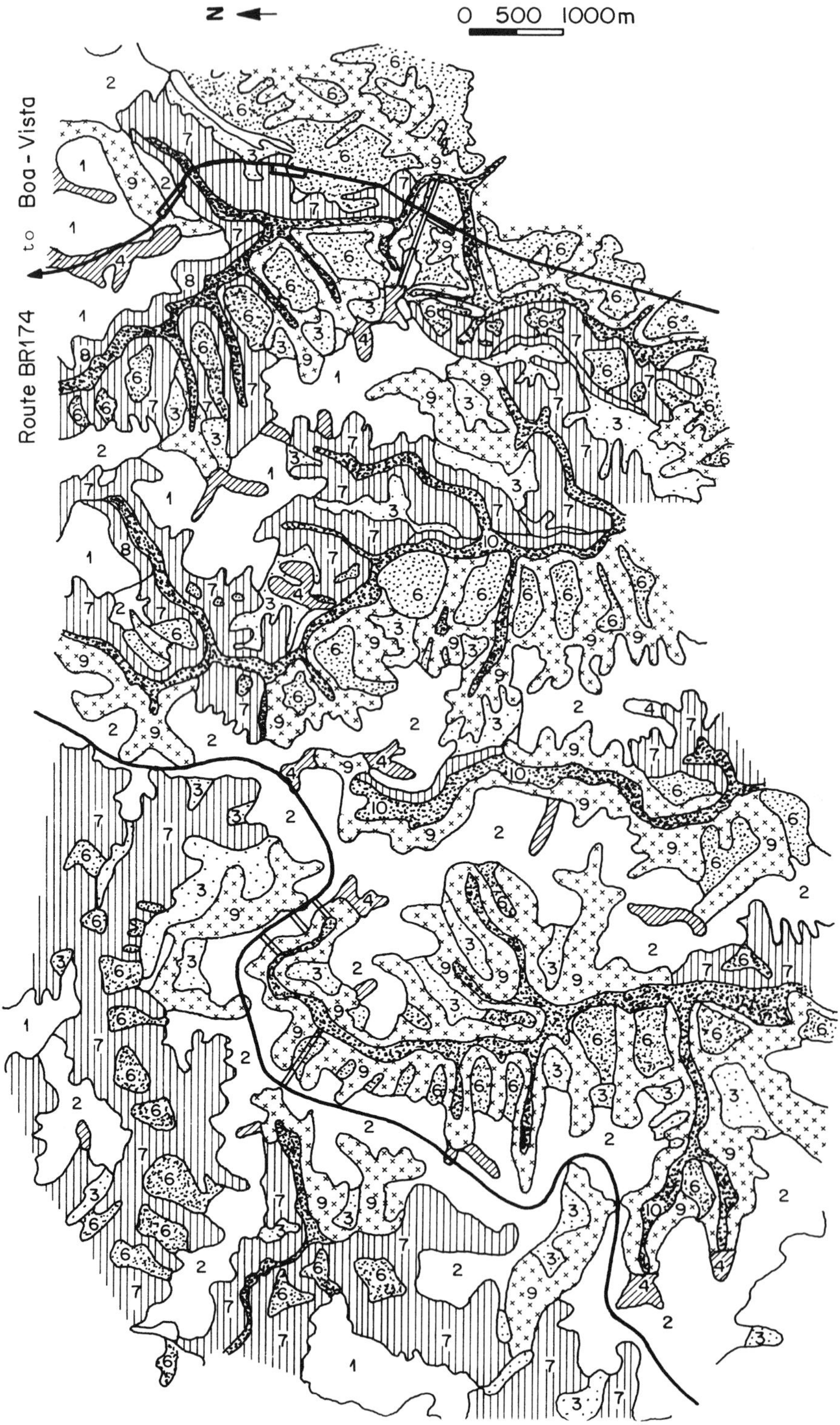
N
0 500 1000m
Route BR174 to Boa-Vista

Fig. 14. Schematic morphological map of the northern Manaus area, Amazonia. Photo-interpretation from photos at 1:50000 scale. (Lucas 1989). *1* plateaus with planar surfaces; *2* plateaus with gently undulating surfaces; *3* fragment of plateau in continuity with the arched preceding one, slightly inclining to the drainage axis; *4* plateau inner valley, with gentle slopes and relief; *5* slope between plateau and intermediate surface with regular steep slope and relief of about 20 m; *6* intermediate surface, slightly arched if narrow, saddled if extended. Intermediate surface slightly inclined to the drainage axis and lower than the plateau; *7* irregular hillsides displaying numerous small secondary drainage axes with medium to large relief; *8* regular hillsides with moderate to steep slopes and relief; *9* hillsides with moderate to gentle slopes and relief; *10* flat valley floor. Slopes: steep if >=20%, moderate if between 10 and 20%, gentle if <10%. Relief: large if >=20 m, moderate if betweeen 10 and 20 m, gentle if <10 m

slopes (Fig. 12c). Thus, the landscape of plateaus (northeastern fourth of the map in Fig. 14) is progressively replaced by a zone where only fragments of the initial plateau are preserved and where the major part is occupied by podzolic low slopes (southwestern fourth of the map in Fig. 14).

This transformation corresponds to a mean lowering of the landscape of about 20 m. According to the present state of knowledge, the rate of evolution of such transformation systems is hard to evaluate. This rate is obviously greater than that of the geochemical descent of plateaus. An estimated value can be proposed by extrapolation from data of Turenne (1975) who studied ferrallitic soils–podzols systems in French Guyana and found a rate of 0.5–2 m per 1000 years for lateral progression of the transformation process. Since the length of the podzolic slopes of the Manaus area is about 1 km, their evolution would have necessitated between 500000 and 2000000 years.

3.2.1.2 In French Guyana

Ferrallitic soil–podzol systems are developed in French Guyana, with a mean annual rainfall of 2–3.5 m, at the expense of the basement (Veillon 1990) and of coastal marine sediments (Turenne 1975; Lucas et al. 1987). On both rocks, they result in a planation of the landscape. The soil cover is thicker and the planation is greater over the basement than over the coastal marine sediments. In the basement area, landforms consist of undulating plateaus of large hills. Transformation of sandy-clay ferrallitic soils to sandy podzols starts in the mid part of interfluves (Fig. 15B) and clay removal induces an important collapse (Fig. 15C). This results in an outcropping of the water table, generating swampy and ill-hierarchized drainage axes (Fig. 15D). Transformation progresses from the central part to the margins of interfluves leading to formation of an almost entirely podzolized out-of-slope landform. Only some limited relict zones of ferrallitic soils are observed and they are deeply transformed, since they consist of slightly argillaceous red sands (Fig. 15E). This podzolic landform is lowered by about 20 m with respect to the initial relief.

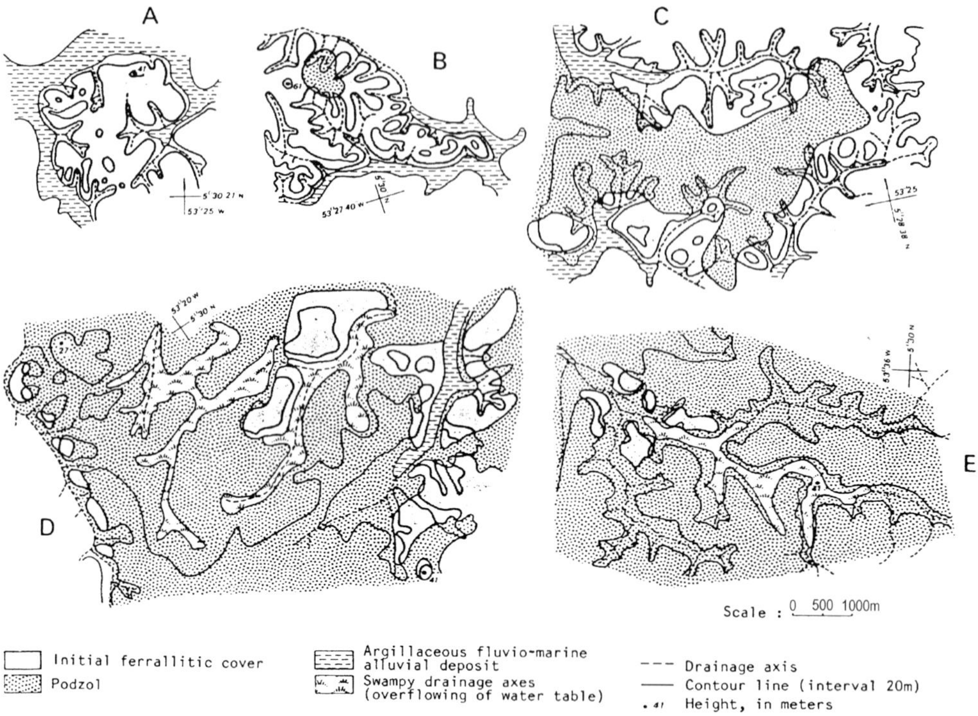

Fig. 15A–E. Examples of landform units displaying different evolutive stages of the ferrallitic-podzol transformation system in French Guyana. (Boulet et al. 1984; reproduced with permission of the Association française pour l'étude du Sol)

3.2.1.3 *In the Basin of the Upper Rio Negro*

According to studies carried out in the basin of the Upper Rio Negro (Dubroeucq and Blancaneaux 1987), with a mean annual rainfall of 3000 mm, sequential soil covers can be related to the different units of basin landforms. From the higher to the lower landform, the following sequence can be observed (Fig. 16): (1) reddish yellow, very thick, ferrallitic soils on the high hills;(2) humic yellow, rather thin, ferrallitic soils on the low-level hills;and (3) podzols and gibbsitic yellow ferrallitic soils in the form of relict patches in the lowlands. Based on cartographic data, this succession of soil covers can be interpreted as an evolutionary sequence. The authors state that the geochemical collapse of landscape corresponds to a lowering of relief of 70 m.

3.2.2 Transformation Systems on Basement with Drainage Inversion

These systems lead to the replacement of a ferrallitic soil cover where water dynamics are vertical, by a ferrallitic soil cover where water dynamics are mainly surficial and lateral. They were identified in French Guyana on "half-orange" landforms developed over schistose or migmatic basements (Boulet 1978, 1983; Boulet et al. 1978; Fritsch 1979), mean annual rainfall being 3 m. The genetic sequence of soil covers observed over schists is presented in Fig. 17.

Stage I corresponds to an initial ferrallitic soil cover in dynamic equilibrium, where water circulation is vertical and supplies a deep groundwater table. Soils display an upper unit of red-brown micro-aggregated argillaceous horizons with large porosity, overlying a lower unit composed of argillaceous horizons with massive structure and fine porosity, which rests on a muscovite-rich red fine saprolite. At stage II, the micro-aggregated upper unit which is located in the upslope part, is similar to that of stage I and is more than 1.5 m thick, but it becomes thinner (around 60 cm) and yellow in the downslope part. Hydrodynamic studies indicated that this lateral transition corresponds rather to impeded vertical water percolation and to establishment of lateral water-flows within the yellow horizon perched over the underlying material with very fine porosity (Humbel 1978; Guehl 1984). Simultaneously, surficial rainwash increases 25 times, although mechanical erosion calculated on patches remains rather weak under forest (about 450 kg ha^{-1} $year^{-1}$; Sarailh 1983). Stages III and IV are characterized by progressive sinking of the yellow horizon with lateral water circulation within the initial soil cover

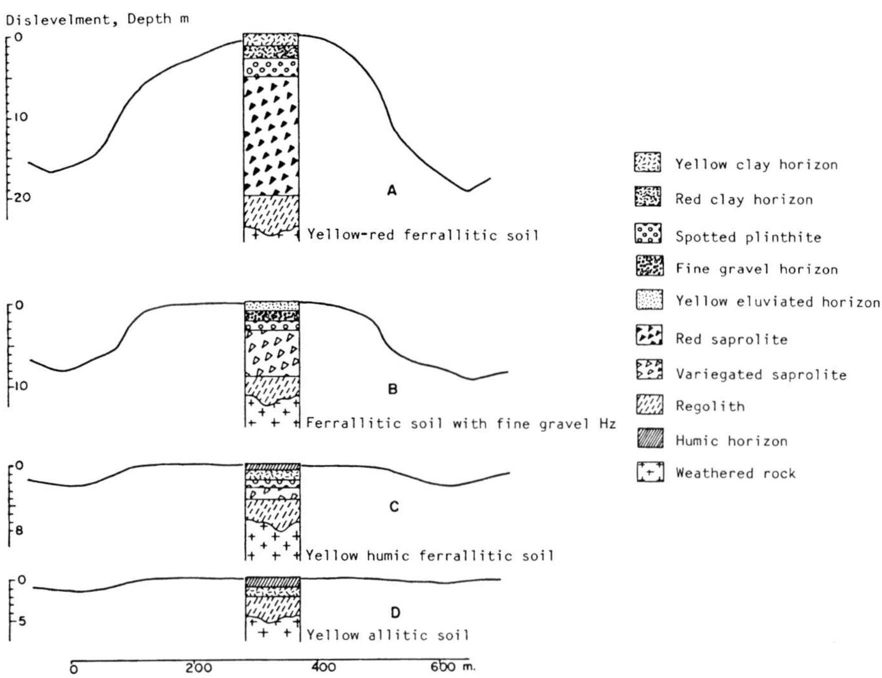

Fig. 16A–D. Basin of the Upper Rio Negro river (Dubroeucq 1991). **A** Convex hill. **B** Levelled hill. **C** Slightly undulating surface. **D** Planar surface. (Reproduced by permission of the Académie des Sciences)

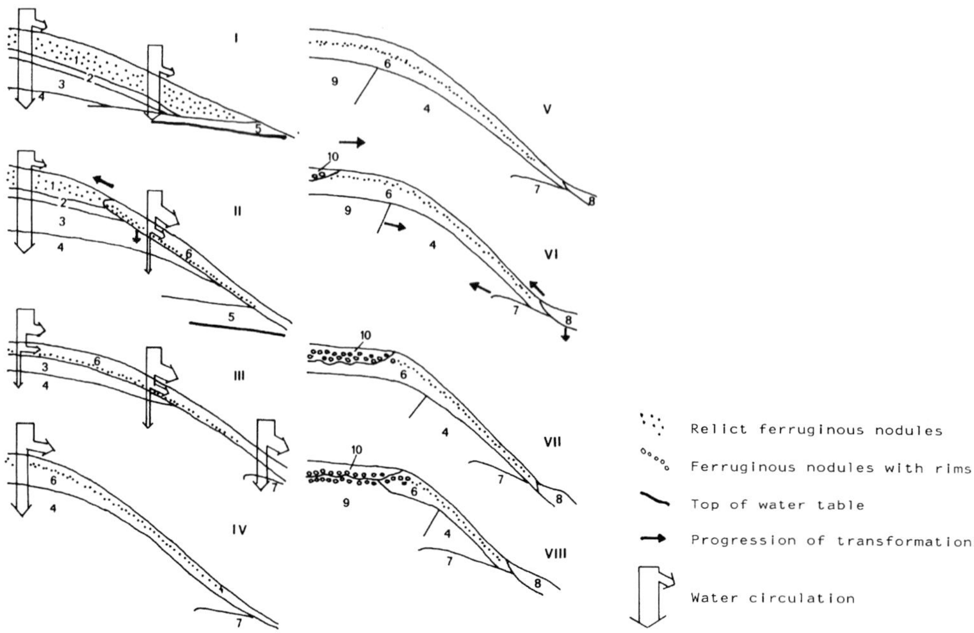

Fig. 17. Evolutionary stages of the soil cover (**I–VIII**) developed over schistose basement in French Guyana. Stages **I–IV** from Boulet (1977) and stages **V–VIII** from Fritsch (1984). *1* red-brown micro-aggregated horizon, rich in relict lithic nodules; *2* transitional horizon; *3* red finely-porous clay horizon; *4* red fine saprolite; *5* water-table reservoir; *6* yellow sandy-clay horizon; *7* white fine saprolite; *8* pale grey, sandy horizon; *9* pale red, locally yellowish fine saprolite; *10* sandy-clay horizon, bright yellow at the bottom, pale greenish yellow at the top

whose horizons progressively disappear, the transformation starting from downslope zone. As early as stage III, in locations where slopes are lower than 10% (summits, passes), large closed depressions of one to several metres across and 50–70 cm in depth, are observed. Water accumulates during rainy periods and then percolates slowly. According to hydrodynamic studies in progress, disappearance of the argillaceous red horizon with fine porosity seems to result in increased vertical drainage, but rainwash remains important too. From stage V, new transformations take place (Fritsch 1984).

Upslope and at depth, the fine red saprolite transforms to a lighter material by partial Fe loss and kaolinization of muscovite. In the uppermost part of the profile, a pair of horizons transforms the yellow horizon to a depth of around 70 cm: the more surficial pale greenish-yellow and sand-clayey horizon is affected by clay and Fe loss; the lower bright yellow, clay-sandy and Fe nodule-rich horizon is affected by absolute Fe accumulation. The presence of light grey sandy mottling in the upper horizon and of ochreous-brown nucleii in the lower one attests to incipient podzolization (Fritsch et al. 1986). The closed depressions observed on the flat part of the landforms during stage III become general on the levelled top of the interfluve. Downslope and at depth, fine red saprolite transforms to a fine white one in which red lithorelicts display ochreous rinds due to complete Fe removing of plasma and partial Fe removing of lithorelicts, and in which kaolinization of muscovite is observed. At the top of the profile, a light-grey sandy horizon has transformed the yellow horizon from the bottom and the underlying fine white saprolite from the top, by complete loss of Fe and by disappearance of most of the kaolinite.

Each of these transformations results in geochemical erosion, which is greater when the transformation is more effective. From stage I to stage IV, evolution of the transformation system results in a steepening of the slopes. From stage V, some planation, lowering of interfluve tops and slope retreat can be observed. Stage VIII corresponds to the part of the landform studied by Fritsch (1984). The progress of transformations ought obviously to generate a more evident planation. Disequilibrium of initial soil cover with vertical water circulation is related to relative lowering of the base level ascribed to slight epeirogenic basement uplift in French Guyana, which is the consequence of subsidence of the adjacent sedimentary basins of the Berbice and Amazon rivers (Boulet et al. 1979). The resulting planation of landforms is induced by evolution of the transformation system due to both geochemical and mechanical erosion; the rate of the latter, measured on the scale of mottling under original forest, and which is less than 70 kg ha^{-1} $year^{-1}$ in the initial soil cover, reaches values between 100 and 440 kg ha^{-1} $year^{-1}$ according to the year considered, in soil covers with principally surficial and lateral water dynamics (Sarrailh 1983). It should be noticed that this increase in erosion, as related to the significant increase in rainwash (from 0.6 to 15% on average; Sarraihl 1983), is a consequence of inversion of drainage, thus of pedological transformations, as observed above.

4 Conclusions

Two major types of soil covers – soil covers in dynamic equilibrium and transformation systems whose inventory is still not exhaustive – have been identified. For both types, a tentative demonstration of the role of pedogenesis in the evolution of landforms has been proposed. It can be emphazised that in tropical zones with contrasting seasons, geochemical erosion plays a predominant role in the planation of landforms, in the case of soil covers in dynamic equilibrium. This role is also important in transformation systems, but mechanical erosion operates more easily because of destruction of the clay-skeleton or of oxihydroxides–clay bonds due to transformation. Recent studies have shown that the direct role of geochemical erosion on planation of landforms is especially evident in humid tropical zones, either by continuous deepening of ferrallitic soil covers in dynamic equilibrium, resulting in slow lowering of the topographic surface, or by lateral transformation of these soil covers.

Soil covers in dynamic equilibrium correspond to zonal soil covers, since they are in dynamic equilibrium with the pedoclimatic conditions to which they were subjected since formation. However, it is not yet possible to provide a continuous latitudinal sequence of these soil covers: the transition from one to the other is difficult to explain except for the transition from indurated Fe soils to iron crusts. The ideal case would consist of a latitudinal sequence developed on the same parent rock (a mesocratic granite for example) without any inherited palaeoclimatic material, if possible. West Africa cannot provide such an ideal example for two reasons. The first is the presence of the Sahara desert between the Mediterranean and the tropical zones, where soil formation is impeded. The second is the palaeoclimatic history of West Africa, which displays an evolution from a humid to a dry climate as early as the Mesozoic (Tardy et al. 1988), which favours inheritance of materials. In contrast, the same history goes from an arid to a humid climate in South America, which tends to obliterate inherited materials. A north-south transect in Brazil is thus fundamentally important, if developing on sufficiently young surfaces. Both north-south transects in West Africa and in South America would shed light on each other.

Very thick and monotonous ferrallitic soil covers in humid areas are usual and have long been related to "half-orange" landforms composed of convex hills, which were considered by Georges Millot (1980a) as the opposite of planation. In fact, the lower parts of slopes can reach 50%. But in humid zones, soil covers which are considered to be in dynamic equilibrium, correspond to plateau-like landforms. A contradiction thus occurs, which cannot actually be answered because of an insufficient number of detailed studies in humid tropical zones with "half-orange" landforms. But the monotony of soil covers in these zones may only be apparent. For example in French Guyana, the half-orange landform which developed on basement rocks in coastal areas, corresponds to transformation systems. In Manaus, the notch of ferrallitic plateaus with convex slopes is related to the appearance of discordant horizons and of transformation systems. In order to complement this synthetic presentation, it thus seems reasonable to wait for more detailed studies on the structure and evolution of soil covers in half-orange landform areas.

Acknowledgments. The authors are deeply indebted to Prof. Larry Frakes for his review and assistance in the English translation, and to Professors Yves Tardy and Adolpho J. Melfi for their constructive and critical remarks.

References

Beauvais A (1991) Paléoclimats et dynamique d'un paysage cuirassé du Centrafrique. Morphologie, pétrologie et géochimie. Thèse, Univ Poitiers, 315 pp

Beauvais A, Tardy Y (1991) Formation et dégradation des cuirasses ferrugineuses sous climat tropical humide à la lisière de la forêt équatoriale. CR Acad Sci Paris 313 (II):1539–1545

Bilong P, Belinga SE, Volkoff B (1992) Séquence d'évolution des paysages cuirassés et des sols ferrallitiques en zone forestière tropicale d'Afrique centrale. Place des sols à horizon d'argile tachetée. CR Acad Sci Paris 314 (II):109–115

Bitom D (1988) Organisation et évolution d'une couverture ferrallitique en zone tropicale humide (Cameroun). Thèse, Univ Poitiers, 164 pp

Bocquier G (1971) Genèse et évolution de deux toposéquences de sols tropicaux du Tchad. Interprétation biogéodynamique. Thèse Sci, Mém ORSTOM, 62, 1973, 350 pp

Bocquier G, Rognon P, Paquet H, Millot G (1977) Géochimie de la surface et formes du relief II Interprétation pédologique des dépressions annulaires entourant les inselbergs. Sci Géol Bull (Strasb) 30 (4):245–253

Boulet R (1974) Toposéquences de sols tropicaux en Haute-Volta. Equilibre et déséquilibre pédobioclimatique. Thèse Sci, Mém ORSTOM, 85, 1978, 272 pp

Boulet R (1978) Existence de systèmes à forte différenciation latérale en milieu ferrallitique guyanais: un nouvel exemple de couverture pédologique en déséquilibre. Sci Sol 2:75–82

Boulet R (1983) Organisation des couvertures pédologiques des bassins versants ECEREX. Hypothèses sur leur dynamique. In: Le projet ECEREX (Guyane), pp 23–52 and in: Sarrailh JM (dir) Mise en valeur de l'écosystème guyanais. Coll Ecologie et Aménagement rural, INRA, 1990

Boulet R, Bocquier G, Millot G (1977) Géochimie de la surface et formes du relief. I. Déséquilibre pédobioclimatique dans les couvertures pédologiques de l'Afrique tropicale de l'Ouest et son rôle dans l'aplanissement des reliefs. Sci Géol Bull (Strasb) 30 (4):235–243

Boulet R, Brugière JM, Humbel FX (1978) Relation entre organisation des sols et dynamique de l'eau en Guyane septentrionale: conséquences agronomiques d'une évolution déterminée par un déséquilibre d'origine principalement tectonique. Sci Sol 1:3–18

Brabant P (1990) Les sols des forêts claires du Cameroun. Exemple d'étude d'un site représentatif en vue de la cartographie des sols et de l'évaluation des terres. ORSTOM, Paris, 2 vols, 530 and 278 pp

Brabant P, Gavaud M (1976) Les sols et les ressources en terre du Nord-Cameroun. ORSTOM, Paris, Notice Expl, 103, 285 pp

Brewer R (1964) Fabric and mineral analysis of soils. Wiley, New York, 470 pp

Carvalho A (1970) Study ofTerra Roxa Estructurada and Latossolo Roxo on a topographic sequence in São Paulo State, Brazil. MS Thesis, Univ Newcastle, 93 pp

Chauvel A (1977) Recherches sur la transformation des sols ferrallitiques dans la zone tropicale à saisons contrastées. Thèse Sci, Trav Doc ORSTOM, 62, 532 pp

Chauvel A, Pédro G (1978) Genèse de sols beiges (ferrugineux tropicaux) par transformation des sols rouges (ferrallitiques) de Casamance (Sénégal). Modalités de leur propagation Cah. ORSTOM, Sér Pédol 16:231–249

Dubroeucq D, Blancaneaux P (1987) Les podzols du Haut Rio Negro, région de Maroa, Vénézuela. Environnement et relations lithologiques. In: Righi D, Chauvel A (eds) Podzols et podzolisation. AFES-INRA, Paris, pp 37–52

Dubroeucq D, Volkoff B, Pédro G (1991) La couverture pédologique du Bouclier du Nord de l'Amazonie (bassin du Haut Rio Negro) Séquence évolutive des sols et son rôle dans l'aplanissement généralisé des zones tropicales perhumides. CR Acad Sci Paris 312 (II):663–671

Eschenbrenner V (1987) Les glébules des sols de Cite d'Ivoire. Thèse Sci, Univ Dijon 1:498 pp; 2:282 pp

Fauck R (1972) Contribution à l'étude des sols des régions tropicales Les sols rouges sur sables et sur grès d'Afrique Occidentale. Thèse Sci, Mém ORSTOM 61:257 pp

Freyssinet P (1990) Géochimie de l'or dans les couvertures latéritiques. Thèse Univ Strasbourg, Doc BRGM, 203, 1991, 279 pp

Fritsch E (1979) Etude des organisations pédologiques et représentation cartographique détaillée de quatre bassins versants expérimentaux sur schiste Bonidoro de Guyane française (piste de Saint-Elie). Rapp, ORSTOM, Cote P183, 30 pp

Fritsch E (1984) Les transformations d'une couverture ferrallitique: analyse minéralogique et structurale d'une toposéquence sur schistes en Guyane Franáaise. Thèse, Univ Paris VII, 138 pp
Fritsch E, Bocquier G, Boulet R, Dosso M, Humbel FX (1986) Les systèmes transformants d'une couverture ferrallitique de Guyane française Analyse structurale d'une formation supergène et mode de représentation. Cah ORSTOM, Sér Pédol 22 (4):361–395
Fritsch E, Valentin C, Morel B, Leblond P (1990a) La couverture pédologique: interactions avec les roches, le modelé et les formes de dégradation superficielles. In: Equipe HYPERBAV (eds) Structure et fonctionnement hydropédologique d'un petit bassin versant de savane humide. Etudes et thèses, ORSTOM, Paris, pp 31–57
Fritsch E, Chevallier P, Janeau JL (1990b) Le fonctionnement hydrodynamique du bas de versant. In: Equipe HYPERBAV (eds) Structure et fonctionnement hydropédologique d'un petit bassin versant de savane humide. Etudes et thèses, ORSTOM, Paris, pp 185–206
Guehl JM (1984) Dynamique de l'eau dans le sol en forêt tropicale humide guyanaise. Influence de la couverture pédologique. Ann Sci For 41:1
Humbel FX (1978) Caractérisation par des méthodes physiques, hydriques et d'enracinement, de sols de Guyane française à dynamique de l'eau superficielle. Sci Sol 2:83–93
Leprun JC (1972) Cuirasses ferrugineuses autochtones et modelés des bas-reliefs des pays cristallins de Haute-Volta orientale. CR Acad Sci Paris 275 (D):1207–1210
Leprun JC (1977) Géochimie de la surface et formes du relief IV La dégradation des cuirasses ferrugineuses. Etude et importance du phénomène pédologique en Afrique de l'Ouest. Sci Géol Bull (Strasb) 30:265–273
Leprun JC (1979) Les cuirasses ferrugineuses des pays cristallins d'Afrique Occidentale sèche. Genèse, transformations, dégradations. Sci Géol Mém (Strasb) 58:224 pp
Lucas Y (1980) Carte pédologique de la région de Paoua. Une carte à 1/20 000 et notice. Centre ORSTOM, Bangui, 127 pp
Lucas Y (1989) Systèmes pédologiques en Amazonie brésilienne. Equilibre Déséquilibre et transformations. Thèse Sci, Univ Poitiers, 157 pp
Lucas Y, Boulet R, Chauvel A, Veillon L (1987) Systèmes sols ferrallitiques-podzols en région amazonienne. In: Righi D, Chauvel A (eds) Podzols et podzolisation. AFES-INRA, Paris, pp 53–65
Lucas Y, Kobilsek B, Chauvel A (1989) Structure, genesis and present evolution of Amazonian bauxites developed on sediments. Travaux ICSOBA 19:81–94
Lucas Y, Nahon D, Cornu S, Eyrolle F (1996) Genèse et fonctionnement des sols en milieu équatorial. CR Acad Sci Paris 322 (IIa):1–16
Melfi A (1968) Contribution à l'étude des «Terras Roxas Legétimas» du Brésil: caractéristiques géochimiques et minéralogiques des formations de Campinas. Bull Assoc Fr Etude Sol 6:31–39
Millot G (1977) Géochimie de la surface et formes du relief. Présentation. Sci Géol Bull (Strasb) 30:229–233
Millot G (1979) Présentation du mémoire «Phénomènes de transport de matière dans l'écorce terrestre». Sci Géol Mém (Strasb) 53:I–II
Millot G (1980a) Géochimie de la surface et formes du relief. In: Exposés de géologie. Acad Sci 290 (Suppl):1–18
Millot G (1980b) Les grands aplanissements des socles continentaux dans les pays subtropicaux et désertiques. Mém h. sér Soc Géol Fr 10:295–305
Millot G (1982) Planation of continents by intertropical weathering and pedogenetic processes. 2nd Int Seminar on Lateritisation processes, São Paulo, pp 53–63

Millot G, Bocquier G, Paquet H (1976) Géochimie et paysages tropicaux. La Recherche 7 (65):236–244

Millot G, Bocquier G, Boulet R, Chauvel A, Leprun JC, Nahon D, Paquet H, Pédro G, Rognon P, Ruellan A, Tardy Y (1979) Géochimie de la surface, pédogenèse, aplanissements et formes du relief dans les pays méditerranéens et tropicaux. Sci Géol Mém (Strasb) 53:39–43

Nahon D (1976) Cuirasses ferrugineuses et encroûtements calcaires au Sénégal occidental et en Mauritanie. Systèmes évolutifs, structures, relais et coexistence. Sci Géol Mém (Strasb) 44:232 pp

Nahon DB (1991) Introduction to the petrology of soils and chemical weathering. Wiley, New York, 313 pp

Nahon D, Millot G (1977) Géochimie de la surface et formes du relief. V. Enfoncement géochimique des cuirasses ferrugineuses par épigénie du manteau d'altération des roches mères gréseuses Influence sur le paysage. Sci Géol Bull (Strasb) 30:275–282

Nahon D, Melfi A, Conte CN (1989) Présence d'un vieux système de cuirasses ferrugineuses latéritiques en Amazonie du Sud, sa transformation in situ en latosol sous la forêt équatoriale actuelle. CR Acad Sci Paris 308 (II):755–760

Pimentel Da Silva L, Hodnett MG, Rocha HR, Cruz Senna R (1992) A comparison of dry season soil water depletion beneath central Amazonian pasture and rainforest. Proc VII Congr Bras Meteorologia, São Paulo, 28/09–02/10 1992, 308–313

Pédro G, Chauvel A, Melfi AJ (1976) Recherche sur la constitution et la genèse des Terra roxa estructurada du Brésil. Ann Agron 27 (3):265–294

Plancon O, Janeau JL (1990) Le fonctionnement hydrodynamique à l'échelle du versant. In: Equipe HYPERBAV (eds) Structure et fonctionnement hydropédologique d'un petit bassin versant de savane humide. Etudes et thèses, ORSTOM, Paris, pp 165–183

Sarrailh JM (1983) Les parcelles élémentaires d'étude du ruissellement et de l'érosion (programme ECEREX). Synthèse après quatre années d'étude. In: Le projet ECEREX (Guyane), pp 394–403

Soubiès F, Chauvel A (1984) Présentation de quelques systèmes de sols observés au Brésil. Cah ORSTOM, Sér Pédol 21:237–251

Tardy Y (1990) Recueil des documents relatifs à l'opération «Latérites d'Afrique» du Programme PIRAT-INSU-ORSTOM, Paris, Strasbourg, I et II

Tardy Y (1993) Pétrologie des latérites et des sols tropicaux. Masson, Paris, 461 pp

Tardy Y (1994) PIRAT Programme interdisciplinaire de recherche de biogéodynamique intertropicale périatlantique. 1. Rapport d'activité scientifique: synthèse et essai prospectif. 2. Climats, paléoclimats et biogéodynamique du paysage tropical. Sci Géol Mém (Strasb) 96:100 pp

Tardy Y, Melfi A, Valeton I (1988) Climats et paléoclimats périatlantiques. Rôle des facteurs climatiques et thermodynamiques: température et activité de l'eau, sur la répartition et la composition minéralogique des bauxites et des cuirasses ferrugineuses au Brésil et en Afrique. CR Acad Sci Paris 306 (II):289–295

Tricart J, Michel P (1965) Morphogenèse et pédogenèse. I. Approche méthodologique: géomorphologie et pédologie. Sci Sol I:69–85

Turenne JF (1975) Mode d'humification et différenciation podzolique dans deux toposéquences guyanaises. Thèse Sci, Mém ORSTOM 84:173 pp

Veillon L (1990) Sols ferrallitiques et podzols en Guyane septentrionale. Relations entre systèmes de transformations pédologiques et évolution historique d'un milieu tropical humide et forestier. Thèse, Univ Paris VI, 194 pp

Volkoff B (1984/1985) Organisations régionales de la couverture pédologique du Brésil. Chronologie des différenciations. Cah ORSTOM Sér Pédol 21:225–236

5 Evolution of Lateritic Manganese Deposits

Francis Weber

1 Introduction

Manganese is a transition element placed just before Fe in the periodic table. It is, after Fe, the second most abundant heavy metal in the lithosphere, at about 0.1%, which is about 50 times less but 12 to 15 times more than Ni and Cu. The rather special electronic configuration of the Mn atom enables it to form numerous compounds of valencies between –III and +VII, although in naturally occurring minerals, usually only +II and +IV valencies are found, and more rarely +III.

In igneous and metamorphic rocks, Mn occurs most often at the valency +II, as a substitution for Fe^{2+} and Mg^{2+} in Fe–Mg silicates; its abundance, of 400 ppm (granites) to 1000 ppm (ultrabasites; Crerar et al. 1980), is roughly proportional to the rocks' mafic origin. The +II valency form can also be found in sediments and supergenic formations, especially as a substitution for Ca^{2+} and Mg^{2+} in carbonates. At the valency +IV (and more rarely +III), Mn forms a great variety of oxides and hydroxides, more or less well crystallized. It is most abundant in carbonate, especially dolomitic rocks (>1000 ppm), not considering sedimentary ore deposits, oceanic nodules and rock varnish.

Geochemically, the behaviour of Mn in the hydrosphere is similar to that of Fe. It is soluble in the reduced state (Mn^{2+}), but almost insoluble when oxidized (Mn^{4+}). Nevertheless, the Mn^{2+} ion is less easily oxidized than the Fe^{2+} ion:

$$Mn^{2+} = Mn^{3+} + e^- \quad (e^o = -1.50\ eV)$$
$$Fe^{2+} = Fe^{3+} + e^- \quad (e^o = -0.77\ eV)$$

which explains to a large extent the distribution of Mn and Fe in sedimentary environments (Wedepohl 1980). However, there are two other important differences between Mn and Fe: (1) the solubility of Fe decreases considerably in a highly reducing environment due to sulphide precipitation (as pyrite), whereas that of Mn remains high because Mn sulphide (alabandite) can only form in extreme reducing conditions that are rare in nature (Force and Cannon 1988); and (2) Mn is much more amphoteric than Fe, which means that it can form manganates with a number of metals (K, Ba, transition metals) scavenging preferentially manganese to iron under certain conditions.

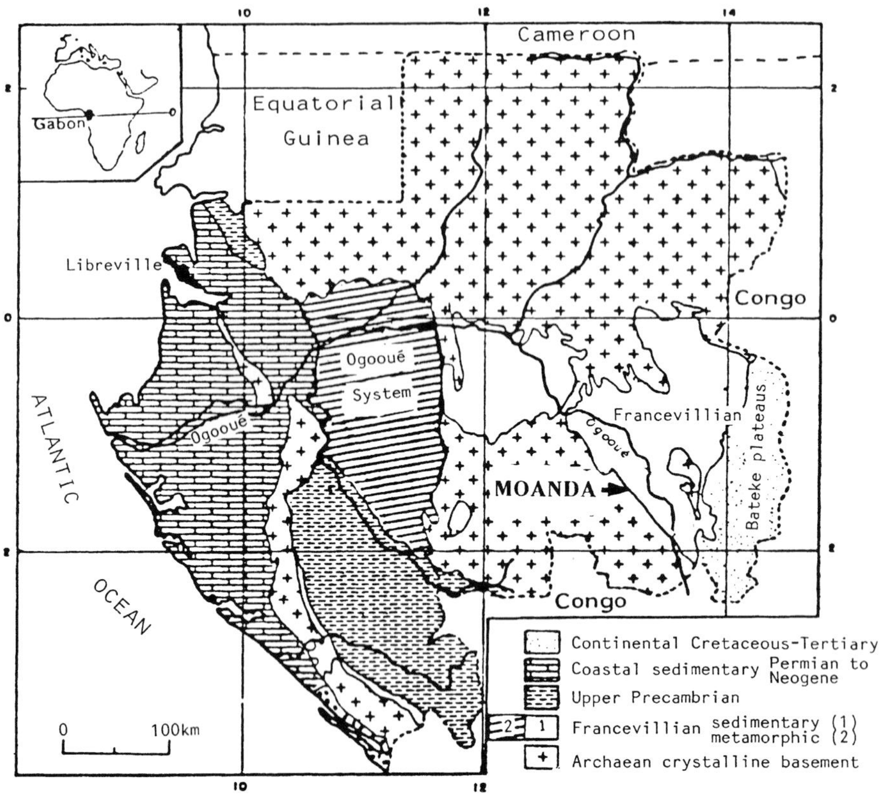

Fig. 1. Location of Gabon, the Francevillian series and the Ogooué System. (After Weber 1968)

The large Mn deposits which are presently mined in the world, are either of sedimentary or volcano-sedimentary origin. They have sometimes been metamorphosed (gondites with oxides and garnets, queluzites with carbonates and/or tephroite), possibly enriched by weathering and are most frequently found in lateritic environments. Such deposits are stratigraphically very unevenly distributed. Apart from oceanic nodules, which are not yet mined, two extensive periods of manganogenesis have been confirmed by recent discoveries. They were already described by Varentsov (1964) in Lower to Middle Proterozoic (Potmasburg and Hotazel in South Africa, Moanda in Gabon, Azul in Brazil, Amapa and Nsuta in Ghana and many small deposits in Brazil, West Africa and India) and in Cretaceous to Oligocene host rocks (Groote Eylandt in Australia, Imini in Morocco and Nikopol in Ukraine).

2 Example of the Moanda (Gabon) Sedimentary Manganese Deposit Enriched by Lateritization

Mining of the Moanda deposit which begun in 1962, made Gabon the world's third-largest Mn producer, about equal to Brazil depending on the year and following the former USSR and South Africa, with 2 million metric tonnes extracted per year.

2.1 Description of the Deposits

The deposits appear as one superficial layer which outcrops at the top of several plateaus and rests on the pelitic FB1 formation of the Francevillian Series (Figs. 1–4) which is a non-metamorphic Proterozoic series dated at about 2000 Ma (Bonhomme et al. 1982; Bros et al. 1992). The main mineral-bearing plateaus are those of Bangombé (40 km^2) being now mined, and Okouma (13 km^2). The mineralized layer comprises four different layers from base to top (Fig. 5; Plate I, 1–2):

1. A compact, very dense, Mn-rich, base layer of 0.2–0.5 m thick, which rests apparently in discordance on pelitic sediments of the FB1 Formation of the

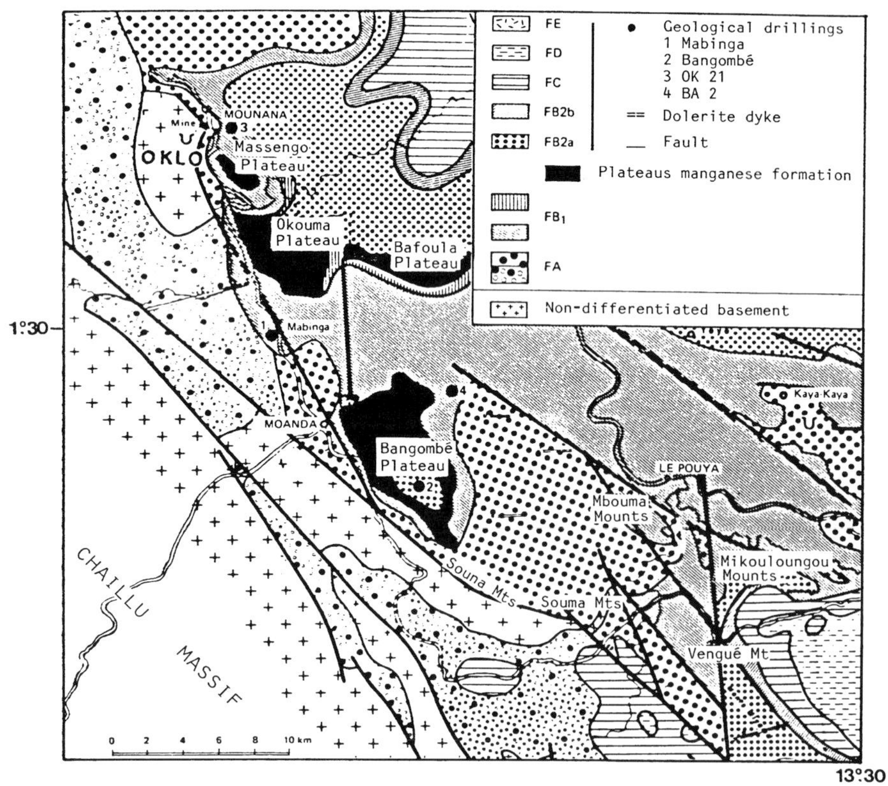

Fig. 2. Geological setting of the Moanda manganiferous plateaus. (After Weber et al. 1979)

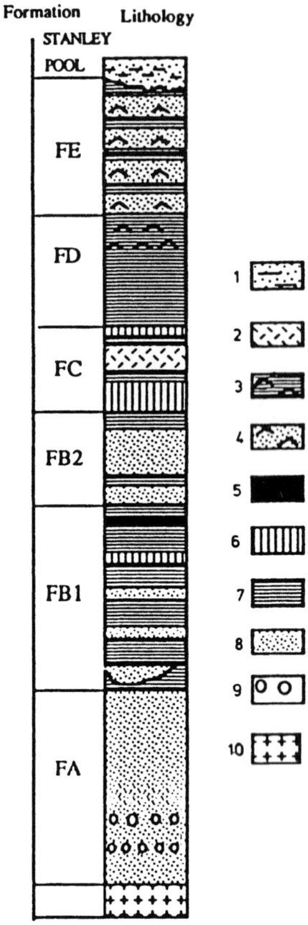

Fig. 3. Schematic stratigraphic section of the Francevillian sequence (after Gauthier-Lafaye and Weber 1989). *1* fine sandstones with kaolinitic cement (Meso, or Cenozoic); *2* cherts; *3* tuffs; *4* pyroclastic sandstones; *5* manganiferous protore; *6* dolomites; *7* shales and black shales; *8* sandstones; *9* conglomerates; *10* granite—gneiss basement (Archaean)

Francevillian Series. Usually a 2–5 cm band of pyrolusite occurs at the base of the ore (Plate I, 3). Locally, some beautiful, well crystallized pink geodes of rhodochrosite occur.

2. A "plaquette horizon" 3–9 m thick (5 m on average) overlies the base layer. The plaquettes (small platelets) of Mn oxides are 1 to a few cm thick and are organized as almost horizontal beds. They are surrounded by an ochre clayey matrix

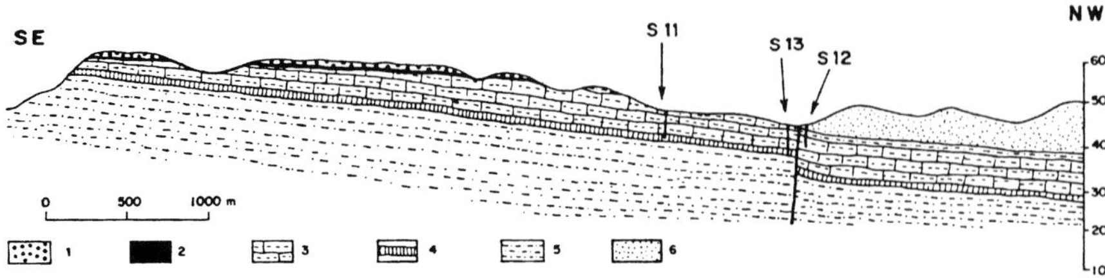

Fig. 4. Geological section of the Okouma plateau (after Leclerc and Weber 1980). *1* pisolites; *2* mined ore layer; *3* carbonate protore; *4* Fe formation; *5* pelitic formation (FB1); *6* sandstones (FB2a)

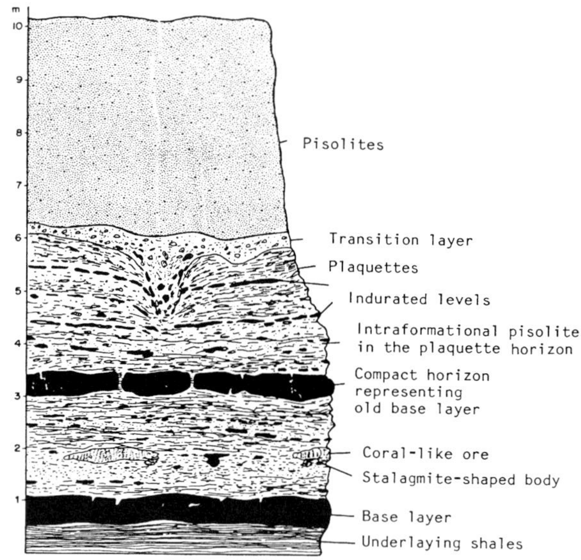

Fig. 5. Schematic section of the mined manganiferous layer (after Leclerc and Weber 1980)

sometimes containing small pisolites. This is the main productive layer of the deposit.

3. A 0.5–1 m thick transition horizon is clearly differentiated from the underlying platelet layers by the lack of bedding and the presence of blocks of very varied composition, together with manganiferous platelets. The large blocks of rough cavernous sandstone are mixed with aggregates of pisolites, while bauxite fragments are cemented by Mn oxides, and by fragments of ferruginous and manganiferous concretions.
4. A pisolitic layer of 5–6 m thick containing pisolites of 2–10 mm diameter in its ochre clayey matrix, is capped by a thin (10–40 cm) humus-bearing layer where some small pisolites are again observed.

Locally, peculiar sinkhole depressions occur at the base of the pisolitic horizon. They are present in the transition layer and penetrate the underlying horizon, where the platelets tend to arrange vertically (Plate I, 2). The horizons 3 and 4 (transition horizon and pisolitic layer), which are too poor in Mn to be mined, are removed before extraction of the platelet horizon and the base layer.

The true manganiferous cuirasses develop locally along the edge of plateaus and in depressions along the rivers: the layer is invaded by concretions of Mn oxides which replace the ochre, clayey matrix and cement plaquettes, pisolites and other fragments. This induration appears at the transition horizon and spreads progressively from the top into the plaquette horizon. On plateaus, margins or along streams that cut the plateaus, the cuirasses can form cliffs of massive ore or even very large boulders, which led to the discovery of the deposits but which often have to be left in place due to inappropriate extraction techniques for this type of ore, which is still exceptional in the deposit. The substratum of the ore-bearing layer can be of three types depending on the location:

Plate I. 1 Lower part of the mined layer. A horizon (*B*) with small plaquettes is noticeable between the present basal layer (*A*) and a degrading fossil base layer (*C*). **2** Upper part of the mined layer. A sinkhole depression (*D*) appears at the contact between pisolitic and transition layers (*B*). It reaches deeply into the plaquette layer (*A*) in which the plaquettes of manganese oxides are arranged vertically. These structures correspond to drainage (paths). **3** Fragment of the base layer. Two layers of well-crystallized pyrolusite (*B* and *D*) record the two successive positions of the oxidation front. The lower ribbon (*B*) cuts the «rhodochrosite shale» (*A*), whose bedding continues across the two pyrolusite bands and the intercalating oxides (*C*). **4** Base of deposit in a zone where the substratum is decarbonated. The «rhodochrosite shale» (*A*) gives way to a decarbonated «black shale» (*B*), which itself alters into a yellow-ochre clay. This clay is penetrated by manganese oxides (*D*) which develop under the old base layer (*B*). A new base layer (not seen on photo) is being formed a little lower down, at the top of the water table.

1. Usually it consists of slightly undulating black shales which are very light and friable and full of organic matter that stains the fingers. However, the shales are depleted in Mn except for some fillings in diaclases by secondary oxides. Rare intercalations of black sandstone occur, sometimes at the wall of the deposit.
2. A yellow-brown argillite identical to argillites at the «lithomargins» can be found at the base of the alteration profiles, which are observed everywhere on the pelitic formations of the Francevillian Series. This «argillite» sometimes bears thin beds of grey-white chert or units with «red shales» consisting essentially of Fe oxides and hydroxides. This lithology appears always at plateau margins, but also, locally, in their central parts.
3. A grey, compact, finely bedded rock, similar in appearance to dolomite, and rich in a rhodochrosite-type carbonate can also be formed. This «rhodochrosite shale» is crumpled at the contact with the «pyrolusite band»: between the beds which become detached from each other, small geodes of well-crystallized rhodochrosite develop. This facies is confined to the highest and most central areas of the plateaus.

2.2 Stratified Deposit or Alterite? The Search for the «Protore»

During the pioneering exploration of the Moanda deposit, Baud (1954, 1956) considered the deposit as residual, resulting from «a process quite close to that of laterite and bauxite formation». This hypothesis, however, raised a problem stated by Routhier (1963): «How can such a concentration of Mn come from a substratum that appears to be extremely poor in this element ?» Starting from the observation that occurrences of Mn which were discovered at several places in the Francevillian basin, seemed to be associated with a stratigraphic horizon well defined in the series, Bouladon et al. (1965) concluded that «lateritization altered initially an enriched banded ore by removing barren interbeds which remain only as residual of hydroxides of Fe and Al easy to eliminate by screening. Nevertheless, beyond the zones of cuirasse formation, it cannot have concentrated the Mn at the expense of Fe and Al, since it is much more soluble than they are in natural conditions of pH and Eh». This idea was supported by the discovery of a banded iron formation with siderite, greenalite, stilpnomelane and pyrite associated with the deposits of the Okouma plateau. Why could this ferruginous deposit not be the precursor to a Mn-bearing one ? It became important, therefore, to find the original ore untouched by lateritic alteration.

A 350 m deep borehole made in 1962 on a sandstone surface layer at the top of the Bangombé plateau, penetrated the whole FB1 formation without encountering the expected mineralized layer. Only several years later, after a systematic geochemical study of the deepest cores of Bangombé, was it possible to show that this borehole did in fact conceal the «protore» of the Moanda deposit (Weber 1968, 1973): a 75 m thick layer with an average Mn content of 13% reaching 25% in some places. Mn was located in a mixed carbonate $(Mn, Ca, Mg)CO_3$ phase which had not been distinguished from dolomite during the preliminary petrographic investigation of the core samples. It was then concluded that the layer being mined resulted

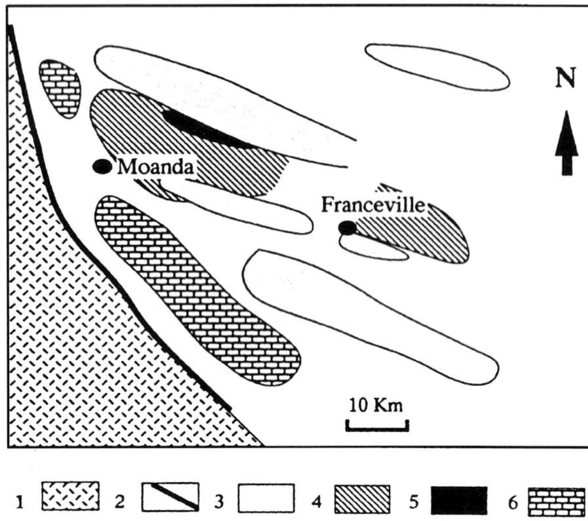

Fig. 6. Outline of the palaeogeography of the Franceville basin at the time of manganese deposition (modified after Azzibrouck 1986). *1* emerged continental area; *2* shoreline; *3* sand bars (FB2a); *4* manganiferous deposit; *5* Fe deposit; *6* stromatolitic, dolomitic and siliceous formations

from lateritic alteration of a carbonate «protore» analogous to that observed in the drilling cores. In fact, the process of modification and the supergene enrichment is still active at the top of the water table. The rhodochrosite or manganiferous mixed carbonate occurred underneath the «friable black shales» that can be considered as the decarbonated protore (Leclerc and Weber 1980). In the following years, deep drilling in the Bangombé plateau for U prospection revealed that the manganiferous carbonate formation continued to a depth of 85 m under the ore deposit, with Mn contents of up to 30% (Azzibrouck 1986). Thus, not only had the alteration front not reached the base of the carbonate protore, but also the latter, in the ore-rich part of the plateau, must have been much thicker than the 70 m found initially. It could also be concluded that the top of the manganiferous formation corresponds laterally to the sandstones, which confirms outcrop observations on the plateau.

The carbonate manganiferous formation was also found by drilling in other parts of the basin, especially on the Okouma plateau where it is clearly associated with preceding ferruginous pyrite units which were also found interbedded with it (Leclerc and Weber 1980). These levels are linked to the greenalite, siderite and pyrite-bearing ferruginous unit already mentioned. The outline of a carbonate manganiferous protore can thus be traced as it develops at the top of the FB1 formation of the Francevillian, which is an epicontinental basin or a filling lagoon divided into compartments by large sand bars (underwater dunes) extending over several kilometres (Azzibrouck 1986; Fig. 6). Towards the open sea, from which the manganiferous lagoon was separated by shoals, the manganiferous deposit gives way to a banded Fe deposit interbedded with silicates, carbonates and Fe sulphides. In the proximal zones, it passes into more or less silicified dolomites that are probably the result of stromatolitic algal constructions (Bertrand-Sarfati and Potin 1994).

2.3 Mineralogy and Geochemistry of the Protore

The manganiferous horizon consists mainly of carbonate-bearing black shales containing 5–30% Mn. Contents vary greatly due both to differences in the amounts of carbonates in the black shales interbedded with sandstone, dolomitic or sandy-dolomitic sediments, which are usually poorer in Mn than the surrounding black shales and to recurrences of the ferruginous formation, especially on the eastern edge of the Okouma plateau (Fig. 7). The ferruginous formation is not well differentiated on the Bangombé plateau, but Fe enrichment is nonetheless often seen there at the base of the manganiferous formation.

The overall average content in the manganiferous formation is 10–16% according to borehole data. The anomaly in Mn is accompanied by a significant anomaly in heavy metals such as Co, Ni, Cu and Zn, whose concentrations in manganiferous ampelites are 20, 5 and 3 times those in the non-manganiferous black shales of the Francevillian sediments, respectively (Table 1), the Mn enrichment amounting from 70 to 80 in the same rocks. Amounts of these elements are positively correlated with those of Mn and negatively with those of Fe (Table 2). Phosphorus is also enriched in manganiferous ampelites, three to four times, but it is

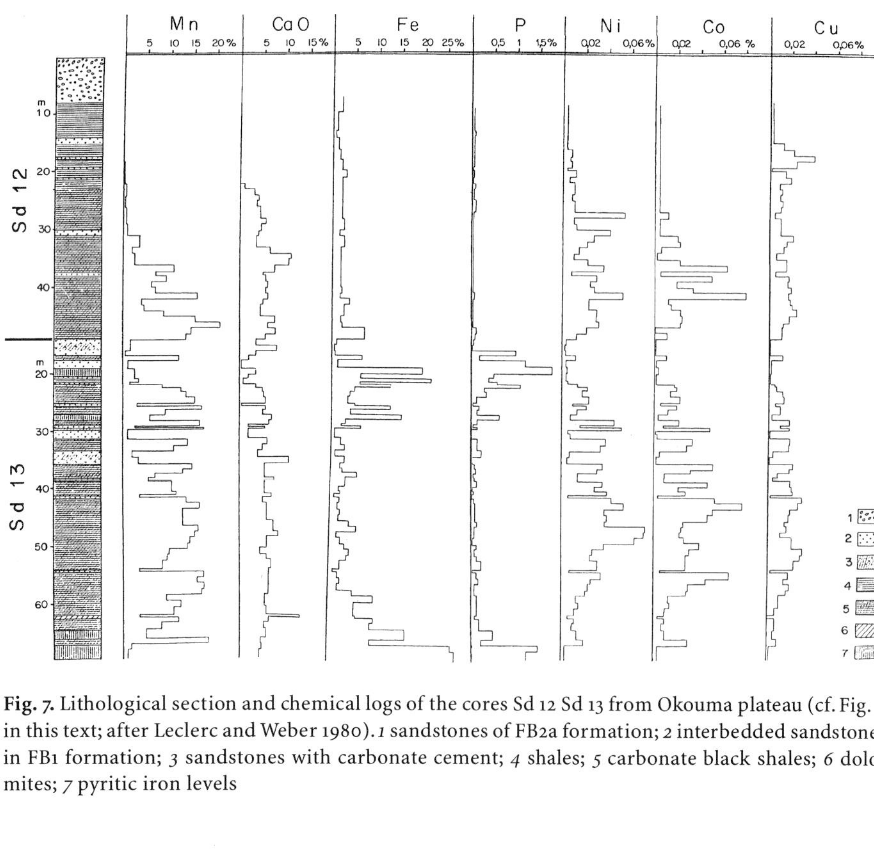

Fig. 7. Lithological section and chemical logs of the cores Sd 12 Sd 13 from Okouma plateau (cf. Fig. 4 in this text; after Leclerc and Weber 1980). *1* sandstones of FB2a formation; *2* interbedded sandstones in FB1 formation; *3* sandstones with carbonate cement; *4* shales; *5* carbonate black shales; *6* dolomites; *7* pyritic iron levels

	(1) Manganiferous black shales (91 samples)		(2) Normal black shales (153 samples)		(3) Ratio (1)/(2)
Mn	16.9	%	0.22	%	76
Fe	2.9		2.75		
SiO_2	25.4		57.2		
Al_2O_3	6.4		20.1		
MgO	1.14		1.2		
CaO	4.0		1.1		
K_2O	1.6		3.0		
P	0.18		0.05		
Ni	297	ppm	62	ppm	4.8
Co	397		17		23.4
Zn	265		92		2.9
Cu	171		53		3.2

Table 1. Chemical composition of manganiferous and normal black shales compared (FB1 formation, Bangombé plateau; after Azzibrouck 1986)

even higher in Fe-bearing layers where it reaches 1.5%. This results in a positive correlation between Mn and P on the Bangombé plateau where ferruginous horizons are rare, and in a negative one among the same Mn and P on the Okouma plateau where drillings showed that such material occurs frequently.

The black shales of the protore consist of a bedded, clayey matrix rich in organic matter, in which the carbonates are scattered. The main clay mineral is illite, chlorite being abundant only in the ferruginous units (ferruginous chlorite or greenalite). Pyrite occurs as aggregates, generally arranged with the stratification planes, sometimes as granules that could represent old recrystallized framboids. Amounts of silt-grade detrital elements (quartz, altered micas and feldspars) are low. Small lenses of chalcedony bordered by rhombohedra of carbonate and cubes of pyrite are present, but they are quite rare.

Petrographic observation, complemented by microprobe analyses (Fig. 8), showed that these carbonates form four groups (Plate II):

1. Dolomite s.s. occurs as large rhombohedral crystals almost always highly corroded (Plate II, 1–2). As a result of this corrosion, some rhombohedra appear

		P	Ni	Co	Cu	Zn
BA[a]	Mn	0.25	0.69	0.80	0.46	0.72
	Fe	0.26	0	−0.04	0	0
OK[b]	Mn	−0.03	0.43	0.58	0.40	n.d.
	Fe	0.54	−0.54	−0.45	−0.55	

[a] BA, on the Bangombé plateau, 451 samples; after Azzibrouck (1986).

[b] OK, on the Okouma plateau, 126 samples (after Leclerc and Weber 1980).

Table 2. Correlation between iron, manganese and trace elements in FB1 formation

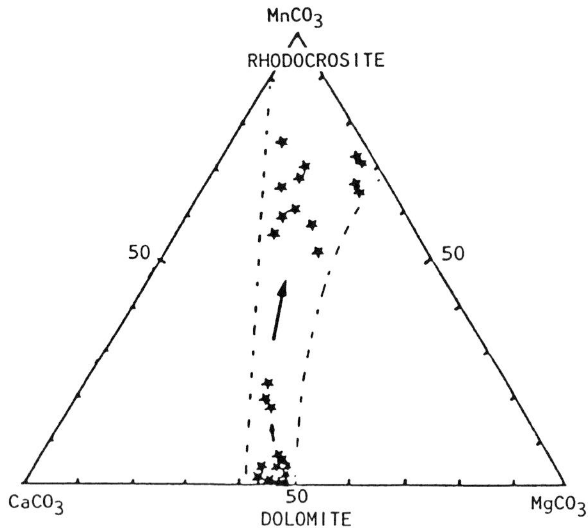

Fig. 8. Molar composition of carbonates of the manganiferous protore. (After Azzibrouck 1986)

emptied: only a thin skin remaining with the interior of the grains filled with newly formed clay and sometimes cryptocrystalline carbonates. This dolomite often contains a small excess of Ca (up to 55 molar%) and sometimes a little Fe (up to 2–3 molar%); it is not manganiferous.

2. Manganiferous dolomite also occurs as large clear rhombohedra sometimes making up the external coating of zoned crystals with dolomite s.s., as the core of the crystals and the different overgrowths are separated by clay films. Manganese, whose amounts do not exceed 20%, occupies the two sites of the dolomite network, but there is usually more Ca than Mg. The type composition of these crystals is: $CaCO_3$=45%, $MgCO_3$=35%, $MnCO_3$=20%. The Fe content does not exceed a few molars per cent.
3. Cryptocrystalline carbonates with very high Mn contents from 50–75 molar% (Plate II, 2–3 and 5) represent the third type of carbonate. The excess of Ca over Mg which is found in Mn-poor carbonates, no longer appears systematically. The reverse is just as likely as such an excess, especially in patches enriched in Mn ($MnCO_3$>65 molar%). These carbonates make up most of the carbonate phase; they are either disseminated in the matrix or grouped together as aggregates that sometimes present a roughly concentric structure, at the centre of which a rhombohedric crystal or a relic of an old, highly corroded rhombohedron can occur.
4. The last group which is much less represented (Plate II, 4), has been encountered only in Fe-rich facies. It is a carbonate of Fe, Mg and Mn or Fe and Mg (Boeglin 1981) which passes into true siderite in Fe-banded chert.

Manganese is found exclusively in the mixed carbonates phases, mainly as $(Mn,Ca,Mg)CO_3$. Fe is distributed between the sulphide (pyrite), carbonate and silicate phases, but occurs predominantly in the pyrite (about 80%), except in the sideritic iron deposits (Plate II, 4 and 6). Phosphorus is obviously localized in a Ca phosphate (apatite) well identified in thin section and by microprobe observations. The transition trace-elements, Co, Ni and Cu, appear to be linked to the sulphides. Leclerc and Weber (1980) observed large anomalies in Ni, Cu and Co in pyrite grains. Azzibrouck (1986) reported several analyses of pyrite containing 1.7% Cu, 0.9% Ni and 1.1% Co. He could not establish whether these elements were present in the form of micro-inclusions of Ni, Co and Cu sulphides, or as substitutions for Fe in the pyrite network, but he noticed that their distribution was very irregular. In the same crystal, one area can show a strong anomaly in one of these elements while these anomalies are totally lacking elsewhere. Although geochemically similar to these elements and strongly correlated with them, Zn has not been localized by microprobe. It could occur equally well in the carbonates as in the sulphides and may not give high-concentration anomalies detectable by this method (Azzibrouck 1986).

The ability of Mn oxides to scavenge transition elements, especially Co, Ni and Cu, is well known, especially in present-day oceanic nodules. The apparent paradox that Co, Ni and Cu are strongly correlated with Mn but mineralogically dissociated, could be explained by an initial precipitation of all these elements as oxides. In addition, the fact that the first carbonate precipitated in the manganiferous sediment is dolomite, suggests that the initial deposit contained a carbonate of Ca and Mg associated with Mn-oxide carriers of Co, Ni and Zn. Early in diagenesis, as the environment became strongly reducing and more acid due to abundant decomposing organic matter, the oxides of Mn, Fe and other transition elements, as well as the dolomite, were destabilized in favour of phases in equilibrium with the new environment: Mn, Ca and Mg (more rarely Fe) carbonates and Fe, Co, Ni and Cu sulphides. This mechanism of returning Mn into solution during early diagenesis explains its enrichment at the top of the FB1 Formation, when an emersion occurred. Before precipitation of Mn, the mechanism concentrated it in the interstitial water of the freshly deposited sediments or even in the deep water of the basin, until excess concentration caused precipitation as carbonates. Fe tends to precipitate earlier, as its sulphide is highly insoluble and its carbonate less soluble than that of Mn. Solutions then become richer in Mn than in Fe (Force and Cannon 1988). At each stage, as Azzibrouck (1986) emphasized, microorganisms must have played an important role in the precipitation of the different mineral phases. Fossil traces of bacterial activity have been described by Cortial (1985), Cortial et al. (1990) and Azzibrouck (1986). Scattered isotopic values for carbon of ferruginous and manganiferous carbonates ($\delta^{13}C=-14$ to $+1$ per mil), which contrast with the higher and narrowly ones of the dolomites ($\delta^{13}C=+2$ to $+6$ per mil), likely represent the contribution of organic C in carbonate diagenetic neoformations. The organic C has a much-lighter isotopic compositions ($\delta^{13}C=-20$ to $-46‰$) in the Francevillian sediments (Weber et al. 1983).

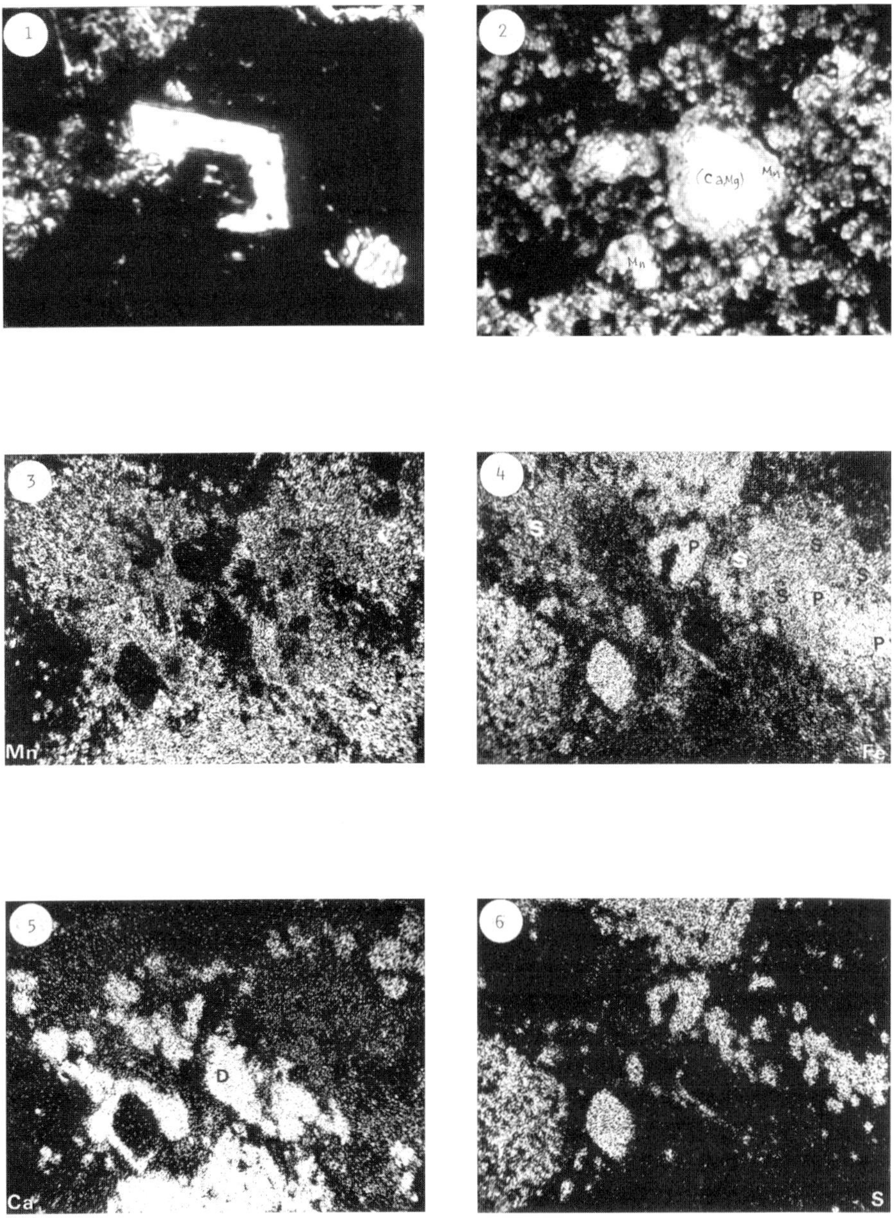

Plate II. **1** Corroded crystal (100μm large) of rhombohedral dolomite in the protore of the deposit. **2** Corroded crystal of dolomite (100μm large) replaced and surrounded by manganese carbonates. **3—6** Distribution of Mn (**3**), Fe (**4**), Ca (**5**) and S (**6**) in the same slice of a protore sample from deposit (x 300). Besides manganese carbonates which also contain Ca, other crystals can be seen: non-manganiferous dolomite (*D*), pyrite (*P*) and iron carbonates (*S*) near the siderite pole. The latter are, however, rare in the protore of this deposit

2.4 Supergene Weathering: Formation and Evolution of the Ore

The ore enrichment corresponds to a three- or a fourfold increase in Mn relative to the content in the protore. However, it is more informative to consider the ratios of the main elements with each other. The Fe/Al ratio, at about 1 in manganiferous shales (0.7–1.3), does not change significantly in the ore or in the pisolites (Table 3). In contrast, the Si/Al ratio shifts from about 3.8 in the protore to 0.4 in the ore and to 0.1 in the pisolites. These changes signify leaching of Si in a lateritic alteration profile tending towards bauxitization. Soluble elements such as Ca, Mg and Na are completely removed. Only K and Ba, which are fixed in cryptomelane-type oxides remain present in small amounts. On the other hand, the Mn/Al and Mn/Fe ratios increase from 4 to 5 in the protore to 8 to 10 in the «raw mined ore», and to 15 to 18 in the commercial-grade ore, and they decrease to about 1 in the pisolites. Mn enrichment in the mined unit corresponds, therefore, to an absolute accumulation. Since Fe and Al have virtually remained in place, it can be deduced that about half the Mn of the mined layer results from this accumulation. The relative paucity of Mn in the pisolites suggests a migration from top to base in the profile, which is confirmed by the relative enrichment of the base layer compared to the plaquette horizon. This does not exclude lateral migration, responsible in particular for the formation of cuirasses, signs of which are much more localized at the scale of the whole deposit. The successive stages of supergene alterations of the deposit can be followed through the different horizons of the profile.

	Mn	Fe	CaO	MgO	SiO_2	Al_2O_3	K_2O	PF	Fe/Al	Si/Al	Mn/Al	d	$\frac{Mn(gr.)}{100cm^3}$
1[a]	16.2	3.5	7.4	3.6	26.6	6.3	1.7	24.0	1.1	3.8	4.9	2.8	45.4
2[b]	25.9	2.6	0.4	0.4	27.3	3.1	1.0	28.3	0.8	4.1	8.4	2.9	75.1
3[c]	57.4	0.8	0.2	0.1	2.5	1.6	0.1	13.4	1.0	1.4	67.5	4.4	253
4[d]	43.6	4.7	–	–	5.1	10.5	–	–	0.8	0.4	7.8	2.1	92.9
5[e]	50.6	2.7	0.2	0.1	3.2	6.6	0.3	–	0.8	0.4	14.5	3.2	162
6[f]	13.7	12.5	0.05	0.03	4.5	31.3	0.13	21.0	0.8	0.1	0.8	–	–

[a] Intact protore: average for the Sd11 manganiferous black shales (Okouma), 9—57 m, 43 samples; after Weber et al. (1979).

[b] Carbonate substratum of cut deposit B8 (Bangombé), 3 samples; after Weber et al. (1979).

[c] Basal layer, cut B8 (Bangombé), after Weber et al. (1979).

[d] »raw mined ore», Bangombé 04/76, after Weber et al. (1979).

[e] Ore >16 mm 04/76 (commercial-grade ore), after Weber et al. (1979).

[f] Pisolites, 155 samples from Okouma (COMILOG analyses).

Table 3. Changes in concentrations of major elements in the main facies of the parent rock and the ore

2.4.1 The Parent Rock and its Development Towards the Surface

In the deep horizons of the weathering profile, about 10–20 m below the present topographic surface, triple carbonates (Mn, Mg, Ca)CO_3 are subject to epigenesis into rhodochrosite. This epigenesis is produced in a reducing environment, below the zone of pyrite oxidation. It induces formation of «rhodochrosite shales» in the substratum which constitute the true parent rock of the deposit, whose appearance is little different from the carbonate black shales of the protore. Particularly noticeable is a recrystallization of carbonates as wide patches, but these do not obscure the microstructure which is visible in natural light.

It is not difficult to describe the mechanism of this epigenesis. The Mn-loaded surface waters, acidic because of pyrite oxidation and themselves oxidizing, dissolve the carbonates. As they percolate through the sediment, which is rich in organic matter and sulphides, their oxidizing power is attenuated and their pH rises following reactions with the carbonates until the solubility product of one of the carbonates is reached, after which the solution will be buffered by reprecipitation of this carbonate. Generally this will be rhodochrosite, as Mn is abundant in the solution and the solubility product of its carbonate is much lower than that of Mg or Ca carbonates. Only Fe carbonate has a solubility product of the same order of magnitude as that of Mn carbonate. However, as Fe is present mainly as pyrite, it is released only in the oxidation zone, where it is fixed at a valency of 3 in hydroxides, its concentration being much lower than those of Mn in the percolating solutions.

Above the water table, the parent rock is depleted in carbonate, whatever its prior development (Plate I, 4). The decarbonatation front appears in distinct zones: a grey «rhodochrosite shale» is followed abruptly by a very light «black shale» which, in turn, gives way to a yellow-brown bedded clay towards the surface. There are no manganese oxide deposits at the decarbonatation front, but gypsum is observed lining the diaclases under it.

2.4.2 The Base Layer

According to Boeglin's (1981) description, the base layer often appears «made up of very hard manganiferous plaquettes, a few millimetres to one or two centimetres thick, with metallic blue-black or blue-grey fracture surface. These plaquettes are bound together, often discontinuously which leaves flat spaces between them». It usually rests on the substratum through an interlayered «band» of largely crystallized pyrolusite. Mamellated concretions sometimes develop and penetrate the substratum.

Besides the well crystallized Mn^{4+} minerals, mainly pyrolusite associated with a little nsutite and lithiophorite, the base layer contains large amounts of cryptocrystalline phases close to cryptomelane. Minerals with Mn^{3+} such as manganite and groutite, and Mn^{2+} such as rhodochrosite, also occur (J. Bouladon 1963, Le gisement de manganèse de Moanda [Gabon] – étude de la zone de première exploitation rapport BRGM 5313/MJMG, unpubl. work). However, distinction must be made here between two extremely different situations depending on whether the base layer is above or at the groundwater level. In the first case, it rests on a more or less altered decarbonated substratum and itself appears to be undergoing alteration. The $Mn^{2+}$and Mn^{3+}minerals are absent and Fe oxide and gibbsite coatings develop in the hollows between platelets, which tend to separate. In the second case, the base layer rests on the «rhodochrosite shale» substratum and indications are that it represents a developing base layer. The contact is always at the top of the water table or in the zone where it fluctuates, the variable conditions leading to precipitation of different minerals. Thus, one can see side-by-side, sometimes inside the same geodes, magnificent neoformations of manganiferous minerals at different valency states: bundles of pyrolusite with a metallic sheen, beautiful scalenohedra of pink or bright red rhodochrosite and fine black acicular crystals of manganite. Whoever has had the chance to look at these mineral bodies just after scouring by the bulldozers, has no doubt from their brilliance, that these neoformations are fresh. In fact, exposed to the weather, these minerals are oxidized and tarnish rapidly.

In addition, observed epigenesis of parent rocks by Mn oxides occurs in two ways:

1. Epigenesis, at the bottom of the layer, of rhodochrosite by manganite which gives way to pyrolusite by pseudomorphosis. This epigenesis was first described by J. Bouladon (1963, Le gisement de manganèse de Moanda [Gabon] – étude de la zone de première exploitation rapport BRGM 5313/MJMG, unpubl. work) and later by all authors who have dealt with the Moanda deposit (Bouladon et al. 1965; Bricker 1965; Weber 1968, 1973; Bouma 1970; Perseil and Bouladon 1971; Weber et al. 1979; Leclerc and Weber 1980; Boeglin 1981; Nziengui-Mapangou 1981; Nahon et al. 1983; Beauvais 1984). It corresponds to the advance of the oxidation front whose intermediate manganite stage is short lived (a few millimetres). The manganite/pyrolusite band is usually horizontal, but because of the slight undulations of the substratum, it is often secant on the stratification which it cuts sharply. Fine bedding can frequently be seen within the pyrolusite band, parallel to its base, which superimposes on the stratification without completely obliterating it. This bedding appears to mark the oxidation front. Under the microscope, it is expressed by alternation of zones of well-crystallized «palisade» pyrolusite in which the protore microstructures are largely obliterated, with zones of microcrystalline structures where the microstructures are much better preserved.
2. Epigenesis of the rhodochrosite schist by cryptocrystalline oxides has been described by Perseil and Bouladon (1971) who have shown that oxidation of the rhodochrosite rock could take place differently from the process just described. «It changes overall, but is keeping its structure, into a grey material, close to cryptomelane, which gives rise to MnO_2g, MnO_2r and ramsdellite». In this case, the pyrolusite band is, of course, absent.

It should be emphasized that, even when the first type of epigenesis with the pyrolusite band at the bottom of the layer is observed (which is far from being the most common case) the base layer consists essentially of a cryptocrystalline assemblage in which the microstructures of the protore are well preserved. Not only the relicts of carbonate micro-concretions are visible (fish roe structures; J. Bouladon 1963, Le gisement de manganèse de Moanda [Gabon] – étude de la zone de première exploitation rapport BRGM 5313/MJMG, unpubl. work), but also relics of detrital grains, mainly phyllosilicates subjected to epigenesis by Mn oxides. It is, therefore, unlikely that the cryptocrystalline matrix results from «retromorphosis» of well-crystallized basal-band pyrolusites in which these structures have been largely obliterated (Nziengui-Mapangou 1981; Beauvais 1984).

The base layer appears to be the horizon where most of the Mn accumulated in the deposit. This accumulation took place at the top of the water table (or more likely in the zone where it fluctuates). It appears that the «pyrolusite bands» fossilize a stabilization surface of the oxidation front. Although the active front with the complete sequence of rhodochrosite, manganite, pyrolusite, can only occur very near the bottom, fossil bands can be seen at various levels of the base layer (Plate II, 3). The second type of epigenesis corresponds, therefore, to a drop of the water table during which crystals would not have had time to grow and oxidation would occur faster than the removal of impurities, especially K which would be held in the cryptomelane-type constituents. These rapidly changing fronts correspond to short-lived situations which are seen much less frequently in the field than the stabilized fronts with pyrolusite bands. However, no quantitative appraisal can be made on the relative importance of the two phenomena in the formation of the ore.

2.4.3 The Plaquette Horizon

Different types of facies frequently appear immediately above the base layer. They are sometimes difficult to interpret, especially the «laminated», «vacuolar» and coral-like ores consisting of nsusite and lithiophorite associated with gibbsite (Nziengui-Mapangou 1981; Azzibrouck 1986). Also seen quite frequently at the base of the plaquette horizon, are large lenses of yellow-brown clay. These lenses are soft with a finely bedded structure and are made up of kaolinite, goethite, quartz and illite and are relics of altered parent rocks saved by epigenetic processes of the base layer. Nziengui-Mapangou (1981) described how these clays passed to highly indurated manganiferous plaquettes. The transition happens over a few centimetres, with gradual appearance of the black colour and a hardening process leading to the plaquettes. It involves the epigenesis of the yellow-brown clays by a mixture of lithiophorite and cryptomelane which subsequently evolves towards pyrolusite. This epigenesis which often develops from fractures or diaclases cutting the parent rock (Azzibrouck 1986), corresponds to the invasion of a completely leached parent rock by Mn oxides, following an abrupt drop of the water table (Weber et al. 1979; Leclerc and Weber 1980). Above these facies, broken-up relics of old base layers can be seen, whose structures disappear upwards in the plaquette horizon (Plate I, 1).

J. Bouladon (1963, Le gisement de manganèse de Moanda [Gabon] – étude de la zone de première exploitation rapport BRGM 5313/MJMG, unpubl. work), Nziengui-Mapangou (1981) and Beauvais (1984) have given detailed descriptions of the mineralogical evolution of the plaquettes along the profile, starting from relicts of a decomposing base layer. Initially, the pyrolusite of the base layer is replaced by cryptomelane which penetrates between crystals along cracks and joints. This process increases the microporosity, while fissures and horizontally elongated voids develop. Up in the horizon, nsutite gradually takes up from cryptomelane and becomes the essential constituent of the plaquettes beyond 1 m above the base layer. Strips of nsutite recrystallized into ramsdellite appear next and around empty spaces, and true well-crystallized ramsdellite can even be seen. In the upper part of the plaquette horizon, pyrolusite which almost disappeared from the middle part, appears again in subhorizontal bands or lenses as «palisade» crystals. Lastly, in the top few tens of centimetres of the horizon, the plaquettes are fragmented and a matrix with lithiophorite and aluminous goethite progressively develops at the expense of all oxidized manganiferous phases.

2.4.4 The Upper Horizons: Transition and Pisolite Levels

The transition and pisolite levels indicate a tendency towards bauxitization at the top of the profile. Large gibbsite nodules appear in the transition layer. The pisolites consist of concentric envelopes of gibbsite, aluminous goethite and lithiophorite around nuclei, usually a plaquette fragment (although sometimes it can be a quartz grain, a ferruginous fragment or even a broken, older pisolite). However, although the pisolites fit well the continuity of the evolution of the top of the plaquette horizon, the transition level, in spite of its name, looks more like a break than a transition. With a chaotic appearance and well defined boundaries with the neighbouring horizons, it appears to be the result of a reworking process. The profile would, therefore, be truncated and the upper two levels allochthonous. If that were the case, one would expect these two levels to be somewhere other than at the top of the deposit, which is not the case. Also, the transition horizon may reflect the original variations of the protore. The frequent sandstone fragments in this horizon lend support to this hypothesis.

2.5 Conclusion

The enrichment of the ore in the lateritic profile takes place at the same time by elimination of the elements that are most mobile in tropical alteration (Ca, Mg, Si) and by absolute accumulation at the bottom of the profile. The cases of K and Ba are noteworthy: fixed by cryptomelane, they are much less mobile than in a normal lateritic profile. One of the original features of the Moanda alteration profile is that the enrichment process begins in the carbonate-bearing parent rock by epigenesis of triple carbonates (Ca, Mg, Mn)CO_3 by $MnCO_3$. This epigenesis leads to an enrichment of 70–100% over the initial protore (Table 3).

The second stage of enrichment is the most spectacular. It occurs in the basal part of the intermediate zone of the water table where the supply of Mn is about 250% higher than in the rhodochrosite substratum and more than 450% higher than that in the protore. However, in the raw ore, the enrichment is only about 100% relative to the protore, hardly more than in the rhodochrosite country rock. Should it be concluded that the ore degradation starts from the plaquette horizon ? Mn certainly undergoes leaching at this level and it accumulates in the base layer, although some of it is lost. We have seen that internal rearrangements also take place in the plaquette horizon itself, and that part of the Mn leached from higher levels is reprecipitated lower down.

The real cause of this apparent impoverishment of the plaquette horizon compared with the base layer must be sought elsewhere. The base layer is not regularly embedded in the substratum, but randomly follows abrupt variations of the water table level. This can be a few decimetres, or even metres lower, the base layer degrading and breaking up into plaquettes when no longer in water. The underlying substratum is leached and the released Mn moves to the top of the displaced water table where a new base layer will start to develop (Fig. 9). The parent rock between the old base layer and the newly forming one, has leached and altered to a yellow-brown clay. It is invaded by remobilized Mn from higher levels, but the ochre clays largely remain in place and are incorporated in the raw mineral deposit. These clays constitute the matrix of the plaquettes which are eliminated from the commercial-grade ore by screening. The result is that the final absolute accumulation of Mn between the rhodochrosite parent rock and the raw mineral deposit is low. The process of absolute accumulation that occurs initially in the carbonate parent rock (epigenesis of triple carbonates by rhodochrosite) is, therefore, vital for the subsequent development of the ore.

Abrupt drops in the water table level result from the way plateaus are drained by kinds of microkarsts. Drainage courses, guided by a fracture network, can be observed in the mine and appear to initiate the sinkhole structures observed in the transition layer and in the plaquette horizon (Plate I, 2). Not only is there leaching of soluble elements but suspended clay particles are also mechanically transported, which explains these sinkhole depressions and the additional enrichment factor, its contribution being difficult to estimate. Capture of one drainage system by another can cause an abrupt drop in the water table level in any given part of the deposit. Such a phenomenon occurred some years ago on the Okouma plateau where a permanent lake, known to exist in human memory, suddenly and definitively emp-

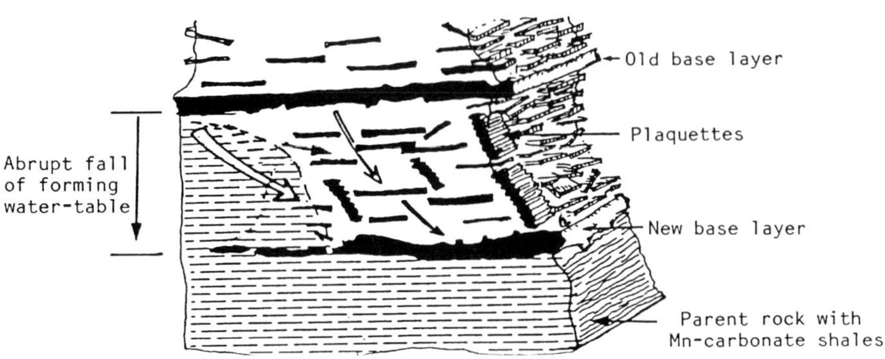

Fig. 9. Sketch showing the formation of a new base layer after an abrupt fall in the water table. (After Azzibrouck 1986)

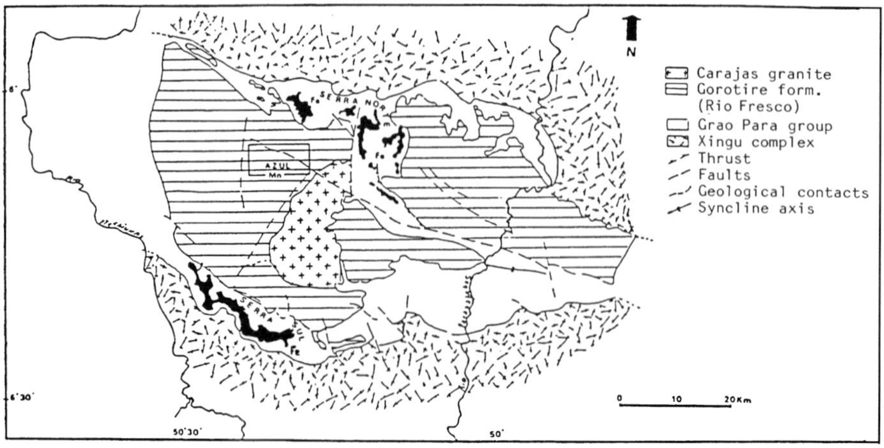

Fig. 10. Geological setting of the Azul deposit in the Serra dos Carajas. (After Beauvais 1984)

tied. In the deposit, similar events are emphasized by significant slidings of the base layer by as much as several metres, which in spite of appearances and misinterpretations, are not the result of fault throws.

3 Comparison With Other Manganese Deposits

3.1 The Azul Deposit

The Azul deposit, near the Fe fields of the Serra dos Carajas in the state of Para in Brazil (Fig. 10), looks like the Moanda deposit. It results from lateritic alteration of a Lower Proterozoic carbonate protore, as at Moanda, which belongs to the Rio Fresco formation (Bernardelli and Beisiegel 1978) which is equivalent to the Francevillian. Beauvais (1984) conducted a comparative study of the two deposits. The protore consists of two manganiferous units separated by a 60–100 m thick Mn-poor unit (4% on average). The lower unit, which is 20–40 m thick, is the richer (21–26% Mn), and the Mn content in the upper unit, which is 30–50 m thick, is no greater than 14%. The same correlation between Mn and base metals (Ni, Co, Cu, Zn and Pb) can be seen, as already observed on the Moanda protore (Figs. 11–12). The parent rock is made up of carbonate-bearing black shales and, as at Moanda, the carbonates sometimes have micronodular micritic structures, the carbonate being here close to the rhodochrosite end member (93% $MnCO_3$). The parent rocks, in fact, underwent alteration which is represented by kaolinization of feldspars and phyllosilicates and recrystallization of carbonates eventually replacing the detrital minerals epigenetically. Valarelli et al. (1978) attributed these changes to a «deep diagenetic alteration» (burial to 5000 m at 200 °C). The analogy of this parent rock with the surface carbonate-bearing rhodochrosite facies of Moanda is striking. But

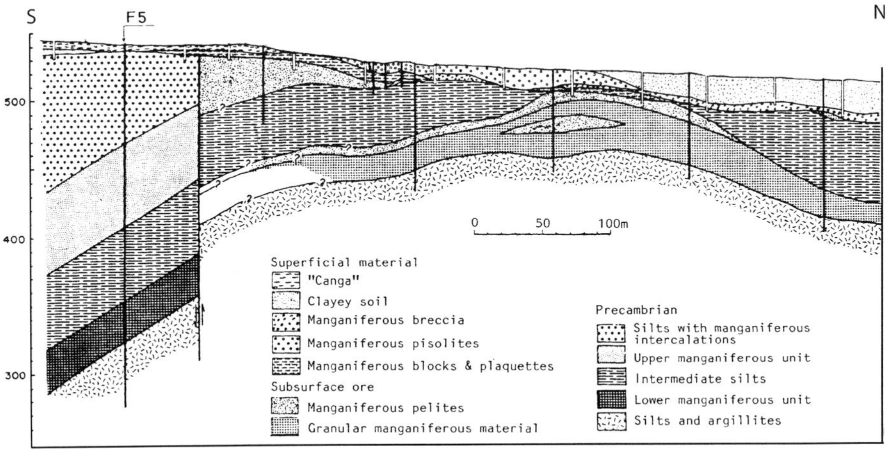

Fig. 11. Geological section of the Azul deposit. (After Bernardelli and Beisiegel 1978)

although we have here, as at Moanda, surface transformation of a mixed carbonate to rhodochrosite, this «alteration» took place to a depth of 200 m at least, instead of only 20 m.

The present climate is similar in the two regions, both being in the equatorial zone, but with a transitional tropical climate having only one marked dry season. The Azul deposit is covered by forest, whereas the Moanda plateaus are covered by scrub savanna. However, the dry season at Azul is longer and more marked, which certainly explains the abundance of haematite colouring the deposit red. The geomorphological context is very different from that of Moanda where the protore outcrops at the structural surface of vast plateaus. The Azul deposit rests on the slopes of the great ferruginous Carajas plateaus and the manganiferous formation, far from being in structural conformity, outcrops only because of an anticlinal structure (Fig. 11). The hydrodynamic situation in which the alteration profiles develop is very different from the one at Moanda: the permanent water table at the base of the deposit is lacking and oxidation of the carbonates is much more pervasive, to several tens of metres beneath the topographic surface.

At the surface, however, the two ore-bearing lithologies are comparable, apart from the colour which is red at Azul and yellow-brown at Moanda. Here too, there is a plaquette horizon, a «lateritic breccia horizon» similar to the «transition level» of Moanda and a pisolitic layer. The vertical arrangement of the profiles is, however, much less regular than at Moanda, especially in the pisolitic level which is extremely variable in thickness and extends well beyond the extent of the deposit. However, no exact equivalent of the Moanda «base layer» is observed. Instead, massive blocks of ore occur in a bedded structure at the bottom of the plaquette horizon. A «subsurface» ore of variable facies (lumpy, massive, pelitic or silty) develops between this «blocky horizon» and the bed rock.

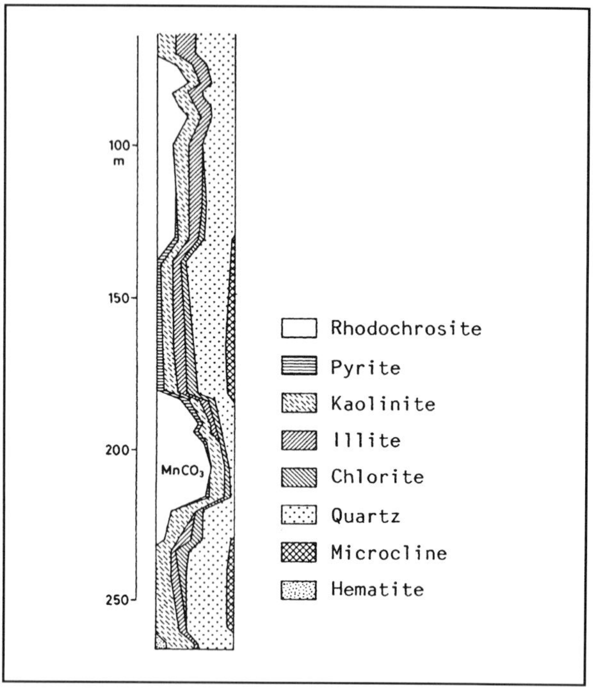

Fig. 12. Mineralogical section of core F5

The structure of the deposit allows the variations of the surface facies to be linked to those of the protore, which is not possible at Moanda. According to Beauvais (1984), «the lower series is changed into a granular ore with very dense plaquettes enriched to a mean Mn amount of 41.50%, whereas the intermediate series gives a reddish clay horizon whose Mn contents do not exceed 7 to 8% on average». It seems that the most enriched Mn^{2+} carbonates become the richest in Mn^{3+} and Mn^{4+} oxy-hydroxides. The oxidized ore formed along three different paths. The first starts from epigenesis of a reddish clay, the other two from rhodochrosite (Beauvais 1984):

1. Reddish clay → lithiophorite (weakly indurated plaquettes) → cryptomelane (indurated plaquettes).
2. Rhodochrosite → cryptomelane → nsutite → ramsdellite → pyrolusite → lithiophorite.
3. Rhodochrosite → todorockite → nsutite → cryptomelane.

The first two paths were already recognized at Moanda, but not the third. However, the transformation of rhodochrosite to manganite and pyrolusite, which is very typical for the base layer in Moanda, has not been observed here. This might result from different conditions, as those prevailing in the Moanda base layer seem not to have occurred anywhere at Azul.

3.2 Sites with Lateritic Alteration on Metamorphic Protore (Queluzites)

About 200 km NW of Moanda, the Francevillian platform series enter the mobile zone of the Ogooué System where they have been intensely folded and metamorphosed about 1900 Ma ago (Azzibrouck 1982; Bonhomme et al. 1982; Ledru et al. 1989).

The different units of the Francevillian can still be identified in some parts of the Ogooué System, for instance in a railway cutting near the Mingooué bridge where the manganese formation was found to be associated with micaschists, graphite schists, quartzites and amphibolites. This represents a Barrovian metamorphism reaching the amphibolite facies (Azzibrouck 1982). The only Mn mineral which was identified at this location is garnet spessartite. Besides this, many Mn deposits were found in metamorphic formations of comparable age in the old shields of Africa, India and Brazil. They attest to the importance of the Lower Proterozoic for metallogenesis potentials of Mn. In most cases, lateritic enrichment was necessary to produce economically workable concentrations.

3.2.1 The West African Deposits

Many Mn ore deposits were found in the West African shield, but only that of Nsuta (Ghana) was of major economic interest. In the context of research prompted by Georges Millot, a study of the supergenic enrichment of these deposits was undertaken by Grandin (1976) who ordered the different stages of supergenic concentration of Mn in the framework of the evolution of the landshape induced by lateritic alteration. Simultaneously and later, studies of the mineralogical evolution of these deposits during weathering were made by G. Grandin himself, E. A. Perseil, D. Nahon and P. Nziengui-Mapangou. The overall conclusion of these studies is that the mineralogical evolution depends on the mineralogical assemblages of the parent rocks which are more complex in metamorphic protores than in sedimentary ones. Many Mn minerals, such as rhodochrosite, garnets (spessartite), tephroite, rhodonite, braunite; hausmanite, etc, were found. At Nsuta (Ghana) and at Tambao (Burkina Faso), the main Mn-bearing mineral is rhodochrosite, at Mokta (Côte d'Ivoire) it is spessartite (Perseil and Grandin 1978) and at Ziemougoula (Côte d'Ivoire) it is tephroite (Nziengui-Mapangou 1981). The main process of enrichment is often the epigenetic replacement of the protore minerals by Mn oxides at the base of the alteration profiles, as is the case at Moanda.

The sequence observed in the carbonate protore of Nsuta, is similar to that seen at the base layer of Moanda: rhodochrosite → manganite → pyrolusite. Cryptomelane develops later, followed by nsutite and ramsdellite which replace pyrolusite (Perseil and Grandin 1978). In the deposits of Tambao (Perseil and Grandin 1978) and Ziemougoula (Nziengui-Mapangou 1981), birnesite appears after, or instead of manganite, and it is replaced by cryptomelane and nsutite. Tephroite, the main carrier of Mn in the protore of the Ziemougoula deposits, changes into manganocalcite and smectite which develop towards a mixture of rhodochrosite and amorphous silica. This rhodochrosite evolves further like the other Mn carbonates of the protore. The complete alteration sequence in this deposit is: carbonates → manganite (or todorokite) → birnesite → cryptomelane + nsutite (+ lithiophorite) → pyrolusite → lithiophorite + aluminous goethite (Nziengui-Mapangou 1981).

Alteration of garnet spessartite often leads directly to the association of the lithiophorite and aluminous goethite (Grandin and Perseil 1977; Perseil and Grandin 1978; Nziengui-Mapangou 1981), but in many cases the garnet quartzite protore is epigenized by cryptomelane, which affects quartz and non-manganiferous silicates, as well as Mn-bearing minerals (Perseil and Grandin 1985).

The Mn concentrated in the lower part of the profile, comes from the superficial horizons from which it has been removed. Therefore, a Mn-depleted horizon with Mn and Fe concretions and pisolites appears near the topographic surface. Sometimes this Mn-depleted horizon can be cemented again by secondary Mn oxides which can also invade soils and colluvia near the protore outcrop, the true Mn cuirasses being formed this way (Grandin 1976; Perseil and Grandin 1978). These cuirasses which can have a high Mn content (>50%), constitute an important part (15 to 20%) of the ore of the Mokta deposit (Grandin and Perseil 1977). The main Mn-minerals of cuirasses are cryptomelane, nsutite and ramsdellite, but pyrolusite and lithiophorite can be found locally (Perseil and Grandin 1978).

In summary, these deposits show evolutions that are similar to that of the Moanda and Azul deposits, in spite of very different composition of the protore. Epigenetic replacements of the primary minerals by Mn oxides at the base of the alteration profile are followed by a cementation by concretionary Mn oxides in the median and upper parts of the profiles, which leads to the formation of cuirasses. The more stable mineralogical phases in the ores are cryptomelane, nsutite and ramsdellite; manganite, birnesite, todorokite and even pyrolusite being only transitional phases of protore oxidation (Perseil and Grandin 1978). Lithiophorite is, as it is at Moanda, an ultimate alteration phase of Mn oxides (Nziengui-Mapangou 1981), but it appears earlier on garnet alteration profiles (Perseil and Grandin 1978; Nziengui-Mapangou 1981).

3.2.2 The Lafaiete Deposit of Brazil

Boeglin (1981) conducted a study with the same objectives on the alterations of the Morro da Mina deposit in the Lafaiete district (Minas Gerais, Brazil). Unlike at Ziemougoula, the triple carbonates are the protore's main carrier of Mn; their composition approaches that of the Moanda protore: $(Mn_{0.54}, Ca_{0.4}, Mg_{0.11}, Fe_{0.01})CO_3$. The various events which followed the metamorphism–retromorphosis, the hydrothermal episode and the meteoric weathering, lead to the progressive evolution of these carbonates towards an almost unsubstituted rhodochrosite (Boeglin et al. 1980). By oxidation, this rhodochrosite evolved into manganite (or groutite) which then developed towards nsutite, cryptomelane and pyrolusite (Bittencourt 1973).

4 Conclusion

The Paleoproterozoic Eon was especially favourable for concentration of sedimentary manganese. In addition to the deposits discussed here, those of the Indian shield and those of the Potmasburg-Hotazel area of the Transvaal, which are probably the largest reserves of Mn ore in the world, should be mentioned. Most of these deposits have undergone supergenic enrichment which has concentrated Mn to an economical ore. These alterations are usually of the laterite type.

The worldwide simultaneous – geologically speaking – sedimentation of a considerable tonnage of Mn is undoubtedly related to great climatic changes taking place during this period, probably caused by the appearance of oxygen in the atmosphere (Holland 1962; Chandler 1978). During the Archaean and the early Proterozoic, virtual absence of oxygen and high concentration of CO_2 in the atmosphere bore upon the composition and properties of the ocean waters in which divalent Mn, along with Fe, could remain in solution at much higher levels than at present. These elements were undoubtedly provided to the ocean waters mainly by volcanism and submarine hydrothermalism, as they are today. For the base of the Proterozoic ocean, Mn contents of the ocean waters can be estimated to be similar to those found in enclosed euxinic environments (like the Black Sea), i.e. 200–500 ppb, which is over 100 times the average concentration in present-day oceans. This represents a stock of Mn in the order of 10^{12} tons in solution in the oceans. By comparison, the deposits represent fixed reserves of a few hundred millions of tons. The increase in the oxygen level in the atmosphere, therefore, induced massive precipitation of Mn, following that of Fe which occurred a few hundred million years earlier (in the form of impressive formations of ferruginous quartzite deposited around 2000 Ma, exemplified by the Carajas site).

Mn was concentrated at basin margins, in reducing environments comparable to present-day euxinic basins where Fe precipitated as sulphides, which explains why Mn is enriched more than Fe. It was deposited with the black shales, as oxides, which could either change into carbonates during early diagenesis, or immediately into carbonates. Microorganisms probably played an important role in the Mn precipitation, as they do presently.

Most of these old rocks were metamorphosed and incorporated in the ancient shields (Birrimian) in which are found many metamorphic occurrences of Mn designated as «grondite» (deposits with oxides and garnets) or queluzites (containing carbonates and tephroite); however, some old rocks are still preserved from both metamorphism and erosion. This is the case in the Francevillian series of Gabon, in the Rio Fresco in Brazil and the Transvaal in South Africa, where the major economic deposits occur.

Laterite enrichment leads to mineral sequences that vary slightly depending on the nature of the starting minerals and local conditions, but which show the same broad tendencies. Rhodochrosite s.s. without Ca or Mg substitutions is only seldom a primary mineral of the protore. It develops at the bottom of alteration profiles at the expense of a manganiferous carbonate (manganocalcite or triple carbonate of Ca, Mg and Mn) or other minerals such as tephroite. Oxidation of primary minerals generally proceeds through a transitory Mn^{3+}phase (manganite, groutite, birnesite, torodockite) which usually evolves towards cryptomelane with, sometimes, an intermediate pyrolusite stage. The latter appears less stable than cryptomelane at the base of profiles that are still rich in K. Higher in the profile, nsutite appears, possibly with ramsdellite, followed by a second generation of pyrolusite. However, all can occur together. Lastly, at the top of profiles all species disappear, giving way to lithiophorite which is usually associated with gibbsite and goethite. Also, development of pisolitic concretions is generalized at this level. When manganese oxides invade a previously altered and leached rock, the first mineral formed is usually lithiophorite which can subsequently evolve towards cryptomelane.

The remarkable aspect throughout this whole evolution is a general tendency for Mn oxides to epigenize all pre-existing materials whatever their nature, manganiferous (carbonates, silicates) or non-manganiferous minerals (mainly the phyllosilicates). It is the amphoteric nature of Mn that allows it to become fixed in the form of compounds with other metals such as Ba, K (cryptomelane) or Al (lithiophorite) by epigenetically replacement of pre-existing minerals. This mechanism of epigenesis leads to absolute accumulation of Mn at the base of profiles where it may be enriched in ore concentrations, not only relatively to soluble elements that are evacuated (mainly Si, Ca and Mg), but also relatively to less soluble or mobile elements such as Fe and Al. It is this propensity of Mn oxides for epigenesis that makes lateritic alteration so efficient in the elaboration of the Mn deposits. On the other hand, an obvious depletion in Mn is seen at the top of profiles as opposed to Fe and especially Al which is represented in gibbsite, aluminous goethite and lithiophorite. The very top pisolitic horizons, therefore, typically tend towards bauxitization if the climate is favourable.

References

Azzibrouck AG (1982) Lithostratigraphie, microtectonique et métamorphisme dans la vallée de l'Ogooué entre Booué et N'Djolé, Gabon. Mém DES, Univ Nancy I, 41 pp

Azzibrouck AG (1983) Contribution à l'étude des gîtes métallifères précambriens du Gabon. Rapport DEA, Univ Strasbourg, 51 pp

Azzibrouck AG (1986) Sédimentologie et géochimie du Francevillien B (Protérozoïque inférieur). Métallogénie des gisements de manganèse de Moanda, Gabon. Thèse, Univ Strasbourg, 214 pp

Azzibrouck AG, Weber F, Besnus Y, Edou-Minko A (1990) Relations géochimiques entre le manganèse, les éléments-traces Ni-Co-Cu et la matière organique dans les ampélites francevilliennes (Protérozoïque inférieur, 2000 Ma) du bassin de Franceville, Sud-Est du Gabon. CR 15 Coll Géol Africaine, Nancy

Baud L (1954) Notice explicative de la feuille Franceville-Est, carte géologique de reconnaissance au 1/500 000, Brazzaville. Direction des Mines et de la Géologie de l'Afrique Equatoriale Française, 34 pp

Baud L (1956) Les gisements et indice de manganèse de l'Afrique Equatoriale Française. XX Congr Int Geol, Mexico Colloque sur les gisements de manganèse, vol II, pp 21–30

Beauvais A (1984) Concentrations manganésifères latéritiques. Etude pétrologique de deux gîtes sur roches sédimentaires précambriennes. Gisements de Moanda (Gabon) et Azul (Brésil). Thèse, Univ Poitiers, 155 pp

Bernardelli AL, Beisiegel VR (1978) Geologia economica da jazida de manganes do Azul. Anais XXX Congr Brazil Geol, Recife 4:1431–1444

Bertrand-Sarfati J, Potin B (1994) Microfossiliferous cherty stromatolites in the 2000 Ma Franceville Group, Gabon. Precambrian Res 65:341–356

Bittencourt AVL (1973) Contribucào as estudo genético do minério de manganès de Conselheiro Lafaiete, Minas Gerais. Thèse, Univ São Paulo, 81 pp

Boeglin JL (1981) Minéralogie et géochimie des gisements de manganèse de Conseilherio Lafaïete au Brésil et de Moanda au Gabon. Thèse, Univ Toulouse, 155 pp

Boeglin JL, Melfi AJ, Nahon D, Tardy Y (1980) Solutions solides caractérisant plusieurs générations de rhodochrosites dans les protores des districts de Conselheiro Lafaïete au Brésil, et de Ziemougoula en Côte d'Ivoire. CR Acad Sci Paris 291 (II):3–15

Bonhomme MG, Gauthier-Lafaye F, Weber F (1982) An example of Lower Proterozoic sediments: the Francevillian in Gabon. Precambrian Res 18:87–102

Bouladon J, Weber F, Veysset C, Favre-Mercuret R (1965) Sur la situation géologique et le type métallogénique du gisement de manganèse de Moanda près de Franceville (République gabonaise). Bull Serv Carte Géol Als Lorr (Strasb) 18:253–276

Bouma (1970) Contribution à l'étude des minéralisations du bassin de Franceville (Gabon). Origine, types métallogéniques, évolutions. Thèse, Univ Lille, 73 pp

Bricker O (1965) Some stability relations in the system MnO_2-H_2O at 25 °C and one atmosphere total pressure. Am Mineral 50:1296–1354

Bros R, Stille P, Gauthier-Lafaye F, Weber F, Clauer N (1992) Sm-Nd isotopic dating of Proterozoic clay material: an example from the Francevillian sedimentary series, Gabon. Earth Planet Sci Lett 113:207–218

Caen-Vachette M, Vialette Y, Bassot JP, Vidal P (1988) Apport de la géochronologie isotopique à la connaissance de la géologie gabonaise. Chron Rech Min 441:35–54

Chandler FW (1978) Proterozoic redbed sequences of Canada. Geol Surv Can Bull 311:49 pp

Cortial F (1985) Les bitumes du Francevillien (Protérozoïque inférieur du Gabon, 2000 Ma) et leurs kérogènes. Relations avec les minéralisations uranifères. Thèse, Univ Strasbourg, 83 pp

Cortial F, Gauthier-Lafaye F, Lacrampe-Couloume G, Oberlin A, Weber F (1990) Characterization of organic matter associated with uranium deposits in the Francevillian formation of Gabon (lower Proterozoic). Org Geochem 15:73–85

Crerar DA, Cornick RK, Barnes HL (1980) Geochemistry of manganese: an overview. In: Varentsov IM, Grasselly GY (eds) Geology and geochemistry of manganese, vol 1. Schweitzerbart'sche Verlag, Stuttgart, pp 293–334

Force ER, Cannon WF (1988) Depositional model for shallow-marine manganese deposits around black shale basins. Econ Geol 83:93–117

Gauthier-Lafaye F, Weber F (1989) The Francevillian (Lower Proterozoic) uranium ore deposit of Gabon. Econ Geol 84:2267–2285

Grandin G (1976) Aplanissements cuirassés et enrichissement des gisements de manganèse dans quelques régions d'Afrique de l'Ouest. Thèse, Univ Strasbourg, Mém ORSTOM, 82, 1976, 275 pp

Grandin G, Perseil EA (1977) Le gisement de manganèse de Mokta (Côte d'Ivoire). Transformations minéralogiques des minerals par action météorique. Bull Soc Géol Fr 19(2):309–317

Holland HD (1962) Model for the evolution of the Earth's atmosphere. In: Engel AEJ, James HL, Leonard BF (eds) Petrological studies, a volume to honor A.F. Buddington. Geol Soc Am, Boulder, 660 pp

Holland HD (1984) The chemical evolution of the atmosphere and oceans. Princeton University Press, Princeton, 582 pp

Leclerc J, Weber F (1980) Geology and genesis of the Moanda manganese deposits, Republic of Gabon. In: Varentsov IM, Grasselly GY (eds) Geology and geochemistry of manganese, vol 2. Schweitzerbart'sche Verlag, Stuttgart, pp 89–109

Ledru P, Eko N'Dong J, Johan V, Prian JP, Coste B, Haccard D (1989) Structural and metamorphic evolution of the Gabon orogenic belt: collision tectonic in the lower Proterozoic? Precambrian Res 44:227–241

Nahon D, Beauvais A, Boeglin JL, Ducloux J, Nziengou-Mapangou P (1983) Manganite formation in the first stage at lateritic manganese ores in Africa. Chem Geol 40:25–42

Nziengui-Mapangou P (1981) Pétrologie comparée de deux gîtes supergènes manganésifères. Gisement de Ziemougoula (Côte d'Ivoire) et de Moanda (Gabon). Thèse, Univ Poitiers, 100 pp

Perseil EA, Bouladon J (1971) Microstructures des oxydes de manganèse à la base du gisement de Moanda (Gabon) et leur signification génétique. CR Acad Sci Paris 273 D:278–279

Perseil EA, Grandin G (1978) Évolution minéralogique du manganèse dans trois gisements d'Afrique de l'Ouest: Mokta, Tambao, Nsuta. Mineral Deposita 13:295–311

Perseil EA, Grandin G (1985) Altération supergène des protores à grenats manganésifères dans quelques gisements d'Afrique de l'Ouest. Mineral Deposita 20:211–219

Routhier P (1963) Les gisements métallifères. Masson, Paris, 1282 pp

Valarelli JV, Bernardelli AL, Beisiegel VR (1978) Aspectos geneticos do minerio de manganès do Azul Anais. XXX Congr Brazil Geol (Recife) 4:1670–1679

Valarelli JV, Hypolito A, Weber F (1980) The manganese deposit of Azul Brazil. XXVI Congr Int Géol (Paris) VII:1024

Varentsov IM (1964) Sedimentary manganese ores. Elsevier, Amsterdam, 119 pp

Weber F (1968) Une série précambrienne du Gabon: le Francevillien. Sédimentologie, Géochimie, relation avec les gîtes minéraux associés. Mém Serv Carte Géol Als Lorr (Strasb) 28:328 pp

Weber F (1973) Genesis and supergene evolution of the Precambrian sedimentary manganese deposit at Moanda (Gabon). Earth Sci 9:307–322

Weber F, Leclerc J, Millot G (1979) Epigénies manganésifères successives dans le gisement de Moanda (Gabon). Sci Géol Bull (Strasb) 32:147–164

Weber F, Schidlowski M, Arneth JD, Gauthier-Lafaye F (1983) Carbon isotope geochemistry of the lower Proterozoic Francevillian series of Gabon (Africa). Terra Cognita 3:220

Wedepohl KH (1980) Geochemical behaviour of manganese. In: Varentsov IM, Grasselly GY (eds) Geology and geochemistry of manganese, vol 1. Schweitzerbart'sche Verlag, Stuttgart, pp 335–351

6 The Lateritic Nickel-Ore Deposits

JEAN-JACQUES TRESCASES

1 Geochemical Characteristics of Nickel

1.1 Nickel in Endogenous Conditions

Nickel belongs to the transition metal family and shows close chemical similarities with Fe and Co. Even more than these elements, it is concentrated in silicated Fe–Mg minerals by octahedral substitution with Fe^{2+} ions. Thus, the mean mineral/matrix partition coefficient of Ni is 14.0, 5.0 and 2.6 for basaltic whole rocks, for olivine, and orthopyroxene and clinopyroxene, respectively. The high Ni contents in olivine (3000–4500 ppm), spinel (3000–3500 ppm) and orthopyroxene minerals (650–1000 ppm) emphasize its high affinity for the first formed minerals during fractional crystallization of magmas. This is due to the electronic configuration of the Ni^{2+} ion: the coordinating octahedron of oxygen atoms is more stable with Ni^{2+} than with Fe^{2+} or Mg^{2+} atoms in the centre (Burns 1970). Ni cannot be defined as a rare element in the universe (in decreasing order of abundance, it takes the 14th place, before Na, K or Mn), but the average crustal content of 75 ppm emphasizes its low lithophile characteristics. Ni is more abundant in materials of mantellic origin, especially in ultrabasic rocks with a world average content of 1450 ppm; the content being as high as 3000 ppm in dunites (Turekian 1978).

Ni can also combine with S or eventually other elements (As, Fe, Co) to form sulphides and sulpho-arsenides (pentlandite, bravoite ...). In the presence of S in a silicated Fe–Mg magma, the partition coefficient of Ni is highest in the sulphide phase. The immiscibility of melted sulphides and melted silicates induces a gravitational segregation of the sulphide minerals to the base of magmatic systems. This model is well illustrated in Archaean ultramafic bodies like komatiites (Western Australia) and in stratiform complexes (Sudbury, Canada).

1.2 Nickel in Exogenous Conditions

Most of the endogenous Ni-bearing minerals (olivine and, to a lesser extent, orthopyroxene) are at least partially serpentinized as a result of hydrothermal alteration. Part of the Fe and Ni from these primary silicates forms oxides, such as magnetites, during the serpentinization process. Thus, serpentines are systematically richer in Mg than olivines and pyroxenes.

During weathering, the main Ni-containing primary minerals (olivine or sulphide) are rapidly destroyed in oxidizing conditions. The behaviour of Mg, Fe and Ni is then very different: Mg is highly soluble as a divalent ion and rapidly removed from the weathering profiles, Fe is oxidized to Fe^{3+} which is highly insoluble above pH 3.0 and accumulates in these profiles as goethite FeOOH or haematite Fe_2O_3, and Ni behaves in a unique manner, as the ultrabasic nature of the parent rocks does not favour formation of clay minerals in the weathering profiles, because these rocks are very poor in Al, Ni being partly a substitute. The high stability of the coordinating octahedra around the Ni^{2+} ion, as described above, also favours the formation of nickeliferous trioctahedral phyllosilicates, the so-called garnierites. Very often garnierites consist of a mixture of two series of isomorphous minerals, each series joining a Mg pole and a Ni one: the Mg-kerolite/Ni-pimelite family which are talc-like minerals (d_{001}~10 Å), and the Mg-serpentine/Ni-nepouite family (d_{001}~7 Å; Maksimovic 1966, 1975; Faust 1966; Brindley and Hang 1973; Brindley and Maksimovicz 1974; Brindley and Wang 1975; Brindley 1978; Brindley et al. 1979). Other nickeliferous silicates such as chlorite, vermiculite, talc, smectites, sepiolite have been described, but they are less frequent than 10 Å and 7 Å garnierites. Although Ni substitutes for Mg in the octahedral sites of these silicates (Cervelle and Maquet 1982; Manceau et al. 1985; Manceau and Calas 1986), this only occurs in purely nickeliferous clusters, as shown by Decarreau et al. (1987) for Ni–Fe–Mg smectite.

Ni can also accumulate as oxides and hydroxides in the weathering profiles: (1) Mn–Co and other trace-element oxides (so-called asbolanes with structures mainly of cryptomelane type or lithiophorite type; Chukrov et al. 1983; Manceau et al. 1987); and (2) Fe oxi-hydroxide goethite (Schellman 1978; Kuhnel et al. 1978; Maquet et al. 1981; Gerth 1990). In these phases, Ni is substituted in the oxide or hydroxide structures, although it is always poorly crystallized.

In summary, Ni is expressed in multiple mineralogical forms in exogenous environments in contrast to endogenous conditions. The stability of the Ni-bearing secondary minerals is controlled by the physicochemical conditions of the solutions (Eh, pH, chemical composition and concentration, speed of renewal, ...) which change from bottom to top of the weathering profiles.

2 Nickel-Ore Deposits

Nickel is a metal of a great economic importance as it is used to produce stainless steels and alloys. Two major types of deposits are known: concentrations of sulphides in old ultrabasic rocks (from Archaean to Proterozoic in age), and lateritic weathering of ultrabasic rocks which were mainly formed since the Cenozoic and occur now in tropical zones. In these lateritic covers, we can distinguish silicate from oxide ores, according to the distribution of silicates or oxides among the nickeliferous minerals. The two types of ore are generally associated in the same deposit (Avias 1978). Nodules of ocean deeps composed of Mn and, to a lesser extent, of Ni could also be interesting Ni reserves in the near future, but they are not economically workable today.

Until the end of the 19th century, all Ni produced was from sulphide-type ores. The lateritic type was discovered in New Caledonia by Garnier (1867), and has been more and more observed at the surface of ultrabasic bodies which were subject to tropical climatic conditions. In 1982, 65% of the worldwide production of Ni (630 000 t of metal) was supplied by the sulphide-type ores. But 72% of the worldwide reserve (97 000 000 t of Ni; Deyoung et al. 1985) are of laterite type. Nearly one half of the total reserves of Ni are in New Caledonia, the other half being distributed in a restricted number of countries, because of the rare occurrence of large areas of ultrabasic rock outcrops.

The sulphide-type ores are still predominantly used for Ni production for many reasons: they are often the only deposits occurring in industrial countries outside the tropical area, they can be valorized, provided that they can be concentrated by non expensive physical techniques, and they often contain sub-products, such as Cu and Pt, which are found less commonly in laterite ores, and the recovery of which valorizes the sulphide-type ore. In contrast, the lateritic ores (silicated or oxidized) very often need more expensive processing with a high consumption of energy.

3 Petrology of Lateritic Weathering Profiles on Ultrabasic Rocks

Above fresh ultrabasic rock, the top of which is often highly distorted, the following typical sequence of weathering layers occurs in tropical profiles (Trescases 1975; Lelong et al. 1976; Troly et al. 1979; Golightly 1981; Schellmann 1983; Fig. 1): a weakly weathered rock (saprock), a partially weathered rock with conserved lithological fabric (coarse saprolite), an entirely weathered rock with collapsed fabric (ferruginous saprolite or "yellow laterite"), and a red soil with ferruginous concretions ("red laterite"). The respective thicknesses of these successive layers may vary according to: climate (and palaeoclimate) which controls the weathering process, mineralogical composition and fabric of the bedrock, and topographic position of the profile in the landscape. The two layers which are of economic interest are the coarse saprolite, with a range of thickness from a few cm to 10 m (silicate Ni ore), and the ferruginous saprolite (yellow laterite), the thickness of which can reach several tens of metres (oxide nickel ore). A negative correlation is frequently observed between the thickness of these two layers. The superficial red-soil level is almost always found with a thickness of 1–10 m. The presence of a sub-superficial ferricrete near the top of the red laterite is not systematic. Finally, some profiles have a silcrete layer at the base, or, more rarely, at the top of the profile(Melfi et al. 1980 ; Oliveira et al. 1992).

This sequence of weathering layers is typical for profiles on ultrabasic rocks almost devoid of Al. It can be compared to lateritic profiles developed on silico-aluminous rocks (Tardy and Nahon 1985; Nahon 1987; Table 1). Lateritic weathering

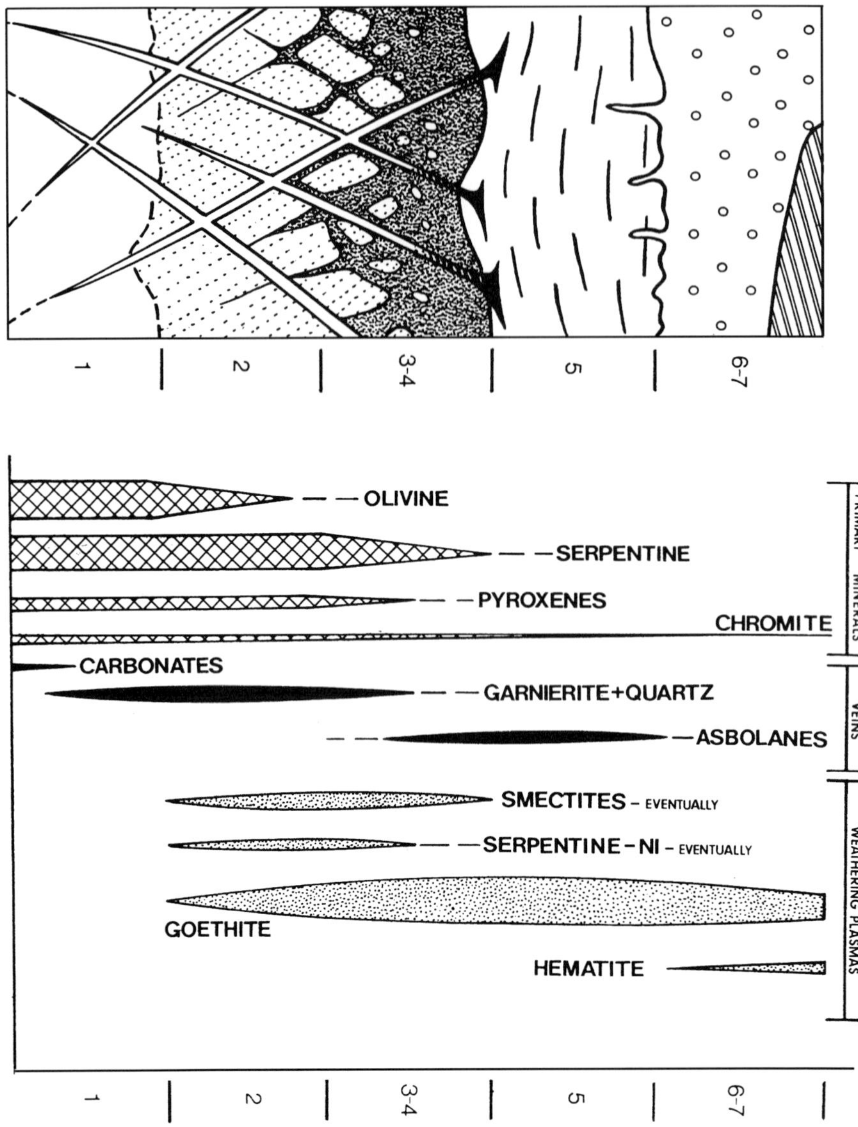

Fig. 1. Ni-laterite profile with the mineralogical composition of the successive layers. Numbers *1—7* indicate the various layers: *1* fresh bed-rock; *2* weakly weathered rock; *3* coarse saprolite (bottom); *4* coarse saprolite (top); *5* ferruginous saprolite (yellow laterite); *6* nodular red soil (red laterite); *7* Fe crust (sometimes present)

leads to a valuable accumulation of Ni ore when the parent rocks are dunites or harzburgites which have been more or less serpentinized. Olivine and serpentine are the two prime minerals having the highest Ni content. The most standard fabric is the so-called mesh fabric, with septa of serpentine isolating olivine nuclei. The

Ultrabasic rocks	Basic to acid rocks
Saprock and coarse saprolite Ferruginous saprolite (yellow laterite)	Isalterite Clay Alloterite (kaolinic)
Red soil with Fe concretions (sometimes ferricrete)	Mottled clay, ferricrete and sandy- clay superficial layer

Table 1. The sequence of weathering layers for profiles on ultrabasic and on silico-aluminous rocks

other primary minerals are less abundant (pyroxene, mostly orthorombic, chromiferous spinels), and normally play an accessory role in the weathering balance and the accumulation of Ni. The situation is completely different for some phases of hydrothermal origin, for instance finely grained magnetite closely associated with serpentine and chlorites, because these two minerals contain Ni. In contrast, talc, amphiboles and carbonates ($CaCO_3$, $CaMg(CO_3)_2$, $MgCO_3$), which are frequent but subordinate hydrothermal phases devoid of Ni, are not important in the Ni balance during the weathering process. However, early dissolution of carbonates in the filling cracks, induces voids where weathering solutions percolate and where neoformed nickeliferous silicates (garnierites) and quartz precipitate (Colin 1985; Colin et al. 1990).

The saprock layer (weakly weathered rock) can be distinguished by fast weathering of the olivine nuclei. The weathering process begins from borders and cracks in the olivine crystals. The weathered areas show a brown material (weathering plasma). This weathering plasma is ferruginous and nickeliferous (up to 25% Fe_2O_3 and 5% NiO). That means a relative high concentration of Ni (Besset 1978; Trescases 1979), a fast leaching of Mg and, to a lesser extent, SiO_2. At least in the primary stages the weathering plasma is amorphous to cryptocrystalline; it transforms later into goethite, with a smaller content of Ni than the initial plasma. Sometimes, weathering of olivine leads to ferruginous and nickeliferous smectites. At this point, the serpentinous septa are still intact and will continue to exist long after the disappearance of the last olivine crystals. However, the yellow-brown colour of this serpentine mesh emphasizes the oxidation of the associated magnetite granules. It can also result from filling of voids by ferruginous and nickeliferous products coming from weathering of olivines.

After disappearance of the last remains of olivine, progressive weathering of serpentines starts in order to form a saprolite layer with a conserved lithological fabric. This process is initiated from cracks in the fresh rock, where solutions percolate: on both sides the rock is transformed into a clay-like material with blocks of weakly weathered wall rock. Then, close to the top of the layer, the blocks

	SiO_2	MgO	$FeO+Fe_2O_3$	Al_2O_3	Cr_2O_3	NiO	CoO	H_2O
7	0.1-1	0.1	72-75	4.4-6.0	5.0-5.5	0.2-0.4	0.01-0.14	10-14
6	1-2.8	0.2-1.6	65-74	4.0-8.0	2.0-6.2	0.1-1.2	0.08-0.14	8-13
5	3-8	0.5-2	60-74	4.0-6.4	1.0-4	0.4-2.5	0.04-0.2	10-13
4	26-38	3.0-20	20-50	2.0-4.6	0.6-2.0	0.4-4.0	0.04-0.2	10-13
3	33-41	23-29	15-25.5	0.7-2.9	0.6-1.3	0.4-3.5	0.03-0.08	10-14
2	34-43	30-33	10.5-13	0.5-1.3	0.4-0.7	0.3-3.8	0.02-0.03	12-14
1	37-44	36-45	8.0-12	0.4-1.2	0.4-0.6	0.25-0.40	0.01-0.02	1-14

1, fresh bed-rock;
2, weakly weathered rock;
3, coarse saprolite (bottom);
4, coarse saprolite (top);
5, ferruginous saprolite (yellow laterite);
6, nodular red soil (red laterite);
7, Fe crust. The ore corresponds to the layers 2—4 (silicated ore) and 5 (oxidized ore).

Table 2. Chemical composition (ranges of weight in %) of nickeliferous laterite layers

become rounded and their size and proportion decrease. The weathering of serpentinous septa can be achieved in different ways. It can be close to dissolution, leaving a small amount of nickeliferous goethite with residues organized as a replica of the previous network of serpentine; in this case, the coarse saprolite layer is not very thick. Alternatively and more frequently, a second generation of nickeliferous serpentine (from 4–8% NiO) is neoformed in the space created by the hydrolysis of the first generation, and this second generation is further transformed into nickeliferous goethite. When the weathering plasma of olivine is smectitic, the last alternative possibility is that serpentine is also transformed into ferruginous and nickeliferous smectite; in this case the coarse saprolite layer will be developed to the greatest extent. During this stage, pyroxenes disappear (sometimes weathered into talc) frequently leaving goethite or smectite relicts. Garnierite and quartz in the cracks of the rocks also disappear and the cracks are covered with black asbolane coatings which are Mn–Ni–Co oxides.

In the layer of the deeply weathered rock with a collapsed fabric (yellow laterite), weathering of serpentine always leads to nickeliferous goethite (with only 1–2% NiO). This goethite results from replacement of the serpentinous network. But the goethite relicts collapse as soon as serpentine has completely disappeared, and the initial fabric of the rock vanishes. Within the brown–yellow matrix, mainly composed of Fe_2O_3 (Table 2), crystals of chromiferous spinels are progressively corroded, but talc and chlorite inherited from the rock may possibly persist. The cracks are still covered with black coatings of asbolane.

The top of the previous layer shows a change of colour, from brown–yellow to brown–red, which characterizes the upper layer of the red soil with ferruginous concretions (red laterite) and represents a mineralogical and chemical evolution. Most of the Ni is removed from nickeliferous goethite, asbolanes coatings disap-

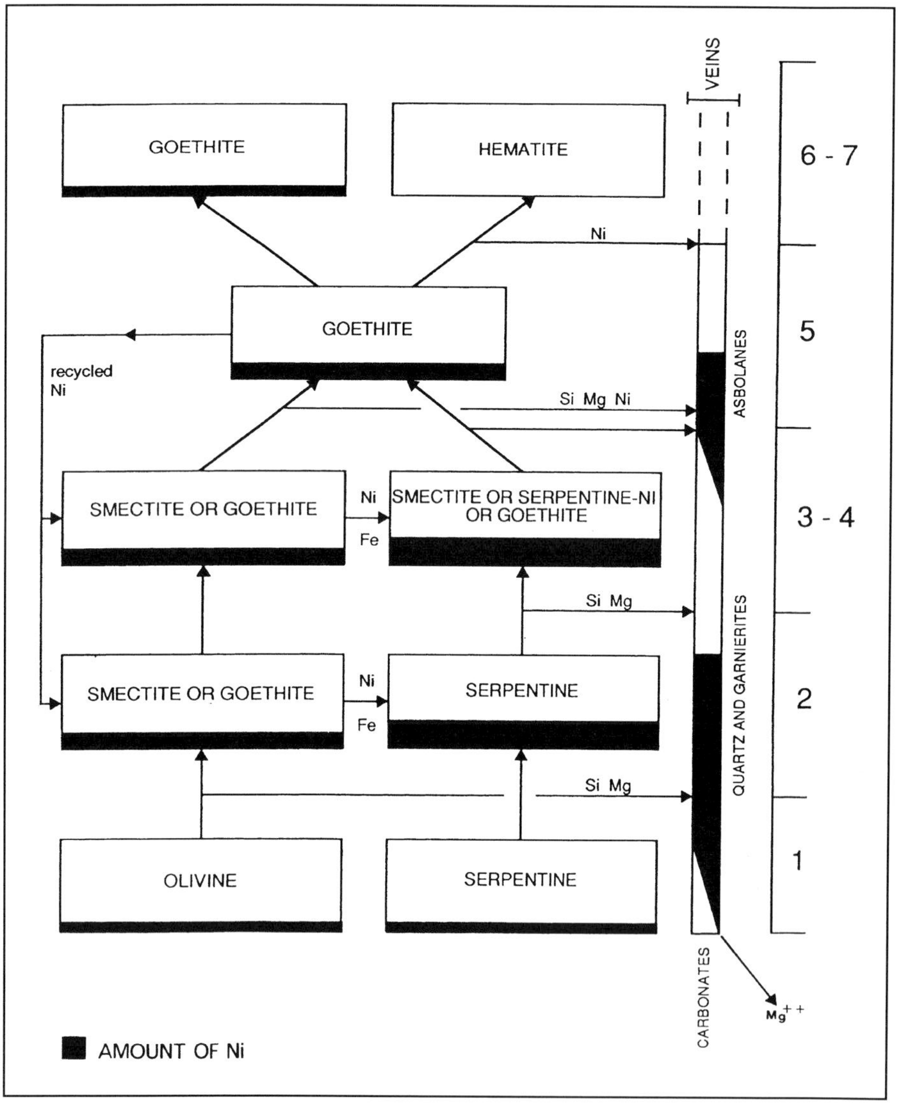

Fig. 2. Main mineralogical sequences occurring during the laterite weathering of ultrabasic rocks and chemical-element transfers (from one mineral to another and from one layer to another). The *numbers* on the *right side* identify the layers (see Fig. 1)

pear and haematite nodules are formed within the ferruginous matrix (Schwertmann and Latham 1986). Haematite precipitates are concomitant with Ni leaching (Ouangrawa 1990; Ouangrawa et al. 1996). With the extension of the process, the nodules increase in size and are cemented, which may lead to the formation of haematite and goethite ferricrete with a thickness generally limited from 0–1 m.

4 Geochemical Interpretation of the Evolution of Weathering Profiles

4.1 Geochemical Balance of Vertical Profiles

Table 2 summarizes the chemical compositions of the successive layers of a vertical profile, which have been described above and are represented with their mineralogical compositions in Fig. 1. All the mass-balance calculations (Millot and Bonifas 1955) applied to various profiles show the same geochemical behaviour of the elements. (1) The removal of Mg is total and instantaneous, proportionally to the weathering of Fe–Mg minerals (first olivine, then pyroxene and finally serpentine). (2) The leaching of Si is less pronounced in the saprolitic layers, which can contribute to neoformation of quartz, garnierites, nickeliferous serpentine of second generation and Ni–Fe–smectites, but the removal of Si from lateritic layers is almost complete. (3) Fe is relatively accumulated, as well as small quantities of Al, Cr, Mn present in the rock; Fe represents most of the superior part of the profiles as oxide and oxihydroxide. (4) The accumulation of Ni is absolute, sometimes important, at the base of the profiles (silicate ore), and relative in the intermediate part (yellow laterite forming the oxide ore), but Ni is removed in the red laterite and ferricrete. The Ni lost from top of the profile seems to supply the accumulation at the base, with a turnover of this element.

4.2 Control of the Geochemical Evolution

The main mineralogical filiations with exportation of products and successive recycling of Ni are shown in Fig. 2. The geochemical evolution of the ultrabasic rock components seems to be controlled by three parameters:

1. The delay in the weathering of serpentine compared to olivine, which results from their standard Gibbs free-energy of formation DG°f (Trescases 1975). Serpentine is stable in the concentrated solutions maintained at the base of the profiles by the weathering of olivine. However, in presence of large amounts of Ni, serpentine is transformed into a more nickeliferous serpentine-like mineral because of the crystal field stabilization energy induced by Ni^{2+} ions in octahedral sites. Higher in the profile where olivine is absent, the solutions are more dilute and serpentine is destabilized. In the open cracks where diluted solutions percolate rapidly, Si and eventually Ni can precipitate (quartz, garnierite).
2. The oxidizing conditions which transform all of the Fe released from Fe–Mg minerals into Fe^{3+}. Fe precipitates early as goethite which is an efficient trap for the released Ni. Higher in the profile, the oxidation of Mn^{2+} into Mn^{4+} and that of Co^{2+} into Co^{3+} induce the formation of asbolane precipitates, another scavenger of Ni.
3. The pH conditions which are alkaline at the bottom as a result of the hydrolysis of olivine, favour the immobilization of Ni. After disappearance of the last

silicate minerals, the pH becomes acidic (pH 5), which favours mobility of Ni towards the bottom of the profile, although Fe is not mobilized.

4.3 Evolution of the Landscapes

The geochemical evolution is vertical, provided the drainage of the profiles is vertical. As the weathering front lowers progressively, time will favour a higher absolute accumulation of Ni at the base (silicate ore). This will also favour the relative accumulation of a thicker layer (oxide ore), provided the latter is not dismantled too rapidly at its top. Finally, the tendency is towards "climatic levelling" (Millot 1982), as shown by all laterite Ni deposits which are systematically associated with "levelled surfaces". In addition to vertical drainage, a lateral circulation of the soil solutions can start and it can even become major in very plane surfaces. The upper part of the landscape then plays an equivalent role at the top of a profile (leached environment); the downstream part being equivalent to the base of a profile (confined environment; Millot 1964; Tardy 1969; Lelong et al. 1976). Ni which accumulates at the base of the downstream profiles, is not only imported from the top of the profile located above, but also from the upstream profile.

Finally, and often with the influence of tectonic uplift, different levelled surfaces can be affected in sequences along the slopes, the oldest being upstream and the most recent downstream. The weathering features of these surfaces can differ, the smectite pathway being more characteristic of the lower surfaces as in New Caledonia (Trescases 1975). In this case, the upper parts of the landscape with their old profiles, can induce preconcentration of Ni (protore), faster accumulation in the lowest parts having been described in Brazil (Melfi et al. 1980; Oliveira and Trescases 1980, 1985; Trescases et al. 1987; Oliveira et al. 1992).

5 Discussion: Role of the Weathering Factors in the Formation of Deposits

Weathering of ultrabasic rocks under tropical climates induces accumulation of Ni ore according to the interactions of different factors (Trescases 1975; Troly et al. 1979; Oliveira et al. 1992).

5.1 Role of the Bedrock and the Geological Context

The Ni content of the parent rocks depends on their petrographic nature, essentially their initial Ni content in olivine. Only dunites and harzburgites with Ni >2000 ppm allow a valuable accumulation. However, a smectite weathering of pyroxenites at Niquelàndia (Brazil) traps the Ni issued from dunitic profiles located upstream (Paquet et al. 1987; Colin et al. 1990).

The grade of serpentinization (hydrothermal alteration) affecting olivine, governs the relative development of the different layers. A nearly monomineral rock (a fresh dunite with 100% olivine or alternatively a serpentinite with 100% serpentine) is transformed nearly directly into goethite (yellow laterite layer i.e. oxidized ore at lower concentration). In contrast, the delay of the weathering of serpentine over olivine allows the maintenance of one silicate layer (with inherited serpentine) at the base of the profile (coarse saprolite, richer silicated ore). An intensively serpentinized area within an ultrabasic massif, may avoid percolation of the solutions and may induce an absolute accumulation of Ni in the upper part.

The tectonic setting is very important for many reasons: it governs the grade of serpentinization, speed and modalities of weathering being indirectly controlled; the groundwater preferentially percolates in faulted areas, where the weathering starts and where garnierites accumulate; on the landscape scale, when a levelled and lateritizated landscape is tectonically uplifted, this first surface is transformed into a plateau and a new plane surface forms downstream. Ni will be mobilized from plateau profile to the lower surface profile. According to the degree of erosion and the age of the uplift, Ni can be essentially accumulated in the high surfaces (as in New Caledonia) or it might have already been transferred downstream, as in most of the Brazilian deposits.

The size of the massif is also important: if the ultramafic body is small, Ni is scattered in the environment after lateral dispersion. If, in contrast, the massif is large, the same lateral movements from high levelled surface towards low surface can lead to an improvement of the concentration.

5.2 Climate Dependence

A wet tropical climate is required to dissolve primary silicates, to release Ni and to remove SiO_2 and Mg^{2+} in the soluble phase. Weathering is slowed down in a dry climate, the evacuation being inadequate, silicification becoming predominant and leading to the formation of silcrete as reported by Melfi et al. (1980) in Brazil at the beginning of Cenozoic time, by de Waal (1971) in South Africa and by Zeissink (1969) and Elias et al. (1981) in Australia.

The difference in the rate of weathering of the two major framework minerals containing Ni (olivine and serpentine) decreases in very humid climatic conditions. The poor oxide ore prevails, as in the Brazilian Amazonia (Bernardelli et al. 1983).

A tropical climate with contrasted seasons allows the transitory conservation of the inherited serpentine (and/or the formation of nickeliferous serpentine or smectite). The longer the dry season, the more important will be the basal horizon of coarse saprolite (silicate ore), and the more reduced will be the oxide ore. The normally richer silicate ore is almost the equivalent of the cementation layer of the gossans developed on the sulphide deposits.

5.3 Time Dependence

Complete development of a laterite profile is a slow process. Despite the fact that the rate of weathering of ultrabasic rocks is higher than that of the Al rocks (Trescases 1975; Nahon 1985), lateritic profiles develop in metres per million years. All lateritic deposits of Ni in the world correspond to levelled surfaces which were initiated during the Tertiary. When an old pre-concentration upstream supplies a recent downstream surface, the enrichment of this latter is more rapid.

6 Conclusion

Understanding of the ultrabasic bed-rock/nickeliferous lateritic cover system requires a global approach, from localization of Ni in the secondary minerals (crystal scale) to landscape scale (geomorphological analyses). Efficient guides for research of the most favourable sites of Ni concentration will be defined from these global observations of the results of the interactions between different weathering factors, as they are recorded in the landscape.

References

Avias J (1978) L'évolution des idées et des connaissances sur la genèse et sur la nature des minerais de nickel, en particulier latéritiques, de leur découverte à nos jours. Bull BRGM Paris Sect II:162–165

Bernardelli AL, Melfi AJ, Oliveira SMB, Trescases JJ (1983) The Carajas nickel deposit. Proc 2nd Int Seminar on Lateritisation Processes, São Paulo 1982. In: Melfi AJ, Carvalho A (eds) Laterisation processes. Instituto Astronômico e Geofisico, Univ São Paulo, Brazil, pp 53–63

Besset F (1978) Localisations et répartitions successives du nickel au cours de l'altération latéritique des péridotites de Nouvelle-Calédonie. Thèse, Univ Montpellier, 129 pp

Brindley GW (1978) The structure and chemistry of hydrous nickel-containing silicate and aluminate minerals. Bull BRGM Paris Sect II:233–245

Brindley GW, Hang PT (1973) The nature of garnierites I. Clay Min 21:17–40

Brindley GW, Maksimovicz Z (1974) The nature and nomenclature of hydrous nickel-containing silicates. Clay Min 10:271–277

Brindley GW, Wan MM (1975) Compositions, structures and thermal behavior of nickel-containing minerals in the lizardite-nepouite series. Am Mineral 60:863–871

Brindley GW, Bish DL, Wan HM (1979) Compositions, structures, and properties of nickel-containing minerals in the kerolite-pimelite series. Am Mineral 64:615–625

Burns RG (1970) Mineralogical applications of Crystal Field Theory. Cambridge Univ Press, Cambridge

Cervelle BD, Maquet M (1982) Cristallochimie des lizardites substituées Mg-Fe-Ni par spectrométrie visible et infra-rouge proche. Clay Min 17:377–392

Chukrov FV, Gorshov AI, Sivtsov AV, Baresovskaya VV (1983) On the manganese mineralogy, in the laterite weathering crusts of ultrabasic rocks. 2nd Int Seminar on Lateritisation Processes, São Paulo. In: Melfi AJ, Carvalho A (eds) Lateritisation processes. Instituto Astronômico e Geofisico, Univ São Paulo, Brazil, pp 147–158

Colin F (1985) Etude pétrologique des altérations de pyroxénite du gisement nickélifère de Niquelandia (Brésil). Paris, Trav Doc Mém ORSTOM, Paris, 137 pp

Colin F, Nahon D, Trescases JJ, Melfi AJ (1990) Lateritic weathering of pyroxenites at Niquelandia, Goias, Brazil: the supergene behavior of nickel. Econ Geol 85:1010–1023

Decarreau A, Colin F, Herbillon A, Manceau A, Nahon D, Paquet H, Trauth-Badaut D, Trescases JJ (1987) Domain segregations in Ni-Fe-Mg smectites. Clays Clay Min 35:1–10

De Waal SA (1971) South African nickeliferous serpentinites. Min Sci Eng Pretoria 3 (2):32–45

Deyoung JR, Sutphin DM, Werner ABT, Foose MP (1985) International strategic minerals inventory: sumary report. Nickel Denver, CO, US Geol Surv, 62 pp (circular, 930-D)

Elias M, Ronaldson MJ, Giorgetta N (1981) Geology, mineralogy and chemistry of lateritic Ni-Co deposits near Karlgoorlie, Western Australia. Econ Geol 76:1675–1683

Faust GT (1966) The hydrous nickel-magnesium silicates. The garnierite group. Am Mineral 51:279–298

Garnier J (1867) Essai sur la géologie et les ressources minérales de la Nouvelle- Calédonie. Ann Mines Paris 6:1–92

Gerth J (1990) Unit-cell dimensions of pure and trace metal associated goethites. Geochim Cosmochim Acta 54 (2):363–371

Golightly JP (1981) Nickeliferous laterite deposits. Econ Geol 75:710–735

Kuhnel RA, Roorda HJ, Steensma JJ (1978) Distribution and partitioning of elements in nickeliferous laterites. Bull BRGM Paris Sect II:191–206

Lelong F, Tardy Y, Grandin G, Trescases JJ, Boulangé B (1976) Pedogenesis, chemical weathering, and processes of formation of some supergene ore deposits. In: Wolf KH (ed) Handbook of strata-bound and stratiform ore deposits, vol 3. Elsevier, Amsterdam, pp 93–173

Maksimovic Z (1966) B-kerolite-pimelite series from Goles Mountain, Yougoslavia. Proc 2nd Int Clay Conf, Jerusalem, 1966, pp 97–105

Maksimovic Z (1975) The isomorphous series lizardite-nepouite. Int Geol Rev 17:1035–1040

Manceau A, Calas G (1986) Nickel-bearing clay minerals 2. X-ray absorption study of Ni–Mg distribution. Clay Min 21:341–360

Manceau A, Calas G, Decarreau A (1985) Nickel-bearing clay minerals. I. Optical study of nickel crystal chemistry. Clay Min 20:367–387

Manceau A, Llorca S, Calas G (1987) Crystal chemistry of Co and Ni in lithiophorite and asbolane from New Caledonia. Geochim Cosmochim Acta 51:105–113

Maquet M, Cervelle BD, Gouet G (1981) Signatures of Ni^2 and Fe^3 in the optical spectra of limonitic ore from New Caledonia: application to the determination of the nickel content. Miner Deposita 16:357–373

Melfi AJ, Trescases JJ, Oliveira SMB (1980) Les latérites nickélifères du Brésil. Cahi ORSTOM Sér Géol 11:15–42

Melfi AJ, Trescases JJ, Carvalho A, Oliveira SMB, Ribeiro Filho E, Formoso ML (1988) The lateritic ore deposits of Brazil. Sci Géol Bull (Strasb) 41:5–36

Millot G (1964) Géologie des argiles. Masson, Paris, 499 pp

Millot G (1982) Weathering sequences. "Climatic" planations. Leveled surfaces and paleosurfaces. Proc 7th Int Clay Conf AIPEA, Bologne-Pavia. Dev Sedimentol 35:585–593

Millot G, Bonifas M (1955) Transformations isovolumétriques dans les phénomènes de latéritisation et bauxitisation. Bull Serv Carte Géol Als Lorr (Strasb) 8:3–20

Nahon D (1985) Evolution of iron crusts in tropical landscapes. In: Colman M, Dethiver X (eds) Rates of chemical weathering of rocks and minerals. Academic Press, New York, pp 168–191

Nahon D (1987) Microgeochemical environments in lateritic weathering. In: Rodriguez-Clemente R, Tardy Y (eds) Geochemistry and mineral formation in the earth Surface. Consejo Superior de Investigaciones Cientificas, Madrid, pp 141–156

Oliveira SMB, Trescases JJ (1980) Geoquimica da alteração supergena das rochas ultramaficas de Santa Fé (Goias-Brazil). Revista Brasileira de Geociências, São Paulo 15:243–257

Oliveira SMB, Trescases JJ (1985) O deposito de niquel de Jacupiranga (SP): evolução mineralogica e geoquimica. Revista Brasileira de Geociências, São Paulo 15:249–254

Oliveira SMB, Trescases JJ, Melfi AJ (1992) Lateritic nickel deposits of Brazil. Miner Deposita 27:137–146

Ouangrawa M (1990) Etude des composés du fer dans l'altération latéritique de roches ultrabasiques. Exemples de Nouvelle-Calédonie et du Burkina Faso (Ton-Brédié). Thèse Univ Poitiers, 148 pp

Ouangrawa M, Trescases JJ, Ambrosi JP (1996) Evolution des oxydes de fer au cours de l'altération supergène de roche ultrabasique de Nouvelle Calédonie. CR Acad Sci Paris 323(IIa):243–249

Paquet H, Colin F, Duplay J, Nahon D, Millot G (1987) Ni, Mn, Zn, Cr- smectites, early and effective traps for transition elements in supergene ore deposits. In: Rodriguez-Clemente R, Tardy Y (eds) Geochemistry and mineral formation in the Earth surface. Consejo Superior de Investigaciones Cientificas. Madrid, pp 221–229

Schellmann W (1978) Behavior of nickel, cobalt and chromium in ferruginous lateritic nickel ores. Bull BRGM Paris Sect II:275–282

Schellmann W (1983) Geochemical principles of lateritic nickel ore formation. 2nd Int Seminar of Lateritisation Processes, São Paulo. In: Melfi AJ, Carvalho A (eds) Lateritisation processes. Instituto Astronômico e Geofisico, Univ São Paulo, Brazil, pp 119–135

Schwertmann U, Latham M (1986) Properties of iron oxides in some New Caledonian oxisols. Geoderma 39:105–123

Tardy Y (1969) Géochimie des altérations. Etude des arènes et des eaux de quelques massifs cristallins d'Europe et d'Afrique. Mém Serv Carte Géol Als Lorr (Strasb) 31:199

Tardy Y, Nahon D (1985) Stability of Al-goethite, Al-Hematite, Fe^3-kaolinite in bauxites, ferricretes and laterites. An approach of the mechanism of the concretion formation. Am J Sci 285:865–903

Trescases JJ (1975) L'évolution supergène des roches ultrabasiques en zone tropicale. Mém ORSTOM Paris 78:259 pp

Trescases JJ (1979) Remplacement progressif des silicates par les hydroxydes de fer et de nickel dans les profils d'altération tropicale des roches ultrabasiques Accumulation résiduelle et épigénie. Sci Géol Bull (Strasb) 32:181–188

Trescases JJ, Dino R, Oliveira SMB (1987) Un gisement de nickel supergène en zone semi-aride: São João do Piaui (Brésil). In: Rodriguez-Clemente R, Tardy Y (eds) Geochemistry and mineral formation in the earth surface. Consejo Superior de Investigaciones Cientificas, Madrid, pp 273–288

Troly G, Esterle M, Pelletier B, Reibell W (1979) Nickel deposits in New Caledonia: some factors influencing their formation. Proc Int Symp on Lateritisation Processes, New-Orleans, 1979, pp 81–119

Turekian KK (1978) Nickel (section B-O). In: Wedepohl KH (ed) Handbook of geochemistry. Springer, Berlin Heidelberg New York, p 37

Zeissink HE (1969) The mineralogy and geochemistry of a nickeliferous laterite profile (Greenvale, Queensland, Australia). Miner Deposita Berlin 4:132–152

7 The Behavior of Gold in the Lateritic Alterosphere

Fabrice Colin

1 Introduction

Man has been searching for Au at the surface of the earth for more than 6000 years, and has used it as a symbolic, social and economic reference. Since the beginning of this century, many studies have dealt with the geology of primary Au deposits of hydrothermal origin, and with secondary Au deposits which result from sedimentary, mainly alluvial processes. During the last decade, research has focused on the weathering of Au protores which generates the so-called lateritic Au in tropical climatic zones.

The two aims of this chapter are: to present a synthesis of the main results which have been found on the behavior of Au in laterites located in intertropical climatic zones, and to discuss the mechanisms of the physical and chemical dispersion, which control the mobility of Au within the lateritic alterosphere developed under equatorial lateritic rainforest conditions.

2 Chemical and Mineralogical Properties of Gold

Gold is a member of the IB group of the periodic table of the elements with Cu and Ag. Its atomic number is 79 and its atomic weight is 196.967 g. The density of Au is 19.32 g/cm^3 at 20 °C. The pure metal is extremely malleable and can be marked with a fingernail. The crystalline cell contains 4 atoms with a=4.07825 Å at 25 °C. Ag and Au have the same atomic radius of 1.44 Å. The main oxidation states of Au are Au (I) and Au (III). ^{197}Au is the only natural isotope yet known and its half-life is 3×10^{16} years.

Au is mainly found naturally in its native form or as a major component of alloys containing Ag, Cu and platinoids. Its average concentration in the upper lithosphere is about 0.005 ppm and the average Au/Ag ratio is about 0.1. According to Boyle (1979), the average Au content is about 0.004 ppm in ultramafic rocks, 0.007 ppm in gabbros and basalts, 0.005 ppm in diorites–andesites and 0.003 ppm in granites–rhyolites. The average level in soils is about 0.005 ppm, and it is about 0.00003 ppm in natural river waters. The world ocean contains an average Au level of 0.000012 ppm. Au exists also as a trace element in various plants and animals.

3 The Precambrian Gold Lithosphere and Its Weathering

An economic Au level in deposits is reached by natural concentration mechanisms from an initial material with a very low Au content. The scientific understanding of these concentration mechanisms in the lithosphere is, thus, very important and it provides information about the interactions between large volumes of rocks and the fluids carrying Au (Fyfe and Kerrich 1984). These authors calculated that the formation of an ore deposit with 10^6 to 10^8 g of total Au, located in a material containing 5–10 ppm, requires a volume of initial rock of about 2.5×10^8 to 2.5×10^{10} m^3. The volume of the parent rock ranges from 10–100 km^3 if the output of the concentration processes is 100%. For instance, a deposit with a potential of 100 tonnes of Au, half of which consisted of a material with an average content of 10 g/t and half of a material without economic mineralization, would result in the transformation of 100 km^3 of material if the efficiency of the transformation processes is about 50%. This 100 km^3 would represent a volume defined by a surface of 4 km^2 multiplied by the distance to the Moho's discontinuity. Such processes of Au concentration are very remarkable and imply a great efficiency of extraction and a succession of several enrichment stages. They also show that even a small Au surficial anomaly is of great importance (Fyfe and Kerrich 1984).

During Archean time, magmatic intrusive and effusive processes transferred Au towards the surface in the oldest levels of the crust which are known today and which form the cratons of the earth. The main type of deposit is represented by veins of Au-bearing quartz and silicified zones (Boyle 1984). The oldest Au mines known are the Nubie mines located in the Precambrian belts; they have been mined for more than 4000 years. Au is found in geological formations of all ages, however, 55% is located in Archean greenstone belts, or in sedimentary basins whose hinterlands consisted of Archean greenstones (Keays 1984). Most of the Archean Au deposits are related to the emplacement of komatiic magmas favored by the separation of Archean cratons and high geothermic gradients (Keays 1984). Many studies (see reviews by Boyle 1979, 1984; Groves et al. 1984; Hutchinson and Burlington 1984; Keays 1984; Pretorius 1984; Saager and Meyer 1984; Seward 1984; Viljoen 1984) deal with the processes of Au concentration which result from combination of tectonic and hydrothermal events in the initial lithosphere, followed by processes of erosion and sedimentation in the Archean intracratonic basins. For example, the Witwatersrand basin in South Africa has provided until now 58% of the estimated Au reserves of the world. In the Precambrian deposits and, generally speaking, in the deposits which have formed during the slow evolution of the earth, Au appears as a marker of tectonic, hydrothermal, volcanic and sedimentary events in the vein-type deposits (epigenetic) and in the sulfide bodies (syngenetic), as well as a physical marker in the recent and old alluvial placers. According to Bardossy (1982), the continental masses (including the Au deposits) were intensively weathered during the Siluro-Devonian and the mid Cretaceous–early Tertiary.

The BLAG model proposed by Berner et al. (1983) to calculate the CO_2 variation during the last 100 million years with its potential effects on the temperature, suggested that the CO_2 levels in the atmosphere were 13 times higher during the mid

Cretaceous (ca. 90 Ma) than today, and that the CO_2 level was also very important during the mid-Eocene. These high levels are generally ascribed to an intense volcanic activity. The resulting greenhouse effect led to warm and humid climates (Frakes 1976) which favored weathering processes and biological activity (Nahon 1991; Colin 1992). This biological activity implies physical and chemical mechanisms, and enhances the weathering of the lithosphere.

The structural and geomorphological controls and the low rates of isostatic uplift are important factors which determine the drainage conditions. The rate of a few mm/1000 years which has been recently found for the African cratons seems to be reasonable (Brown et al. 1994). The tectonic stability and the long exposure of the Precambrian craton allowed the development of thick regoliths under high rainfall and temperature conditions, which are characteristic of tropical humid climates. Combination of these various conditions favors laterite weathering leading to the formation of residual material or regoliths which cover as much as one-third of the landmass (Nahon 1986; Fig. 1). Therefore, because of its pre-, syn-, and post-lateritic weathering history, the «transcontinental» surface which consists of the Au Archean belts in South America, Africa, India and Australia, is an ideal location for studying the behavior of Au in the lateritic alterosphere. In addition, the understanding of Au behavior in surficial environments is of major importance for mining companies and consequently of critical concern for the economic development of tropical countries (Colin 1992).

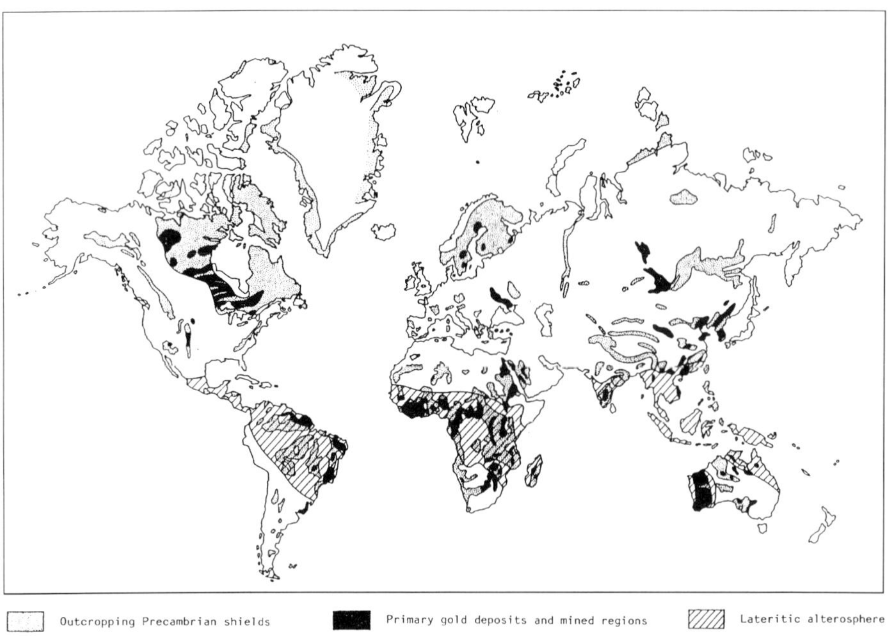

Fig. 1. Distribution and spatial relationships of the Precambrian cratons, the primary gold deposits and the tropical lateritic alterosphere

4 Gold Behavior in Natural Supergene Systems

In supergene systems, the conditions of mobility, transport and precipitation of Au depend on the nature and the degree of fracturation of the parent rock, the nature of the mineralogical phase which contains Au, the nature of the Au itself, the past and present climatic regimes (rainfall, composition of the rain water), the location and composition of the water table (high and low), the local and regional geomorphology, the proximity of the sea and ocean, the quantity of organic matter and the biological activity. In other words, Au mobility is controlled by chemical parameters (Eh, pH, fO_2, T, presence of oxidizing or reducing ions) and their relative variation at the water/Au-bearing rock interface, by the nature and stability of the transport agents (ligands, fauna and vegetation) in a weathering system, and by the chemical composition and size of the Au particles (primary particles and particles formed by weathering processes), which range from nugget size to colloid size.

4.1 The Mobility of Gold Issued From Parental Sulfides

The mobility of Au is shown by the comparative petrographic study of fresh and weathered Au-bearing rocks, and by the chemical reactions which take place during weathering of the primary Au phases into the supergene phases, e.g. the transformation of pyrite (FeS_2) into goethite (FeO(OH)). The studies generally conclude that the chemical mobility of Au is controlled by the stability of the Au–thiosulfate complexes $Au(S_2O_3)_2^{3-}$ in an oxidizing environment with a moderately acid to moderately basic pH, and by the stability of Au bisulfide complexes $Au(HS)_2^-$ in a more reducing environment. Supergene Au is generally pure and forms millimetric dendritic assemblages, micrometric cubic and octahedral crystals, micrometric spherules, or coatings of various dimensions on the surface of Fe oxi-hydroxides (Machairas 1967; Petrovskaya 1971; Boyle 1979; Nesterov et al. 1979; Roslyakov and Roslyakova 1986; Stoffregen 1986; Webster 1986; Webster and Mann 1984; Butt 1987a). The presence of visible Au in regoliths, by comparison with the so-called invisible Au in primary sulfides, prompted authors to believe that weathering is a concentration process of Au. A good example of such a process is given by Cottard et al. (1994) at the Al Hajar Au deposit.

4.2 The Mobility of Gold Under s.s. Lateritic Environments

The behavior of Au in s.s. lateritic systems, which result from surficial weathering under tropical conditions, is not very well documented. The lower part of deep regoliths which developed on sulfide bodies (gossans/ironstones), even if they were formed in the intertropical climatic zone, are not s.s. laterite. The behavior of Au during lateritic weathering is, thus, better approached by the study of past and present weathering of protores which do not contain sulfide mineralizations, or by the study of very surficial auriferous lateritic rocks. However, the scientific community started to be interested in the laterites developed from this type of deposit (mainly Precambrian epigenic quartz deposits) only recently (in the early 1980s) and this new interest was essentially driven by economic reasons (oil crashes in 1973–1974, and 1979–1980, and strong inflation) and by the advancement of developing countries.

The search for Au thus became a priority. With the help of new analytical techniques (chemical analyses with very low detection limits of soils, water, plants), mining companies undertook prospection in Africa, India, South America, Indonesia. They targeted lateritic soils and river sediments which often showed anomalous Au values, lower than 1 g/t (Balasundaram 1972; Boyle 1979; Le Count Evans 1981; Watters 1983; Schiller 1985; Rao et al. 1987).

Exploration in the intertropical belt initiated programs of applied and fundamental studies of the Au behavior in latosols and by extension in lateritic systems. Australian and French scientists already have a good knowledge in geochemical exploration: the Australians because of their mining experience of laterites and consequently because of their knowledge of lateritic environments, especially in Western Australia (Butt 1980, 1987a,b; Smith 1987), and the French because of historical relationships with West and Central Africa and mining experience in French Guyana (Wilhem 1975; Zeegers 1987).

Evidence of Au mobility has been revealed by exploration surveys. Au anomalies outside the limits of the expected or known mineralizations have been found in soils: in the clay lateritic formations of Mont Flotouo d'Ity in the Ivory Coast (Granier et al. 1963), in latosols at Liptako in Niger (Schoeder and Mikhailoff 1977), in sandy-clay lateritic soils at Esperance in French Guyana (Zeegers 1979), in latosols of the Afema region in the Ivory Coast, in surface lateritic formations in Guinea and British Guyana, in the Marlu District in Ghana (Boyle 1979), in latosols at St-Elie and Adieu Vat in French Guyana (Le Count Evans 1981), in latosols at Kalana in Mali (Bassot and Traore 1980), in pisolitic laterites of the Golden Grove District in Western Australia (Smith and Perdrix 1982), in latosols of the Samoa, Tahiti and Oahu Pacific islands (Roslyakov and Roliaskova 1986), in lateritic bauxites of Boddington in Western Australia (Davy and El Ansary 1986), and in latosols of eastern Cameroon (Freyssinet et al. 1989a). Hypotheses for the physical and chemical mobility of gold have been suggested from studies of its distribution on the landscape scale, or locally inside the regoliths, and from morphological and chemical studies of Au particles.

Minatidis (1979) suggested a physical mobility of Au particles by vertical transfer through bauxites in India and Greece. According to Le Count Evans (1980), placers could result from a lateral chemical dispersion of Au particles from laterites. Smith and Perdrix (1982) suggested that the lateral dispersion of Au at the surface of the Golden Grove latosols is due to a physical dispersion of Au-bearing pisolites. For Wilson (1983), lateral and vertical transfers explain the presence of nuggets found in latosols and laterites of Coolgardie in Western Australia. Zeegers et al. (1985) noticed that the decrease in the Au concentration in soils at Esperance in French Guyana is more pronounced in the fine fractions, and thus concluded that the fine Au particles are mechanically dispersed laterally on the surface. Michel (1987) observed high concentrations of Au particles in the lower part of the ferruginous crusts at Cuiaba, Mato Gross (Brazil) and suggested a gravity transfer of Au from the soils to the bottom of the crusts. Carver et al. (1987) believe that termites help the physical dispersion of fine Au particles towards the surface in West Australian soils. According to Butt (1987a), lateral mechanical dispersion of Au at the surface results from mechanisms of washing during erosion of the lateritic surfaces. This author also concluded that it is difficult to estimate the importance of the physical transfer of Au during the formation and the evolution of lateritic systems.

Machairas (1963) demonstrated the chemical mobility of Au in ferruginous crusts and in lateritic regoliths at Ipoucin in French Guyana and at Ity in the Ivory Coast. This author showed the presence of supergene micrometric crystals which are well crystallized, euhedral, and contain low Ag amounts in ferruginous crusts, in Fe-rich crusts and «at the foot of the hills». For Bonnemaison (1986), such precipitations are caused by upwards capillary circulation of Au solutions which evaporate at the surface in dry tropical climates. Freyssinet et al. (1989b) showed evidence for precipitation of Au (dendrites of pure Au at the surface of ferruginous nodules) in crusts, in Mali. The studies carried out in Western Australia, in the Yilgarn Block and in Queensland have supplied evidence for the chemical mobility of Au (Wilson 1983; Mann 1984; Webster and Mann 1984; Davy and El Ansary 1986; Lawrance 1988). The Archean West-Australian craton has been intensively lateritized during the warm and humid Tertiary time, inducing formation of thick ferruginous crusts and bauxites. Then under dry conditions (200–600 mm rainfall/year), which have continued until the present, the old formations were dismantled and lateritic systems have developed locally in valleys (Mann 1983). In most lateritic formations, and especially in Kalgoorlie and Coolgardie, two types of Au particles have been observed: residual particles from primary veins with dissolution features and high Ag contents sometimes surrounded by a cortex with a low Ag content, and dendrites, octahedral micrometric crystals, pure Au filaments occasionally associated with Fe or Al oxi-hydroxides. Mann (1984) suggested that acid solutions resulting from oxidation and hydrolysis of Fe, are responsible for the dissolution of Au and Ag in quartz Au-bearing veins. The high chloride levels (10^{-1} mol/l) required to form $AuC1_{4-}$ and AgCl were supplied during Tertiary time by winds enriched in NaCl and blowing from inland, such as the present-day winds. The Au–chloride complexes migrated from veins to the parent rock/saprolite interface and were reduced in the presence of Fe^{2+} and H_2O to give supergene Au which is syngenetic with goethite. The Ag–chloride complexes are not reduced by Fe^{2+} because the Ag/AgCl redox potential is lower that of the ferrous–ferric couple. Butt (1987b) concluded that this dissolution–precipitation mechanism cannot be applied to laterites evolving under a tropical humid climate, because free Au is resistant to weathering in such conditions because adequate ligands are lacking.

These studies first suggested and showed that Au which was previously considered as resistant to weathering, could be dissolved, migrate and reprecipitate in supergene laterite conditions, but they remain qualitative. A model of the mechanisms of mechanical dispersion has not yet been proposed. The chemical dispersion of Au in lateritic conditions has been qualitatively shown by the presence of partially dissolved particles and supergene particles extracted from profiles, but the proportion of the residual stock compared with the precipitated stock is unknown. In addition, most of the observations have not always been carried out in complete profiles from parent rocks to surface, and they were not combined with more general studies including chemistry, mineralogy, structure and petrology of the local and regional laterites. A model of the dissolution and precipitation of Au in natural lateritic systems has only been proposed by Australian authors, but this model can only be applied to the Tertiary paleolaterites and laterites of the Yilgarn Block exposed to fluids enriched in chlorides.

Many questions remain about the supergene chemical and physical mobility of Au, the distances of transport and the enrichment or/and the depletion rate in laterites. These questions result from difficulty in accurately locating fresh mineralization, the heterogeneous distribution of Au in regoliths and fresh rocks, from lack of knowledge of the chemical composition of the rain waters, the water tables, the springs, the river waters, and from lack of quantitative calculations made on detailed petrologic studies of Au laterites which are very often polygenic. In an attempt to answer some of these questions, the next section presents results concerning the behavior of Au in an equatorial lateritic environment.

4.3 Laterites in a Humid Equatorial Climate: the Gabon Example

Due to an increase in budget resulting from oil resources, the Gabonese government decided in 1979 to undertake a systematic inventory of the main mineral resources (Diouly-Osso 1988). This research program was carried out with various methods such as teledetection, geophysics, geology, aerogeophysics. It allowed discovery and identification of many Au anomalies in suitable geological contexts (Fig. 2). Four sites were investigated in detail among all the Au sites which have been identified: Mebaga in the north-central part of the country (Colin and Lecomte 1988), Dondo Mobi (Colin et al. 1989; Lecomte and Colin 1989; Colin and Vieillard 1991; Colin et al. 1992, 1993a), Ovala (Colin et al. 1989; Minko et al. 1992) and Pounga (Colin et al. 1993b) in the Eteke Au district, in south-central Gabon.

The landscape of the equatorial forest consists of hills with convex slopes. Veins and lenses of Au-bearing quartz cut the granito–gneissic bedrock. The weathering profiles which are located directly above the mineralizations, are the thickest and can reach 100 m (Fig. 3). From parent rock to top, the weathering profile presents: (1) a saprolite characterized by a matrix rich in goethite, kaolinite and gibbsite; (2) a nodular level a few meters thick and enriched in gibbsitic, goethitic and hematitic nodules surrounded by a sandy-clay matrix; (3) a sandy-clay level with a micro-aggregated matrix, which can be about 10 m thick; and (4) a humus horizon characterized by intense biological activity.

Au bounded by the primary silicate minerals in fresh mineralizations (Plate Ia) is exposed to intense geochemical alteration in the regoliths. This autochthonous alteration proceeds progressively right above the mineralizations, from fresh rock towards the soil throughout the saprolite, nodular and sandy-clay levels (Fig. 3a). It comprises: (1) the detachment of the Au particles from the matrix during the early stages of weathering within the saprolite; (2) a decrease in the size of the particles

Fig. 2. Location of the gold sites found during the Mineral Inventory of Gabon (*black dots*) and the sites which have been studied in detail (*black stars*)

and of the total content of Au (Fig. 4); (3) a chemical smoothing of the particles (Plate Ib) with more pronounced surface changes as the dissolution voids become larger and more connected (Plate Ic,d); and (4) a loss of Ag, except in the core of the particle which memorizes the parental Au content, resulting in a natural refinement of the Au (Plate Ie,f; Fig. 5).

In addition to the progressive alteration right above the Au veins and lenses, a para-autochthonous and mechanical dispersion of the Au particles takes place at the surface (Fig. 3a). As the weathering proceeds, this lateral dispersion of Au particles creates a halo which spreads over several hundred meters and generally reaches the hydrographic system. The Au particles which migrate downhill are then located in a «pre-alluvial» environment. The sandy-clay and nodular levels which are located at the rims of the halos on the slopes of the half-spherical hills, are also polluted by Au particles migrating downwards. This dispersion occasionally reaches the upper part of the saprolite which was not initially mineralized. In the upper levels characterized by a high density of roots and intense microfauna activity, the Au particles which are usually less than 100 mm in size, migrate with other minerals such as goethite, rutile and xenotime. These transfers use the vesicular and interconnected voids and also the active and fossil root axes. The dispersion of Au has been demonstrated qualitatively (alteration gradient of the residual particles) and quantitatively (Fig. 6).

The geochemical alteration is the vector of the mechanical dispersion. The dissolution which proceeds from the inter-granular joints leads to a micro-division of multi-granular particles into mono-granular particles. This micro-division increases the specific surface area of the particles in contact with the fluids and facilitates their mechanical transport. Moreover, the geochemical alteration continues during and after the lateral and vertical mechanical transfers. The alteration gradients (chemical smoothing, decrease in the particle size, presence and size of the dissolution voids, depletion of Ag) characterizing the particles located at the rims of the dispersion halos are the opposite of the alteration gradient of the particles resulting from in situ alteration of the mineralizations. The dissolution of Au is promoted by highly oxidizing and acidic to moderately acidic conditions which prevail within the surface alterosphere. In the equatorial forest, the humic sub-

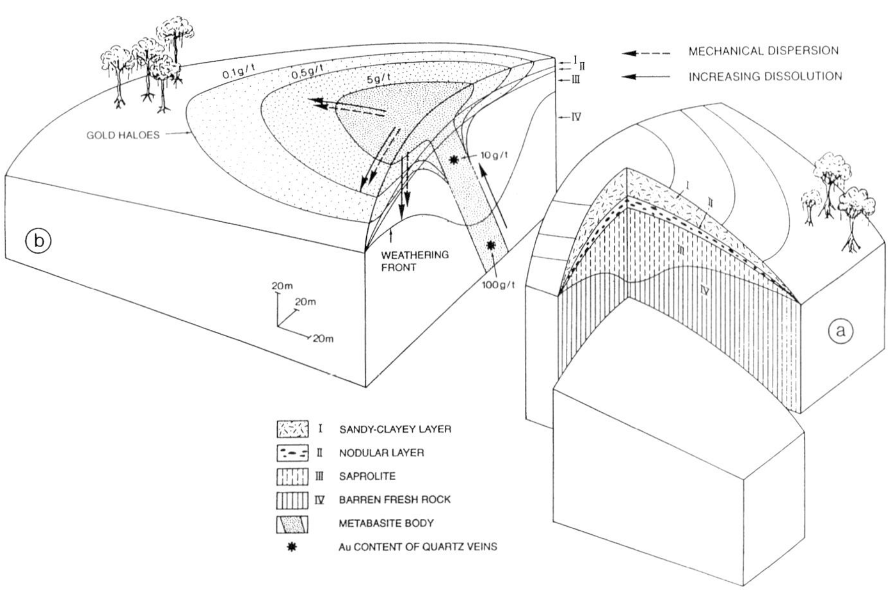

Fig. 3.a Schematic section through a dome-shaped hill (half-orange) characteristic of the lateritic landscape in the Gabonese equatorial forest. **b** Spatial distribution of the Au mineralizations and dispersion haloes resulting from the chemical and mechanical surficial dispersion

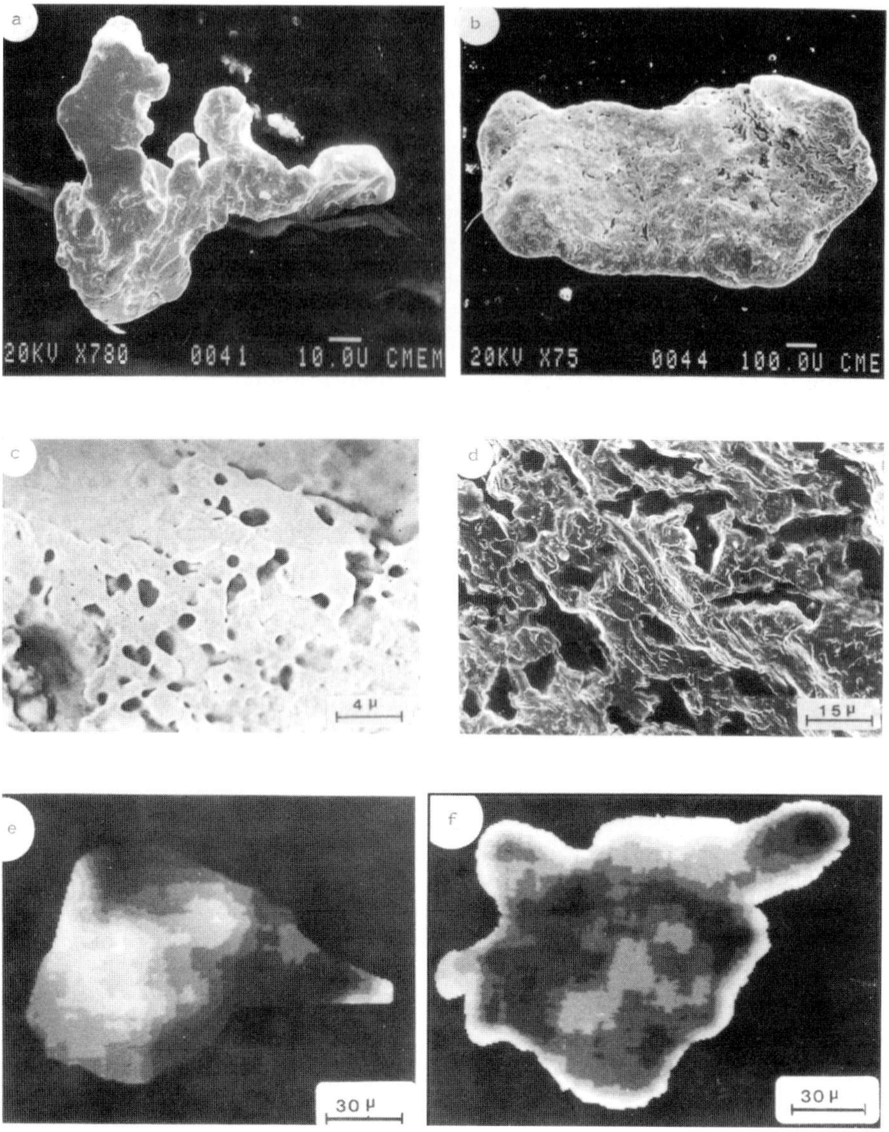

Plate I.a Euhedral and xenomorph particle of primary gold with a non-porous surface and well-defined edges, and the quartz matrix in the fresh mineralization. **b** Gold particle characteristic of the sandy-clayey level. The surface is smooth and porous, and the initial euhedral shape can no longer be recognized. **c, d** The dissolution porosity is characteristic of the gold particles in the sandy-clay level. As the weathering proceeds (from **c** to **d**), there is an increase in the number and size of the pores which become connected and follow the preferential dissolution axes (**d**). **e, f** Distribution of silver in the particles during weathering. The lowest silver content is represented by the *white areas*. **e** Fresh particle. The distribution of silver is homogeneous (5—8%) and is not related to the shape of the particle. **f** Weathered particle. The distribution of silver is heterogeneous (0—8%) and depends on the shape of the particle. The depletion of silver is centrifugal in the grains which compose the particle, the core of each grain retaining the high silver content of the grains in the protore

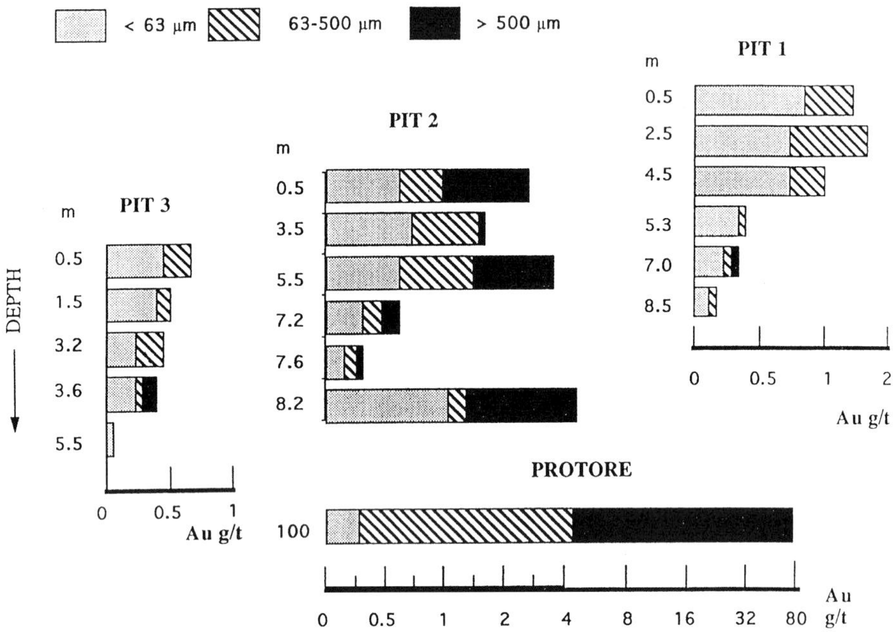

Fig. 4. Distribution of total gold contents and size variation at the Dondo Mobi site: *P1*, *P2* and *P3* are located respectively at the top of the hill, directly above the mineralization and downslope (see Fig. 3b). The protore represents the fresh mineralization which is at about 100 m depth. The weathering gradient results in a decrease in the total gold content and the particle size, from fresh rock to the surface (*P2*) and in the dispersion haloes (*P1* and *P3*; after Colin and Vieillard 1991)

stances and especially the fulvic acids resulting from intense biological activity, provide ligands able to form complexes with Au. In addition, the rain in areas close to the Atlantic Ocean in Gabon and especially in the Eteke region, contains chlorides and it is well known that chlorine is a potential ligand able to form complexes with Au. By considering the hydrolyses reactions also, three main types of complexes responsible for the chemical mobility of Au can form: $Au(OH)_3$ (H_2O), AuClOH

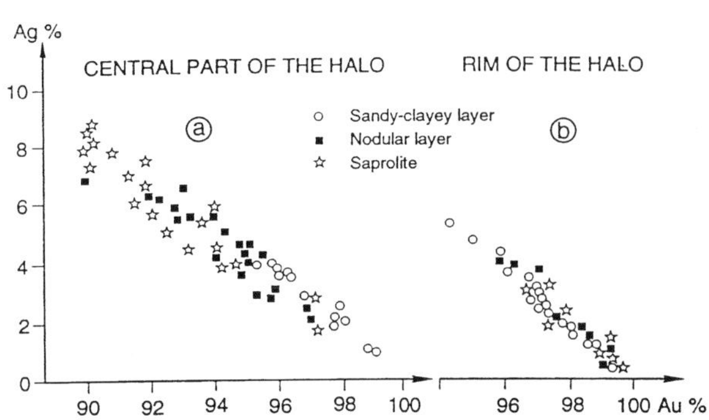

Fig. 5. Distribution of the silver content in gold particles of the Ovala site: in the central part of the halo (see Fig. 3b), the silver content ranges from 10% (weight) in the saprolite to 2% on the surface. On the rim of the halo, the content ranges from 4% on the surface to 0.5% in the upper saprolite. (After Edou Minko et al. 1992)

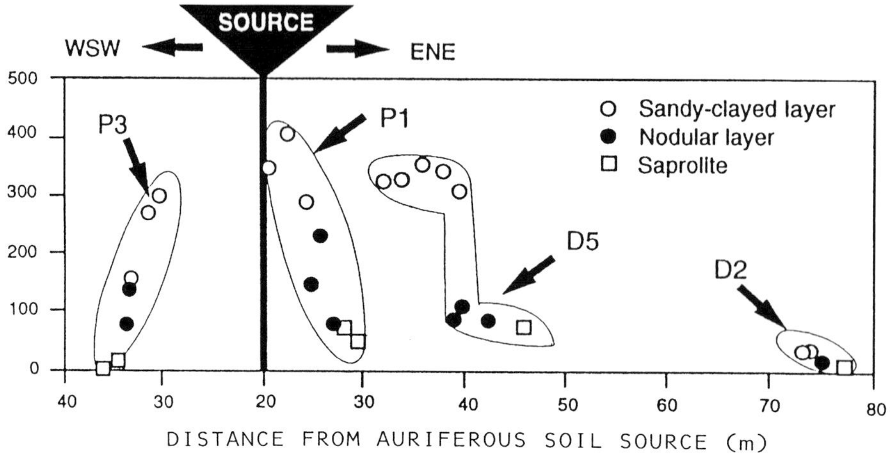

Fig. 6. Quantification of the mechanical dispersion of gold with the distance from the auriferous soil source directly above the mineralization, at Dondo Mobi site. The drill holes *1*, *5* and *2* are located uphill and the drill core *3* is located downhill (see Fig. 3b). The net gains (t%) are calculated for each weathered level and refer to the fresh rock; they decrease at the surface from source to the rims of the halos, and from the surface to the upper part of the saprolite in the halo. The dispersion limit of the gold particles is controlled by physical barriers. *D2* and *D5* are drill cores; *P1* and *P3* are pits. (After Colin et al. 1993)

and $Au(OH)_2$ FA (FA=fulvic acid). The solubility of Au increases with the Ag content and pure Au seems to dissolve only in environments rich in organic acids (Figs. 7–8).

The Au complexes seem to be stable in the conditions prevailing in the alterosphere and also in the hydrosphere of the equatorial forest. They can leave the regoliths and join the river systems which drain the sites, as has been described in the Congo basin by Benedetti and Boulègue (1991). The reprecipitation of Au within the regolith has been very rarely observed and it occurs mostly at the alteration front in the fluctuation zone of the water table. In this case, the common association of residual particles with supergene particles shows that the precipitation occurs close to the site of dissolution. When supergene particles are present, they statistically represent a very low to negligible amount compared to the residual Au particles. The quantification of the Au masses which are transferred during alteration of the Au mineralizations and during the formation of the dispersion halos shows a significant negative balance on the scale of the study. The quantification is based on tridimensional integration of a mass-transport function which includes the enrichment factors, the porosity and the changes in volume of the different weathering levels using the parent rock as a reference. It shows that the depletion of Au related to the geochemical alteration of the mineralized bodies is far from being compensated by the mechanical enrichment of the neighboring surface formations; the laterite weathering results in a loss of 80–90% of the initial Au. The time required to form a dispersion halo of approximately 100 m wide on both sides of

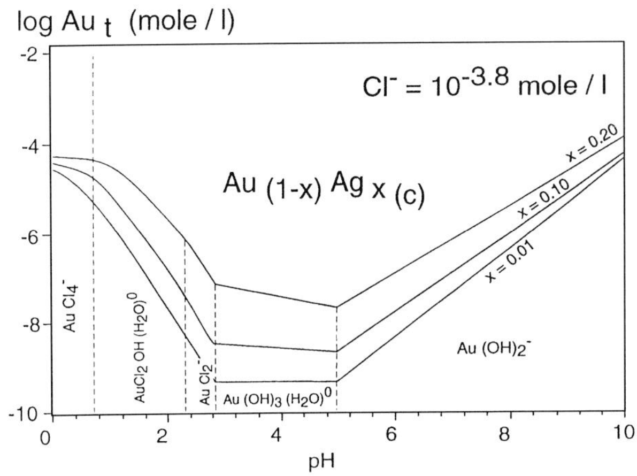

Fig. 7. Solubility of Au–Ag alloys as a function of pH. Chloride concentration=$10^{-3.8}$ mol/l; $fO_2=10^{-0.68}$ (natural conditions at Dondo Mobi). Hypothetical concentration of humic substances<0.2 mg/l. The pH measured in soils and waters at Dondo Mobi ranges from 4–6. For a total Au concentration of 10^{-9} mol/l (average of the river waters circulating in the Au deposits in inter-tropical regions), the solubility of gold increases with the silver content, and pure gold is stable. The dominant form of gold is $Au(OH)_3(H_2O)$. If the chloride concentration exceeds $10^{-3.5}$ mol/l, $AuClOH^-$ forms and is more stable than the hydroxyl complex, under natural pH conditions. (After Colin et al. 1993)

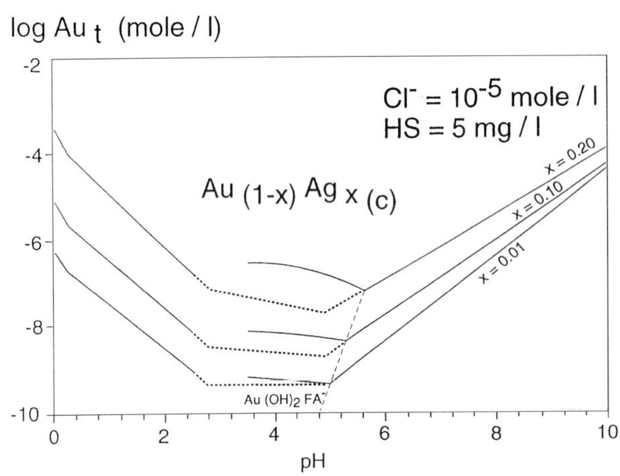

Fig. 8. Solubility of Au-Ag alloys as a function of pH. Chloride concentration=10^{-5} mol/l; $fO_2=10^{-0.68}$ (natural conditions at Dondo Mobi). Hypothetical concentration of humic substances = 5 mg/l (average concentration on the surface in tropical systems). The solubility of gold increases with the Ag content, and the nearly pure gold can be dissolved. Au-fulvic acids complexes $Au(OH)FA^-$ form and are much more abundant than the non-organic complexes. If the concentration of humic substances increases by only a few mg/l, pure gold becomes soluble. (After Colin et al. 1993)

the mineralized body, and 10 m deep, has been estimated to be about 1 million years, assuming that the humid equatorial conditions prevailing today were globally the same during this period in Gabon.

Thus, the mobility of Au is controlled by low-energy and low-amplitude dissolution processes and surface mechanical transfers taking place during the evolution of the weathering systems, which are geochemically opened and developed from an Au Archean crust characterized by a tectonic stability. The geochemical balance is strongly negative in the conditions prevailing in equatorial forests and the Au-hydroxyl, chloride and organo-metallic complexes are released into the local and regional drainage system.

5 Conclusions and Perspectives

The mobility of Au is controlled by geochemical, biochemical and mechanical processes in the lateritic alterosphere. This has been demonstrated in the regoliths developed on the dome-shaped hills of the Gabonese equatorial forest. However, this environment represents optimal conditions for the dissolution of Au particles, such dissolution being the first impetus (through size reduction of the Au particles) of the mechanical dispersion on the surface and in the upper part of the local weathering systems. The dispersion which is concomitant with the in situ digestion of nodular crusts, is subactual.

In lateritic systems which are evolving since their formation towards more arid conditions with a shorter but still intense wet season (e.g. Burkina Faso and Mali), Au dispersion on the foothill and peripheral depressions should be mainly related to mechanical erosion resulting from running of rain water (Sanfo et al. 1993).

The three remarkable properties of Au, namely its high density, its sensitivity to geochemical weathering and its extreme malleability which allows the preservation of mechanical marks, make Au an excellent recorder of the present-day conditions of genesis and transformation of the lateritic alterosphere.

Under equatorial rainforest conditions, the relation of Au to biota should be of future concern for improving Au exploration methods and providing complementary information on its supergene behavior (Bowell et al. 1993; Machesky et al. 1993).

Although the evolution of Au after formation of the old lateritic crust systems is better understood today, its behavior during the genesis of these systems remains poorly known. For example, the behavior of Au as a function of time and climate changes should be investigated within the complex lateritic polygenetic systems of the sub-Sahelian African area. In continuation of the work reported by Machesky et al. (1991) and Gréffié et al. (1993), experimental syntheses associating Au with Fe oxi-hydroxides and Fe-rich phases, may also help to answer some of the unknowns.

References

Balasundaram MS (1972) Applied geochemistry in mineral exploration. Geological Survey of India, Misc Publ 21, pp 1–50

Bardossy G (1982) Karst Bauxite. Elsevier, Amsterdam, 441 pp

Bassot JP, Traoré H (1980) Le gisement d'or de Kalana (République du Mali). Chron Rech Min 457:5–16

Benedetti M (1991) Géochimie de l'or: mecanismes de transport et de dépôt. Sci Géol Mém (Strasb) 91:148 pp

Benedetti M, Boulègue J (1991) Transfer and deposition of gold in the Congo Watershed. Earth Planet Sci Lett 100:107–115

Berner RA, Lasaga AC, Garrels RM (1983) The carbonate-silicate geochemical cycle and its effect on atmospheric carbon dioxide over the past 100 million years. Am J Sci 283:641-643

Bonnemaison M (1986) Les filons de quartz aurifères: un cas particulier de shear zone aurifère. Chron Rech Min 458:5–11

Bowell RJ, Foster RP, Gize A (1993) The mobility of gold in tropical rain forest soils. Econ Geol 88:999–1016

Boyle RW (1979) The geochemistry of gold and its deposits. Geol Surv Can Bull 280:584

Boyle RW (1984) Gold deposits: their geology, geochemistry and origin. In: Foster RP (ed) Gold '82. The geology, geochemistry and genesis of gold deposits. Balkema, Rotterdam, pp 183–190

Brown ET, Bourlès D, Colin F, Sanfo Z, Raisbeck GM, Yiou F (1994) The development of iron crust systems in Burkina Faso, West Africa, examined with in situ produced cosmogenic nuclides. Earth Planet Sci Lett 124:19–33

Bumstead ED(1984) Some comments on the precision and accuracy of gold analysis in exploration. Proc Aust Inst Min Metall 280:71–78

Butt CRM (1980) Weathered bedrock. J Geochem Explor 12:122–126

Butt CRM (1987a) A basis for geochemical exploration models for tropical terrains. Chem Geol 60:5–16

Butt CRM (1987b) The dispersion of gold in the weathered zone, Yilgarn Block, Western Australia. Proc Aust Geol Soc Meet, Perth, Oct 1987, pp 27–53

Carver RN, Chenoweth LM, Mazzucchelli RH, Oates CJ, Robbins TW (1987) «Lag» a geochemical sampling medium for arid region. In: Govett RG (ed) Geochemical Exploration 1985. J Geochem Explor 28:183–199

Colin F (1992) L'or, traceur de l'histoire chimique et physique de l'altérosphère latéritique sous forêt équatoriale. Habilitation à diriger les recherches, Thèse, Univ Aix-Marseille III, 171 pp

Colin F, Vieillard P (1991) Behavior of gold in lateritic equatorial environment: weathering and surface dispersion of residual gold particles at Dondo Mobi, Gabon. Appl Geochem 6:279–290

Colin F, Boulangé B, Lecomte P (1989) Dissolution features of gold particles in lateritic profile from Dondo Mobi, Gabon. Geoderma 45:241–250

Colin F, Brimhall G, Nahon D, Lewis CJ, Baronnet A, Danti K (1992) Equatorial rainforest lateritic soils: a geomembrane filter. Geology 20:523–526

Colin F, Lecomte P (1988) Etude minéralogique et chimique du profil d'altération du prospect aurifère de Mébaga Mvomo (Gabon). Chron Rech Min 491:55–65

Colin F, Vieillard P, Ambrosi JP (1993a) Quantitative approach to physical and chemical gold mobility in equatorial rainforest lateritic environment. Earth Planet Sci Lett 114:269-285

Colin F, Lecomte P, Edou Minko A, Benedetti M (1993b) Regional exploration strategies at Pounga, Gabon, and gold dispersion under rainforest conditions. Chron Rech Min 510:61–68

Cottard F, Abdulhay GJ, Artignan D, Gélot JL, Roubichou P, Trinquard R, Vadala P (1994) The Al Hajar gold deposit (Kingdom of Saudi Arabia): a newly discovered example of supergene enrichment from a massive sulfide deposit of late Proterozoic age. Chron Rech Min 510:13–24

Davy R, El Ansary M (1986) Geochemical patterns in the laterite profile at the Boddington Gold Deposit, Western Australia. J Geochem Explor 26:119–124

Diouly-Osso O (1988) Editorial. Chron Rech Min 491: 2

Frakes LA (1976) Climates throughout geologic times. Elsevier, Amsterdam, 310 pp

Freyssinet Ph, Lecomte P, Edimo E (1989a) Dispersion of gold and base metals in the Mborguéné lateritic profile, east Cameroun. J Geochem Explor 32:99–116

Freyssinet P, Zeegers H, Tardy Y (1989b) Morphology and geochemistry of gold grains in lateritic profiles of southern Mali. J Geochem Explor 32:17–31

Fyfe WS, Kerrich R (1984) Gold: Natural concentration processes. In: Foster RP (ed) Gold '82, The geology, geochemistry and genesis of gold deposits. Balkema, Rotterdam, pp 99–129

Granier C, Lajoinie JP, Vitali C (1963) Géochimie de l'or et du cuivre dans les formations latéritiques argileuses du mont Flotouo, Ity, Côte d'Ivoire. Bull Soc Fr Minér Crist 86:252–258

Gréffié C, Parron C, Benedetti M, Amouric M, Colin F (1993) Experimental study of gold precipitation with synthetic iron hydroxides: HRTM-AEM and Mössbauer spectroscopy investigations. Chem Geol 107:297–300

Groves DI, Phillips GN, Ho SE, Henderson CA, Clark ME, Woad GM (1984) Controls on distribution of Archaean hydrothermal gold deposits in Western Australia. In: Foster RP (ed) Gold '82. The geology, geochemistry and genesis of gold deposits. Balkema, Rotterdam, pp 689–712

Hutchinson RW, Burlington JL (1984) Some broad characteristics of greenstone belt gold lodes. In: Foster RP (ed) Gold '82. The geology, geochemistry and genesis of gold deposits. Balkema, Rotterdam, pp 339–372

Keays RR (1984) Archaean gold deposits and their source rocks: the upper mantle connection. In: Foster RP (ed) Gold '82. The geology, geochemistry and genesis of gold deposits. Balkema, Rotterdam, pp 17–53

Lawrance LM (1988) Behavior of gold within the weathering profile in the Yilgarn Block, Western Australia. Publ Univ Western Australia 12:335–351

Le Count Evans D (1981) Lateritization as a possible contributor to gold placers. Eng Min J:86–91

Lecomte P, Colin F (1989) Gold dispersion and size fraction distribution in a tropical rain forest weathering profile. J Geochem Explor 34:285–301

Machairas G (1963) Etude des phénomènes de migration chimique de l'or. Cas de la Guyane Française et d'Ity en Côte d'Ivoire. Bull Soc Fr Minér Crist 86:78–80

Machairas G (1967) Dissolution et recristallisation de l'or primaire pendant l'oxydo-réduction des sulfures aurifères. Bull BRGM 5:111–121

Machesky ML, Andrade WO, Rose AW (1991) Adsorption of gold (III)-chloride and gold (I)-thiosulfate anions by goethite. Geochim Cosmochim Acta 55:769–776

Machesky ML, Rose AW, Andrade WO, Bliss L, Kato T (1993) Gold speciation and mobility near the Igarapé Bahia deposit, Amapa, Brazil. Chron Rech Min 510:53–62

Mann AW (1983) Hydrogeochemistry and weathering of the Yilgarn Block, Western Australia-ferrolysis and heavy metals in continental brines. Geochim Cosmochim Acta, 47, 181–191

Mann AW (1984) Mobility of gold and silver in lateritic weathering profiles: some observations from Western Australia. Econ Geol 79:38–48

Michel D (1987) Concentration of gold in in situ laterites from Mato Grosso, Miner Deposita 22:185–189

Minatidis DG (1979) The bauxite deposits as a favorable terrain for gold exploration: a theoretical approach. Proc 4th ICSOBA Congr, Athens, Oct 1978, pp 587–596

Minko E, Colin F, Lecomte P, Trescases JJ (1992) Altération latéritique du gite aurifère d'Ovala (Gabon), et formation d'une anomalie superficielle de dispersion. Miner Deposita 25:90–100

Nahon D (1986) Evolution of iron crust in tropical landscapes. In: Colman SM, Dethier DP (eds) Rates of chemical weathering of rocks and minerals. Academic Press, New York, pp 169–191

Nahon D (1991) Introduction to the petrology of soils and chemical weathering, John Wiley, New York, 313 pp

Nesterov VN, Plyusnin AM, Pogrelnyak Yu F (1979) A study of the behavior of gold under the conditions of oxidation zones in gold-sulfide deposit. Geokhimiya 6:878–887

Petrovskaya NV(1971) Growth and subsequent changes in native gold crystals. Miner Soc Japan, Spec Pap 1:116–123

Pretorius DA (1984) The source of Witwatersrand gold: conjecture of the unknown. In: Foster RP (ed) Gold '82. The geology, geochemistry and genesis of gold deposits. Balkema, Rotterdam, 219 pp

Rao BA, Adsumilli MS (1987) Lateritic gold: option for small mine development. In: Small mine development in precious metal. Soc Mining Eng AIME, Littleton, pp 29–36

Roslyakov NA, Roslyakova NV (1986) Evaluation of the economic potential of gold deposit by the analysis of oxidized ore outcrops and exogenic aureols: methods used in the USSR. Appl Geochem 1:451–462

Saager R, Meyer M (1984) Gold distribution in Archaean granotoids and supracrustal rocks from southern Africa: a comparison. In: Foster RP (ed) Gold '82. The geology, geochemistry and genesis of gold deposits. Balkema, Rotterdam, pp 53–70

Sanfo Z, Colin F, Delaune M, Boulangé B, Parisot JC, Bradley R, Bratt J (1993) Gold: a useful tracer in subsahelian laterites. Chem Geol 107:323–326

Schiller EA (1985) Gold in Brazil. Mining Mag 53:313–319

Schoeder G, Mikhailoff N (1977) Geochemical methods in a mineral exploration project in the Niger Republic. Nat Res Dev 5:68–83

Seward TM (1984) The transport and deposition of gold in hydrothermal systems. In: Foster RP (ed) Gold '82. The geology, geochemistry and genesis of gold deposits. Balkema, Rotterdam, pp 165–182

Smith RE (1987) Current research at CSIRO Australia on multielement laterite geochemistry for detecting concealed mineral deposits. Chem Geol 60:205–211

Smith RE, Perdrix J (1982) Pisolitic laterite geochemistry in the Golden Grove massive sulphide district, Western Australia. J Geochim Explor 18:131–164

Stoffregen R (1986) Observation on the behavior of gold during supergene oxidation at Summitville, Colorado, USA, and implications for electrum stability in the weathering environment. Appl Geochem, 1, 549–558

Viljoen MJ (1984) Archaean gold mineralization and komatitites in southern Africa. In: Foster RP (ed) Gold '82. The geology, geochemistry and genesis of gold deposits. Balkema, Rotterdam, pp 595–628

Watters RA (1983) Geochemical exploration for uranium and other metals in tropical and subtropical environments using heavy minerals concentrates. J Geochem Explor 19:103-124
Webster JG(1986) The solubility of gold and silver in the system Au-Ag-SO_2-H_2O at 25 °C and 1 atm. Geochim Cosmochim Acta 50:1837–1845
Webster JG, Mann AW (1984) The influence of climate, geomorphology and primary geology on the supergene migration of gold and silver. J Geochem Explor 22:21–42
Wilhem E (1975) L'or dans le cycle supergène. Rapport BRGM, 75 SGN 077 MET, 20 pp
Wilson AF(1983) The economic significance of non hydrothermal transport of gold, and the accretion of large gold nuggets in laterite and other weathering profiles in Australia. Geol Soc Sci Afr Spec Publ 7:229–234
Wilson AF (1984) Origin of quartz-free gold nuggets and supergene gold found in laterites and soils. A review and some new observations. Aust J Earth Sci 31:303–316
Zeegers H (1979) Regional geochemical prospecting in equatorial areas. An example in French Guiana. In: Watterson JR, Theobald PK (eds) Geochemical exploration. Assoc Explor Geochem, Toronto, pp 209–225
Zeegers H (1987) Remaniements de surface et prospection géochimique de l'or. Chron Rech Min 488:55–61
Zeegers H, Lucas Y, Laville-Timsit L (1985) Dispersion de l'or dans les sols à l'aplomb du gite primaire d'Espérance (Guyane). Rapport BRGM 85 DAM 024 GMX, inédit, 11 pp

8 Comparative Ecology of Two Semi-Arid Regions: the Brazilian Sertão and the African Sahel

Jean-Claude Leprun

1 Introduction

During one of the last field trips of Georges Millot that I had the privilege to share with him in May 1988 through northeastern Brazil and especially through the dry «Sertôes» of Pernambuco and Paraíba, we had the opportunity to study weathered rocks, soil profiles, vegetation covers and rural organization in connection with water. Comparison of these landscapes with those of the African Sahel, already well-known to both of us, was very tempting and it became the purpose of this contribution.

In fact, numerous physical and human factors are similar in both of these semi-arid areas, as follows: (1) tropical climate with well-marked dry and hot seasons and significant annual and inter-annual variation in mean rainfall, resulting in aleatory agricultural yield and/or food crops due to periodical disastrous droughts; (2) a similar Precambrian granito-gneissic basement, Brazil and Africa being separated only since the Jurassic; (3) the same sedimentary borders of the basement; (4) a drought-resistant vegetation consisting of thorny shrubs and trees; (5) a river system without natural water sites, but with predominant temporary and violent runoff; (6) a poor rural population forced to undertake frequent migrations and practising low-level technological subsistance agriculture together with small extensive farming. Besides these similarities, quite different ecological factors and processes can also be evidenced from study of specific natural ecosystems and their evolution into agrosystems, as related to the natural behaviour of soil and water. In the light of the comparison of all available data, a tentative explanation of water–soil–plant relations with resulting effects on the respective environments will be suggested for both areas.

2 Ecological Parameters

The study presented here is based on numerous climatic, agronomic, pedological, hydrological and biological observations and data collected during more than 10 years of prospection and research in the Saharo–Sahelian zone of West Africa and during 12 years of scientific cooperation with SUDENE (Superintendence for

Northeastern Development) and EMBRAPA (Brazilian Agency for Agricultural and Farming Research) in northeastern Sertão.

Most of these data are reported in survey reports and agreements (for the African Sahel: Leprun and Moreau 1968; Boulet and Leprun 1970; Boulet et al. 1971; Leprun 1977, 1983a; for northeastern Brazil: Leprun 1978, 1981, 1983b; Molinier et al. 1989) and in publications (mainly Leprun 1982, 1989, 1992; Leprun and Silveira 1992).

2.1 Geographical Frame

The West-African Sahel and the Brazilian «Nordeste» are located on opposite sides of the Equator but their respective extensions are different (Fig. 1). Extending between 3° and 15° South in latitude and between 34° and 45° West in longitude, the dry Nordeste, also well known under the name of Sertão, is longitudinally smaller and latitudinally larger then the West-African Sahel which is arbitrarily restricted to a zone between 10° and 18° North in latitude and between 4°and 15° West in longitude. Hence, there is a continental factor which differs in the two cases and which governs directly climate and indirectly all the other factors, especially the biological ones.

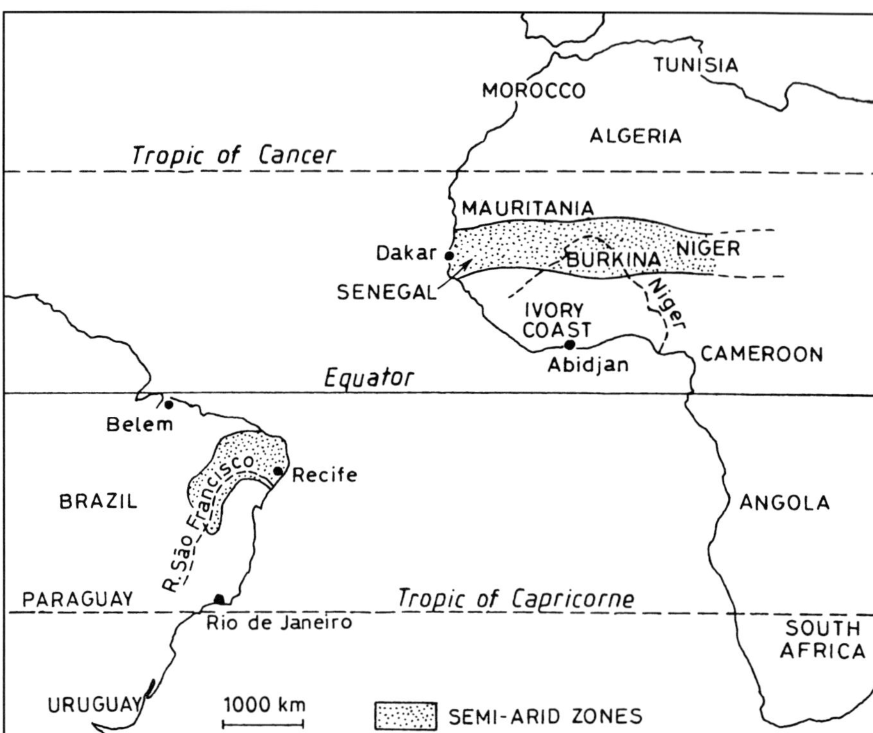

Fig. 1. Location of the study areas

2.2 Climate

Since the purpose of this chapter is to compare two dry areas, only the semi-arid climate will be considered, that is the areas where mean annual rainfall is between 400 and 800 mm. In West Africa, the semi-arid zone, here conveniently called Sahel, corresponds really to the southern part of the true Sahel, i.e. to Sudanian–Sahelian and Sahelian–Sudanian climatic conditions.

In the Brazilian Nordeste, the semi-arid zone is characterized by three different pluviometric regimes, those of the central and northern parts depending, as for the Sahel, on the migration of the inter-tropical front (ITF), and that of the southern part which is under the influence of cold fronts coming from the extreme south of Brazil. Therefore, a latitudinal shift governs the beginning of the rainy season, starting in March in the north and in November in the south (Sertão in Bahia State). This results in a less marked alternation of dry and rainy seasons than in the Sahel. In fact, during a period of 10 years, it has been very difficult to register a complete rainless month in the Sertão.

The pluviometric regime of the Sahel is governed by the ITF, the meeting front of a NE dry continental air mass and a SE humid sea air mass shifting progressively from north to south from January to August. This slow shifting results both in a noteworthy arrangement of parallel isohyets which is not observed in the Sertâo where isohyet values increase regularly from north to south and in a much more contrasted character of the climate, which means a short rainy season of about 2 or 3 months during summer and a long dry season in winter.

Rainfall erosivity is measured by a kinetic energy as a function of intensity. The erosivity index of rainfall is calculated by the formula:

$$R = E \cdot I_{30}$$

STATION	MEAN ANNUAL PLUVIOMETRY (mm)	R	NUMBER OF ANNUAL EROSIVE RAINS
TAPEROA (NE)	515	1648	16
LAGES (NE)	500	1570	15
DORI (B.F)	510	2550	13
OURICURI (NE)	607	1815	17
OUAHIGOYA (B.F.)	600	2953	16
PAU DE FERROS (NE)	751	2609	29
MOGTEDO (B.F.)	753	3708	20

NE, Brazilian Northeast; B.F., Burkina-Faso

Table 1. Mean annual rainfall and rain erosivity index for comparable stations of the Sertão and the Sahel

	STATIONS	MEAN	MAXIMUM	MINIMUM
Temperature (°C)	Quixeramobim	24.7	25.8	23.4
	Matam	37.0	42.6	33.0
	Dori	36.4	41.4	32.0
Air humidity (%)	Quixeramobim	59.4	73.2	50.8
	Matam	47.0	71.0	26.0
	Dori	45.0	77.0	22.0
			MONTHLY MAXIMA	
Evaporation (mm)	Quixeramobim	1764.2	198.8	
	Matam	3291.4	444.8	
	Dori	3817.6	460.6	

Table 2. Mean annual values for some climatic parameters in the Brazilian northeast and the African Sahel

in which E is energy and I_{30} the maximal intensity during 30 min (Wischmeier 1958). If calculated in metric units for periods of more than 10 years, for comparable weather stations, and for a number of days with similar erosivity per year, the values of the mean annual erosivity index are much higher in the case of the Sahelian rains than those of the Sertão ones (Table 1). In a previous publication, Leprun (1989) found evidence for the preponderant role played by instantaneous intensities of rain registered during very short times of 1 or 2 min and by the intensities of exceptional sudden showers, more important in the Sahel.

Like rain erosivity, other climatic factors such as mean annual temperature, humidity, evaporation recorded for the same period (1931–1960) at Quixeramobim (Ceara), Matam (Senegal) and Dori (Burkina Faso) (as well as mean annual insolation and wind frequency and power; Leprun and Silveira 1992) indicate that the climate of the Brazilian Sertão is cooler, wetter and less extreme than that of the Sahel (Table 2).

2.3 Geological Framework

The geological framework is very similar in the two areas. The same Precambrian crystalline basement consists of granites, migmatites, gneisses and basic rocks, the African and South American continents having been separated during the Mesozoic. Also similar are the Mesozoic and Cenozoic sedimentary formations composed of sandstones, shales and limestones, which were generated by degradation of the basement and deposited on its borders. Therefore, the detrital continental formations of the Late Tertiary and of the Early Quaternary, called «Continental terminal» in Africa and Barreiras Formation in Brazil, are lithologically very similar. However, in spite of their similar lithology, the geological formations have evolved differently: from the Tertiary, the internal geodynamic conditions (tectonics) and external geodynamic ones (weathering) have probably been different. The African basement remained stable and was affected by strong and prolonged weathering which resulted in a thick ferruginized lateritic cover (Millot 1964); the cover reaching some hundreds of metres in the Sahel (Leprun 1979; Pion 1979). In contrast, tectonics resulting from elevation of the Andes was active in the NW Brazilian shield during most of the Tertiary; from this period the basement has slowly but continuously been lifted up (Petri and Fulfaro 1983). Otherwise, Quaternary epeirogenic movements were evident in the coastal sedimentary series near Recife (Ponte 1969; Rand 1976) and on the shoreline of Bahia State (Silva and Tricart 1980).

2.4 Landforms and Hydrological Frame

Landforms are quite different in the two areas. The Sahelian valleysides are long and they slope gently (some kilometres in length and with slopes of 1° or less), whereas those of the Sertão (developed on crystalline basement) are short and their slopes are steep (some hundred metres in length with slopes between 5° and 45° or more). The most recent Quaternary formations consisting of sandhills fixed on the southern border of the Sahara desert and covering a large part of the northern Sahel, are lacking in the Sertão. Likewise, there are no equivalents of the colluvial and alluvial deposits of the large Sahelian valleysides in the Brazilian Nordeste. In fact, the alluvial sediments of the river system of the Sertão represent less than 2% of the total area.

Both areas display a hydrographic network composed of two large perennial rivers, i.e. the Rio São Francisco and the Rio Parnaíba for the Brazilian Nordeste, the Senegal and Niger rivers in West Africa, and of many temporary rills with rough and sudden floods which sometimes do not flow during the dry periods. Deep ground water tables with high capacity are found in the sedimentary formations of the Sahel and the deep weathering zone of the basement may also contain less important perched water tables. In the dry Brasilian Nordeste, such water tables are not observed in the crystalline basement and only narrow joints which may contain small volumes of water are found. The only possibility for water to be accumulated is then storage of the runoff water in retention hills called «açudes», whose capacity is between 20000 and 100000000 m^3. The number of «açudes» with more than 1000 m^2 area is about 70000 according to an estimation of Molle (1991).

2.5 Soils

The distinguishing features of the soils of the two areas were recorded and discussed in a recent paper (Leprun and Silveira 1992). The soils of the Sertâo have been extensively studied and carefully mapped (synthesis at 1/5.000.000 carried out in 1981 by the pedologists of the EMBRAPA/SNLCS). Developed from rocks of the slightly or non-weathered crystalline basement, these soils are thin and rich in alterable minerals. They display a high or saturated ion-exchange capacity and a clay fraction composed of 2:1 minerals or mixed layers with little kaolinite. These soils are mainly non-calcic brown soils, planosols, lithosols and regosols, then vertisols, solodized solonetzes with some eutrophic red and yellow podzolic soils (in the American meaning) and less cambisols and brunizems. Latosols and usually dystrophic red and yellow podzolic soils develop preferentially from sedimentary rocks. The rare limestones generate rendzinas, cambisols, indurated soils and red

	SUME (Paraiba)	OURSI (Burkina)
Clay (%)	32.0	33.5
Coarse sands (%)	21.0	18.8
Organic matter (%)	4.6	0.7
Organic C (%)	2.77	0.40
Nitrogen N (%)	0.23	0.024
C/N	12.0	15.9
Ca (me)	18.54	10.0
Na (me)	0.49	0.69
S (me)	29.21	18.06
T (me)	31.99	18.24
S/T (%)	91	99
Log 10 Is[a]	1.30	1.65
Log 10 K[a]	1.47	1.17
Soil thickness (cm)	70	190
Horizon A thickness (cm)	10	22

[a] Structural instability Is and infiltration rate K by Hénin *et al.* (1969).

Table 3. Analytical characteristics of the A horizon in brown vertic soils at Oursi (Burkina Faso, Africa) and Sumé (Paraíba, Brazil)

latosols. The Sahelian soils mapped for the most part over the last 40 years by the pedologists of O.R.S.T.O.M. (synthesis at 1/5 000 000 made for the Food and Agriculture Organization of the United Nations by Boulet et al. 1971) differ in the nature of the parent materials from which they result. Developed from thick, usually indurated or ironcrust-bearing, kaolinitic and lateritic cover or from fixed downsand of ancient ergs, the most frequently occurring soils are little evolved or leached ferruginous soils, lithosols over Fe "carapaces", or ironcrusts. On the crystalline basement, the soils are almost similar to those of the Nordeste; i.e. vertic or non-vertic brown soils, vertisols and solodized solonetzes.

However, some important differences which are ascribed to the dissimilar nature of the parent rocks are now described. The podzolic soils frequently connected with the African ferruginous soils have not been observed in the Sahel. In addition, because of the difficulty of identification, the soils have even been denominated «fersiallitic ferruginous soils» or «fersiallitic ferruginous soil–vertisol intergrades» by a pedologist of Africa who mapped these soils in a semi-arid zone of the Ceara area (Guichard 1970). Similarly, the lithosols and the fine-gravel soils over Fe "carapaces" or ironcrusts on one hand, and the sub-arid brown soils or ferruginous soils developed from fixed dunes on the other, which are widespread in the Sahel, have no equivalents in the Nordeste. Likewise, Brazilian cambisols and brunizems developed from slightly weathered basic rocks have no homologue in the African Sahel.

Another important difference which will be considered further, comes from the frequent occurrence of crusts or thin silt–clay layers in the upper part of Sahelian soils (Leprun 1978; Valentin 1981; Casenave and Valentin 1989); these are very rare in the Sertão. Such crusts plug the soil surface, hindering water infiltration and seedling emergence. Comparison between two vertic brown soils with similar textures, both developed from migmatites under mean annual rainfall between 450 and 550 mm and sampled at Oursi (Burkina Faso) and at Sumé (Paraíba), shows that the soil of Sumé has a much higher content of exchangeable cations and organic matter (Table 3). Physical and chemical data (more than 1500 for the Sertão, about 200 for the Sahel) concerning the surficial horizons of soils from both areas, indicate higher chemical fertility and better physical properties in the Sertão soils (Table 4).

2.6 Vegetation

From data of the F.A.O. (1975), arboreal and shrubby vegetation represents less than 5.3% of the surface of the Sahel, in Senegal, Mali and Burkina Faso of Africa, while it covers 90% of the semi-arid zone of the three States of the Brazilian Nordeste, i.e. Pernambuco, Paraíba and Alagoas (Duque 1980). The natural vegetation of the Sertão consists of a more or less compact thorny bush, or even a true low and dry forest, called «caatinga» whose penetration is difficult. The caatinga is a continuous vegetation formation with trees between 2 and 5 m and sometimes 10 m high, where the arboreal and shrubby strata are rich in leguminous plants, and where the Graminaceae–herbaceous stratum is slightly developed. In a low, mainly shrubby caatinga, Hayashi (1981) identified about 17 200 tree and shrub species resulting in a phytomass of 23 t/ha. In an arboreal caatinga of more humid zones, the tree density is between 500 and 1000 units/ha (Emperaire and Pellerin 1989). The regeneration capacity of a caatinga is remarkable; for example, after a 5-year fallow, a crop of tomatoes on a vertic brown soil of Sumé (Table 3) regenerated a phytomass of about 50 t of dry matter per hectare.

The arboreal and shrubby Sahelian savanna is an open formation composed of an important steppe of gramineous stratum with scattered trees and shrubs, along with leguminous plants. A total of 133 trees and shrubs per hectare resulting in a phytomass of 1.8 t/ha was reported by Lamothe and Boulière (1978) in the case of a loose arboreal savanna of the Ferlo, in Senegal. Another particular vegetation formation, called «brousse tigrée» because of the alternations of dense wooded bands and bare ones, covers a large part of the Sahel.

	SERTAO	SAHEL
Soil texture	clay-sandy	sandy-clay
Organic matter (%)	0.5–4.0 (0.2)[a]	0.3–1.5 (0.7)[a]
Exchangeable calcium (me/100 g)	5.0–10.0	0.5–2.0
Infiltration rate (mm h^{-1})	rapid (50–400)	moderately slow (15–25)
Soil structure	grainy	massive
Surface layer	not crusted	crusted

[a] The number in parentheses indicates the most common value found

Table 4. Average values for soil surficial horizons in the Sertão and the Sahel

2.7 Human Environment

Both areas are essentially rural, the average density of the population being 16 inhabitants per square kilometre in each. In dry Africa, a primitive, manual agriculture has been practiced for 2000 years, and remains largely monocultural. This agriculture develops from villages which move after several years of small-scale farming. The fields are let to lie in fallow during a variable number of years, which tends to decrease because of demographic increase. Cultivation starts by clearing followed by yearly repeated burning. The Sahelian farmer breeds few cattle around his house. Transhumance breeding over long distances is specific to the «Peulh» shepherds. Wood is the only power supply and its consumption contributes greatly to deforestation of the Sahel.

In the Sertão, agricultural activities are only little developed and began less than 400 years ago. The Portuguese and Indian halfcaste population practices non-itinerant agriculture and small-scale breeding in closed and precisely limited and protected properties. Different varieties of beans, maize, shrubby and herbaceous cotton and fodder crops are cultivated manually or with the help of animal traction. *Prosopis juliflora*, introduced 50 years ago from Peru's arid zones, is largely and successfully cultivated as an arboreal fodder crop. The practice of yearly burning following clearing is little developed and the use of charcoal is restricted to bread burning and brick firing, since cooking with gas is known almost everywhere.

3 Discussion

3.1 Similarities and Differences in the Ecological Parameters

All the preceding data indicate that indices of rain erosivity and other climatic parameters result in less aggressive pluviometric and thermic conditions in the Sertão than in the Sahel. Moreover, the differences in landforms governing the working and evolution of ecosystems may be easily explained by the specific geological characteristics. In Nordeste, landforms are undulating, as they are composed of rock outcrops and inselbergs of crystalline basement and of large sandstone mesas on sedimentary rocks, called «chapadas» and «tabuleiros», respectively. In contrast, the Sahelian landforms which developed from a clay or sand–clay iron-crust-bearing cover are remarkably flat. Consequently, external water drainage is always easy in the Sertão, whilst bad drainage and even endoreism prevail in the Sahel. The rock material affected by pedogenesis is also completely different in both cases. In the Sertão, the hard rocks of the crystalline basement as well as thin weathering mantles crop out, whereas thick weathering mantles are observed in the Sahel where they are lixiviated, frequently iron indurated, and hence poorer. In fact, soils of the Sertão are richer in chemical elements, more permeable and display much better aggregation and structural stability. Consequently, they have more satisfactory hydrodynamic and agricultural properties than those of the Sahel. The

FACTOR	BRAZILIAN NORTHEAST	AFRICAN SAHEL
Geographic situation	Southern Amazônia continentality little accentuated	Southern Sahara accentuated continentality
Climate	less accentuated and hot more humid air less erosive rains	more contrasted and hot drier air more erosive rains
Rocks Tectonic	little weathered predominantly crystalline active	very weathered sedimentary predominant stable
Relief	undulated, with rock outcrops, non duned, exoreic	plane, without outcrops, duned, endoreic
Soils	more clayey, less thick, rich, permeable, non indurated	more sandy, thick; kaolinitic insaturated, indurated
Runoff	weak	strong
Vegetation	ligneous low forest	open gramineous savanna contracted "brousse tigrée"
Human activities	recent, fixed culture and herds	old, annual burnings shifting culture and herds

Table 5. Similarities and peculiarities of ecological factors in the Sertão and the Sahel

principal similarities and characteristics of the ecological factors in the dry Nordeste and in the Sahel are reported in Table 5.

3.2 Consequences on Runoff and Infiltration

As rainwater reaches the soil surface, runoff or infiltration will operate according to edaphic conditions. Average surface runoff rates measured over small catchments of the Sertão by the SUDENE (data collected in Cadier 1991 and given in percent of the height of fallen rain) indicate runoff coefficients lower than 5% of the total annual precipitation, even in the case of slopes higher than 5° in little disturbed ecosystems.

Experiments were carried out in similar conditions under simulated rainfall with a mini-simulator conceived by the O.R.S.T.O.M. (Asseline and Valentin 1978) and applied to non-calcic vertic brown soils of the type found at Oursi (Burkina Faso) and Sumé (Paraíba). The results reported by Chevalier (1982) and Audry et al. (1987) show that the height of imbibition rain, which represents the amount of water that infiltrates before the first runoff episode, is eight times higher in the Sertão whether the soil is bare or covered by natural vegetation. At the end of the dry season at Sumé, where the first simulated rainfall is 48 mm, runoff rate is zero, whereas runoff represents 30% of the total rainfall at Oursi, regardless of whether the soil is bare or covered. At Sumé, in the case of a dry soil under 5 years fallow, the first runoff was observed only after simulated rainfall of 63 mm/h, which represents

203 mm in 2 days, whilst mean annual rainfall is 500 mm. At Oursi, where the mean annual rainfall is 400 mm, only one simulated rainfall of 40 mm applied with similar intensity results in a runoff of 13 mm.

The data collected in the Sahel during some decades by hydrologists of the O.R.S.T.O.M. (Rodier 1975; Chevalier 1982) indicate average runoff above 25% in the case of gentle slopes between 1° and 3°, which was corroborated by measurements carried out in similar conditions on plots with bare worked soils (Table 8).

3.3 Consequences for Surface Water Quality

The data reported in Table 6 show differences in the chemical composition of water from both areas. In the Sertão, as infiltration waters percolate through thin soils rich in alterable minerals, thin saprolites and narrow joints of the basement, they are rapidly loaded with chemical elements. After only 1 m infiltration, the contents of Cl and Na increase 100 times and the amounts of the other elements are between 10 and 20 times higher than in rainwater (Molinier et al. 1989; Leprun and Silveira 1992). In contrast, the runoff waters in the Sahel are only slightly mineralized because they originate from lixiviated weathered materials and soils. River waters maintain these differences and ground water tables intensify the process.

3.4 Consequences for Soil Erosion

Runoff water partially releases material extracted from the soil surface, i.e. eroded solid load. The latter depends on different factors (Wischmeier and Smith 1960), among which some are natural, i.e. climatic (such as *R* erosivity index), pedological (like *K* erodibility or susceptibility index) and topographic (like the SL slope parameter). Two other factors which are dependent on man's activities (i.e. the *C* factor as a function of vegetation cover and *P* indicative of anti-erosive practices),

	pH	Ca	Mg	K	Na	Cl	SO_4	SO_3H
		← ...	...	...	$mg.l^{-1}$	...	...	... →
Runoff waters								
Sahelo-sudanian (N = 10)[a]	7.0	4.2	1.2	0.7	1.8	0.5	ND	24.8
Sertão (N = 6)[b]	7.1	6.2	1.5	4.3	0.7	1.2	2.5	24.4
River waters								
Sahelo-Sudanian (N = 29)[a]	7.4	7.3	3.9	2.2	3.1	1.4	ND	50.5
Sertão (N = 27)[b]	7.2	23.8	19.2	5.5	79.7	157.5	8.9	81.3
Reservoir waters								
Sertão Açudes (N = 364)[b]	7.7	25.9	12.7	9.6	53.7	80.9	7.1	117.9
Deep water table waters								
Africa-Sahelian (N = 11)[c]	6.9	19.7	13.7	8.3	20.5	13.1	10.7	75.0
Sahelo-Sudanian (N = 101)[a]	6.8	7.3	3.9	2.2	3.1	1.4	ND	50.5
Sertão (N = 32)[b]	7.6	79.0	41.8	9.2	103.0	96.6	37.3	270.0

Sources: a Blot (1970 and 1980),
b Leprun (1983a),
c Leprun (1979)

Table 6. Comparison of chemical compositions of waters from the studied areas

	FACTORS	NORTHEAST VARIATIONS	SAHEL VARIATIONS
R	(rainfall erosion index)	1 to 5	1 to 10
K	(soil erodibility)	1 to 10	1 to 12
SL	(topographic)	1 to 80	1 to 25
C	(cover and management)	1 to 1000	1 to 1000
P	(practices)	1 to 20	1 to 10

Sources: Sahel (Roose 1977); Northeast (in Leprun 1989)

Table 7. Variation in the values of the hydric erosion factors for the studied areas

can be used. The rain erosivity index established earlier shows that rainwaters are more efficient in the Sahel than in the Sertâo.

If we compare erosion susceptibility of soils in the two areas mostly related to aggregation and permeability, the results obtained by the method of Wischmeier and Smith (1960) on plots, and under natural rainfall, provide erodibility indices higher in the Sahel than in the Sertão for comparable soils (average indices of 0.10 and 0.20 for podzolic and ferruginous soils and of 0.07 and 0.15 for brown soils in the Sertão and in the Sahel, respectively). The range of soil erosivity values is between 0.02 and 0.30 in the Sahel. The *C* factor which represents the effectiveness of vegetation cover, varies concomitantly between 1 and 0.001 in both areas. The values of the *P* factor vary between 1 and 0.05 in the Sertão and between 1 and 0.1 in the Sahel. All the data are recorded for the Sahel in Roose (1977) and for the Sertão in Leprun (1989).

Variations of the above-mentioned erosion factors are given in Table 7. It should be noted that these variations are of the same order of magnitude whatever the area considered, that, as expected, the range of variation of the topographic factor is larger in the Sahel and that the vegetation cover is very efficient since soil losses are reduced by a factor of 1000. The effects of hydric erosion were also compared in both areas from measurements on plots (surface area between 1 and 100 m^2) and on small catchments (surface area between 1 ha and 1 km^2) in which rainfall conditions and soil nature and use were similar.

The first comparison concerns the sites of Linoghim in Burkina Faso (Piot and Milogo 1980) and of Serra Talhada in Pernambuco (Lago 1981), in comparable

Stations	Serra Talhada (Pernambuco)	Linoghin (Burkina)	Sumé (Paraíba)	Allokoto (Niger)
Area (ha)	0,1	0,1	0,42	0,48
Type of soil	non calcic brown		non calcic vertic brown	
Mean annual pluviom. (mm)	850	800	500	440
Slope (%)	4.7	1.5	7.0	3.0
Wischmeier plot				
Soil losses (t/ha)	2.45	13.92	0.36	1.10
Runoff losses (%)	–	35.8	4.4	5.2

Table 8. Comparison of solid and water loss values in similar stations of the studied areas

conditions except for the slope which is 4.7° at Serra Talhada and only 1.5° at Linoghin. The second comparison concerns two drier sites, i.e. Allokoto in Niger (Delwaulle 1973) and Sumé in Paraíba (Cadier et al. 1983) where the slopes are also different. The results reported in Table 8 indicate that soil losses are more important in the Sahel than in the Sertão, although such losses were measured on lower slopes in the Sahel (when other factors are similar , the influence of slope on erosion and runoff becomes relatively important).

This tendency of the Sertão soils to better resist rain aggressivity may be verified from examples observed on larger catchments (surface areas of some thousands of km^2). In fact, the Senegal and Falémé rivers which drain slightly or non-cultivated Sahelian catchments whose surface areas are 218 000 km^2 and 28 900 km^2, respectively, result in mechanical erosion of 10.4 and 16.57 t km^{-2} $year^{-1}$ (Orange 1990). From sedimentometric data collected by the CHESF (São Francisco River Hydroelectric Company) and tests carried out for the Sobradinho Dam Project on the São Francisco River (Hidroservice 1974), mechanical erosion in the 85 000 km^2 of the slightly cultivated lower-middle sub-basin located between Sobradinho and Moxoto dams, is 7.18 t km^{-2} $year^{-1}$. It should be recalled that mechanical erosion was estimated by Kozoun et al. (1974) to be between 12.4 and 30 t km^{-2} $year^{-1}$ for the whole 613 133 km^2 São Francisco River basin, and about 57 t km^{-2} $year^{-1}$ for the 418 000 km^2 upstream part which is more steeply sloping, more watered and more cultivated, and is completely encompassed by the semi-arid zone.

3.5 Consequences for the Functioning Ecosystem

Two ecosystems have been chosen as examples: the caatinga ecosystem which is the most widespread and representative of the Sertão in the Brazilian Nordeste, and the "brousse tigrée" ecosystem studied during some years in the Sahel by Leprun (1983b, 1992). For convenience, each of these ecosystems will be considered as a homogeneous whole, although really composed of different sub-ecosystems. In fact, there are several types of caatinga and of "brousse tigrée".

3.5.1 The Caatinga Ecosystem

It is necessary to refer to the historical setting of the caatinga to understand how it develops. During a long time in the Tertiary, a hot and humid tropical climate prevailed in Brazil, as in the other continents, and resulted in the formation of thick lateritic mantles on the Precambrian crystalline basement. Active tectonics, which has affected the basement several times, provoked significant erosion and removal of material so that the lateritic cover was transported and settled in marginal coastal or pre-coastal basins in the Late Tertiary (Mabesoone et al. 1972; Petri and Fulfaro 1983). Hence, the crystalline basement remained bare, displaying suboutcropping or very slightly weathered rocks. There were no sandy deposits and only a few colluvial-alluvial sediments in the riverbeds. During the Quaternary, tectonics remained active, so that soils developed on the uplifted basement are rather thin with high contents of alterable minerals, low clay amounts and good internal drainage. An efficient exoreic system developed, characterized by strong withdrawal of material, given the absence of storage rocks and of underground reservoirs. Under the dry and contrasted climates of the Quaternary, pedogenesis could not balance the geochemical withdrawal favored by epeirogenesis, so that soils could not develop. Low rainfall and easy infiltration of rainwater in these permeable soils resulted in an enrichment of waters in chemical elements, from contact with weatherable soil minerals and with narrow rock joints. These very mineralized waters were held in deep faults or diaclases, or exported to the sea.

A dense, continuous vegetation formation dependent on edapho-climatic conditions developed on these soils. Because sands were lacking and winds slight, the prevailing species were not Gramineae but trees and shrubs. Since soils have been recently and manually farmed, protected in delimited properties and not subject to yearly burnings, and are free of caatinga, they have retained a good structure due to aggregation favoured by high cation and organic matter content, whose good quality comes from the decay of the abundant phytomass produced by leguminous plants. When let in fallow, this phytomass regenerates easily due to a high germination rate favored by the physical properties of the surficial horizon and by the absence of crusts, allowing water infiltration and retention.

The caatinga ecosystem, specific to the semi-arid Sertão, may thus be considered as a climatic system, i.e. in equilibrium with present-day environmental conditions. Its behaviour with regard to erosion is satisfactory in natural conditions, or in the case of slightly aggressive use; as this ecosystem is degraded, its regeneration rate and efficiency are exceptional.

3.5.2 The "Brousse Tigrée" Ecosystem

As in the preceding case, a reconstruction of the historical setting of this ecosystem is necessary to understand its functioning. The West African crystalline basement was deeply weathered under the Tertiary tropical climatic conditions and probably in the late Mesozoic – a period during which tectonic activity was very slight or absent – this resulted in a very thick cover of Fe-rich lateritic formations. By erosion during the Late Tertiary, a part of these lateritic formations generated detrital formations of the "Continental terminal" which is the African equivalent of the Barreiras Formation in Brazil. In contrast to the Brazilian basement, whose weathering cover was completely removed, the West African basement kept a major part of the latter which continued to be affected by intense weathering. In fact, thick Fe crusts may be observed in the "Continental terminal", whereas they are lacking in the Barreiras Formation. Under Quaternary climatic changes, degradation of the ironcrust-bearing formations of the basement of the "Continental terminal" and of the sedimentary rocks, generated Fe-rich fine-gravel soils, tropical ferruginous soils and mostly sandy, loose material. Raised by wind, this sandy material generated the fixed dune fields covering all the northern zone of the Sahel, as well as the Sahara borders. A plugging is observed, which originates from these sandy formations and from extended colluvial–alluvial sediments. Movement is partly or totally impeded due to inactive tectonics, continental geographical location and decrease of rainfall, so that numerous zones become endoreic. As very aggressive rainwater reaches the soil surface, it penetrates a silt–clay material which is little structured because it is poor in chemical elements, slightly permeable, easily crumbling and likely to be subject to encrustation. Hence, this rainwater will run rather than infiltrate. In both cases, it will be in contact with excessively lixiviated and desaturated materials and thus will have no possibility to become mineralized.

Whereas open savanna or gramineous steppe covers sandy materials or colluvial–alluvial formations or soils on the basement where external drainage is possible, the "brousse tigrée" develops on the endoreic zones of sedimentary or iron-crust-bearing substratum. The vegetation composition, permanence and evolution of the "brousse tigrée" depend on strict edapho-climatic conditions, i.e. mean annual rainfall less than or equal to 500 mm, slope lower than 4°, soil thickness less than l m. The "brousse tigrée" is composed alternatively of wooded and bare bands. Each wooded band consists of dense shrubby and arboreal strata of species belonging to more humid environments but surviving due to the water runoff on bare bands. Like the vegetation, the fauna corresponds to more humid southern environments and is composed of rodents, termites and butterflies of Sudanian type. Under rain impact, bare bands are covered by a pellicular crust preventing water infiltration and seed germination.

By the combined effects of hydric and aeolian erosion, dynamics of rapid lateral displacement between one and several dm per year affects the whole formation composed of alternating wooded and bare bands. Rain erosion applied to bare bands generates a fine-grained material plugging the wooded lower zone, and a sandy material which, under wind action, will build a micro-dune favouring trapping and germination of gramineous seed resulting in a front of settlement. The wooded band will also be displaced in the windward direction, being progressively plugged (Fig. 2).

The Sahelian "brousse tigrée" ecosystem is unstable and can persist only if, as related to its latitudinal position, drought periods are not too long and if fire or farmer machete do not reduce its wooded bands which, then, progressively become dry and disappear. In the Sahelian zone, the distribution area is decreasing since the "brousse tigrée" is an open ecosystem in disequilibrium with present-day environmental conditions. The latter represents a humid ecological niche which is hardly stable in drier regions and which does not regenerate if degraded. Sahelian steppes and savannas are also unstable because of yearly burning, difficult climatic conditions, disastrous influence of man and herds on these vegetation formations. As extended zones are weakened and degraded by overpasturing and badly managed farming, they are denuded and subject to the combined effects of hydric and aeolian erosion.

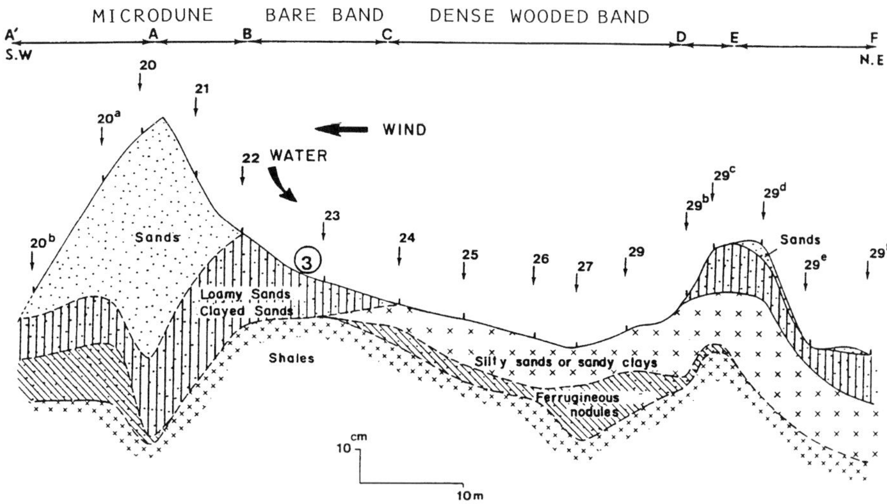

Fig. 2. Cross section of a "brousse tigrée" formation (Gossi, Mali)

4 Conclusion

A comparative study of ecological factors between two semi-arid areas located on both sides of the Atlantic Ocean and of the Equator is presented. Although both areas display similar geological formations and climatic conditions, especially periodical marked dry seasons, some differences are evidenced, which may explain the specific setting and functioning of the different ecosystems under the dependence of some particular factors.

The preponderant influence of internal geodynamics (tectonics and epeirogenic movements) and of external geodynamics (weathering, drainage) on the same Precambrian basement after separation of the African and South American continents results in formation of different materials, soils, waters and vegetation covers. The soil–plant–water relations in the two ecosystems specific to both areas, i.e. the Brazilian caatinga and the Sahelian "brousse tigrée" are compared and their respective functioning effects on runoff and infiltration, on surficial water quality and on erosion are emphasized.

The analysis of the analogies and differences of ecological factors in both areas indicates that the Brazilian Sertâo is much more resistant to natural aggression and less affected by anthropogenic degradation than the African Sahel. However, each advantage has an unfavourable side. In fact, in the Sertão, if soil physical properties and vegetation cover of the caatinga favour infiltration, the high content of chemical elements in soils generates excessively mineralized waters, so that any project of irrigation may be dangerous. In contrast, Sahelian waters are slightly mineralized, since they originate from poorer and less fertile soils. But these poorer soils are deeper and, thus, may store much more water. In the Sertão, retention capacity of soils represent the limits for pluvial yearly farming to be brought to an end and for perennial cultivation to survive. Likewise, the more sloping topography of the Brazilian landforms, although responsible for good drainage and for good cultivation along most steep slopes, leads to severe erosion rates.

In both cases, dangers come from extension of farming to new areas, from acceleration of mechanical cultivation and from overpasturing. This was illustrated by a massive increase in sediments at the mouth of a São Francisco river tributary, as a consequence of farming undertaken in the upstream soils. In fact, specific erosion varies from less than 750 t km^{-2} $year^{-1}$ to around 2000 t km^{-2} $year^{-1}$ (Carvalho 1988). The very high erosion rates of the Rajasthan basins in the inhabited semi-arid area of India (Sharma and Chatterji 1982) represent another case. These examples establish, if it were necessary, that semi-arid and arid zones which cover a third of the emerged areas and which represent a reserve of lands and of natural resources for the future, are particularly sensitive to farming practices and to cultivation changes in relation to climatic variations in time and space. It is, thus, necessary to protect and manage them and to have a better knowledge of the functioning of semi-arid ecosystems and of their processes of transformation into agrosystems.

Acknowledgments. The author is very thankful to Prof. Larry Frakes for his assistance in English translation.

References

Asseline J, Valentin C (1978) Construction et mise au point d'un infiltromètre à aspersion. Cah ORSTOM, Sér Hydrol XV:321–349

Audry P, Cadier E, Leprun JC and Molinier M (1987) Influence à l'échelle régionale des couvertures pédologiques et végétales sur les bilans hydriques et minéraux du sol. Rapport d'avancement des travaux. Rap ATP-PIREN «Dynamique de l'eau et des matières dans un écosystème représentatif du Nordeste brésilien». ORSTOM, Recife, 78 pp

Boulet R, Leprun JC (1970) Etude pédologique de la Haute-Volta. Région Est. Rap ORSTOM, 1 carte au 1/500,000 Dakar, 331 pp

Boulet R, Fauck R, Kaloga B, Leprun JC, Riquier J, Vieillefon J (1971) Carte pédologique au 1/5,000,000 de l'Afrique de l'Ouest avec notice explicative. International West African atlas. Under the auspices of the Organisation of the African Unity ScientificTechnical and Research Commission and with the assistance of the Ford Foundation s.l.:s.n., 1968

Cadier E (1991) Hydrologie des petits bassins du Nordeste brésilien semi-aride. Thèse, Univ Montpellier II, 396 pp

Cadier E, Freitas JB, Leprun JC (1983) Bacia experimental de Sumé (PB). Instalação e primeiros resultados. SUDENE, Ser Hidrologia, Recife, 16, 87 pp

Carvalho N de O (1988) Sediment yield in theVehas River basin (Minas Gerais, Brazil). In: Bordas MP, Walling DE (eds) Sediment budgets. Proc Porto Alegre Symp, Dec 1988, IAHS Publ 174, pp 369–375

Casenave A, Valentin C (1989) Les états de surface de la zone sahélienne. Influence sur l'infiltration. Didactiques, Ed ORSTOM, Paris, 229 pp

Chevalier P (1982) Simulation de pluie sur deux bassins versants sahéliens (Mare d'Oursi-Haute Volta). Cah ORSTOM, Sér Hydrol, XIX:253–297

Delwaulle JC (1973) Résultats de six ans d'observations sur l'érosion au Niger. Bois For Trop (Nogent-sur-Marne) 150:15–36

Duque JG (1980) O Nordeste e as lavouras xerófilas, 3 edn. Fundação Guimarães Duque, Coleção Mossorense, Ed Escola Sup Agric, Mossoro, 238 pp

EMBRAPA/SNLCS (1981) Mapa de solos do Brasil a 1/5,000,000. Rio de Janeiro

Emperaire L, Pellerin J (1989) Le Sud-Est du Piaui: les potentialités d'un écosystème du Nordeste brésilien semi-aride à l'aube d'une transformation. In: Bret B (coord) Les hommes face aux sécheresses. Nordeste brésilien, Sahel africain. Ed EST-IHEAL/Samuel Tastet, Paris, pp 191–199

Food and Agriculture Organization of the United Nations (1975) Formulation of a tropical forest monitoring project, FAO/UNEP, Rome

Guichard E (1970) Les sols du Bassin du Rio Jaguaribe. Mém ORSTOM, Paris, 40, 146 pp

Hayashi I (1981) Plant communities and their environments in the caatinga of northeast Brazil. Latin American studies 2. Univ Tsukuba Skura-Muraz Ibarahi, Japan

Hénin S, Gras R, Monnier G (1969) Le profil cultural. Masson, Paris, 332 pp

HIDROSERVICE (1974) Projeto Sobradinho. Estudo sedimentométrico. Rap REP 26/74, São Paulo, 31 pp

Kouzoun VI, Sokolov AA, Budyko MI, Voskresensky KP, Kalinin GP, Konoplyantsev AA, Korotkevich ES, Kuzin PS, Lvovitch MI (1974) World water balance and water resources of the Earth, USSR. National Committee for the International Hydrological Decade, Leningrad

Lago JC (1981) Erodibilidade de um solo podzólico vermelho-amarelo eutrófico pelos métodos da chuva natural, simulador de chuvas e nomograma no Sertão do Pernambuco. Tese Mestrado, Univ Fed Rural do Pernambuco, 78 pp

Lamothe M, Bourlière F (1978) Problèmes d'écologie: structure et fonctionnement des écosystèmes terrestres. Ed Masson, Paris

Leprun JC (1977) Esquisse pédologique à 1/50,000 des alentours de la Mare d'Oursi (Haute-Volta) avec notice explicative et analyse des sols. ACC DGRST «Lutte contre l'érosion dans l'Oudalan (Haute-Volta)». Rap ORSTOM, Paris, 53 pp, carte annexe

Leprun JC (1978) Compte-rendu de fin d'études sur les sols et leur susceptibilité à l'érosion, les terres de cures salées, les formations de «brousse tigrée» dans le Gourma malien. Rap ORSTOM-DGRST, Dakar, 45 pp

Leprun JC (1979) Les cuirasses ferrugineuses des pays cristallins de l'Afrique occidentale sèche. Genèse, transformations, dégradation. Thèse Sci, Univ Strasbourg, 1979. Sci Géol Mém (Strasb) 59:224 pp

Leprun JC (1981) A erosão, a conservação e o manejo do solo no Nordeste brasileiro. Balanço, diagnóstico e novas linhas de pesquisas. SUDENE, Ser Recursos de solos, Recife, 15, 107 pp, carte annexe

Leprun JC (1982) Comparação dos fatores da erosão hídrica no Nordeste brasileiro seco e na Africa do oeste seco. Consequências. 1e Symp brésilien du tropique semi-aride, Olinda, Pernambuco, Ann, 19 pp

Leprun JC (1983a) Relatório de fim de convénio de manejo e conservação do solo no Nordeste brasileiro (1982–1983). Rap SUDENE-ORSTOM, Recife, 290 pp, 5 cartes annexes

Leprun JC (1983b) Les sols. Les relations sols-végétation. In: Barry JP, Boudet G, Bourgeot A, Celles JC, Manière R (eds) Etude des potentialités pastorales et de leur évolution en milieu sahélien au Mali. ACC – GRIZA-LAT Groupe de Recherches Interdisciplinaires et Zones Arides, Paris, pp 19–56

Leprun JC (1989) Etude comparée des facteurs et des effets de l'érosion dans le Nordeste du Brésil et en Afrique de l'Ouest. In: Bret B (coord) Les hommes face aux sécheresses. Nordeste brésilien, Sahel africain. Ed EST-IHEAL/Samuel Tastet, Paris, pp 139–153

Leprun JC (1992) Etude de quelques brousses tigrées sahéliennes: structure, dynamique, écologie. In: L'aridité, une contrainte au développement. Didactiques, Ed ORSTOM, Paris, pp 221–244

Leprun JC, da Silveira CO (1992) Analogies et particularités des sols et des eaux de deux régions semi-arides: le Sahel de l'Afrique de l'Ouest et le Nordeste brésilien. In: L'aridité, une contrainte au développement. Didactiques, Ed ORSTOM, Paris, pp 131–151

Leprun JC, Moreau R (1968) Etude pédologique de la Haute-Volta. Région Ouest-Nord Rap ORSTOM, 1 carte au 1/500,000, Dakar, 341 pp

Mabesoone JM, Campos e Silva A, Beurlen K (1972) Estratigrafia e origem do Grupo Barreiras em Pernambuco, Paraíba e Rio Grande do Norte. Rev Bras Geoc 2:173–188

Millot G (1964) Géologie des argiles. Ed Masson, Paris, 499 pp

Miranda EE, Oliveira CA (1981) Um método simples para estimar as precipitaçôes anuais em localidades sem pluviométro no trópico semi-árido do Brasil. Docum n 9, CPATSA/EMBRAPA, Petrolina (PE), 36 pp

Molinier M, Audry P, Desconnets JC, Leprun JC (1989) Dynamique de l'eau et des matières dans un écosystème représentatif du Nordeste brésilien. Conditions d'extrapolation spatiale à l'échelle régionale. Rapp final ATP-PIREN, Rap ORSTOM, Recife, 25 pp, annexes 56 pp

Molle F (1991) Caractéristiques et potentialités des «açudes» du Nordeste brésilien. Thèse, Univ Montpellier II, 380 pp

Nimer E (1979) Climatologia do Brasil. Recursos Naturais e Meio ambiante, IBGE, Rio de Janeiro, 4, 421 pp

Orange D (1990) Hydrologie du Fouta-Djalon et dynamique actuelle d'un vieux paysage latéritique (Afrique de l'Ouest). Thèse, Univ Strasbourg. Sci Géol Mém (Strasb) 93 (1992):198 pp

Petri S, Fulfaro VJ (1983) Geologia do Brasil. USP, Ed TA Queiroz, São Paulo, 631 pp

Pion JC (1979) L'altération des massifs cristallins basiques en zone tropicale sèche. Etude de quelques toposéquences en Haute-Volta. Sci Géol Mém (Strasb) 57:220 pp

Piot J, Milogo E (1980) Etude du ruissellement et de l'érosion. Report DGRST/ACC «Lutte contre l'érosion dans l'Oudalan (Haute-Volta)», 16 pp

Ponte FC (1969) Estudo morfo-estrutural da Bacia Alagoas-Sergipe. Bol Tecn Petr 12:439- 473

Rand HM (1976) Estudos geofísicos na faixa litorãnea ao sul de Recife. Tese para Docente livre, Univ Fed Pernambuco, Recife, 101 pp

Rodier JA (1975) Evaluation de l'écoulement annuel dans le Sahel tropical africain. Trav et Doc ORSTOM, Paris, 121 pp

Roose EJ (1977) Erosion et ruissellement en Afrique de l'Ouest; vingt années de mesures en petites parcelles expérimentales. Trav et Doc ORSTOM, Paris, 78, 108 pp

Sharma KD, Chattji PC (1982) Sedimentation in Nadis in the Indian arid zone. Hydrol Sci J 27:345–352

Silva TC, Tricart J (1980) Problemas do Quaternário do litoral sul da Bahia. Ann XXXI Congr Bras Geol, vol I, pp 603–606

Valentin C (1981) Organisations pelliculaires superficielles de quelques sols de région subdésertique (Agadez-Niger). Dynamique et conséquences sur l'économie en l'eau. Thèse Univ Paris VII, Coll Etudes et Thèses, ORSTOM, Paris, 1985, 259 pp

Wischmeier WH (1958) Rainfall energy and its relationship to soil loss. Trans Am Geophys Union Washington 39:285–291

Wischmeier WH, Smith DD (1960) A universal soil loss estimation equation to guide conservation farm planning. Int Congr Soil Sci 7, Madison, vol 1, pp 418–425

9 Importance of the Pore Structures During the Weathering Process of Stones in Monuments

Daniel Jeannette

1 Introduction

After World War II, salvaging monuments which were damaged or even demolished by bombing, justified urgent restoration treatments using techniques that did not always respect the usual constraints for this kind of work. Afterwards, it was decided by those in charge of the monuments to instigate new restoration works which took into greater account the conservation or «preservation» aspect of the monuments. At the same time, research was undertaken to determine the origin of all types of degradation observed in stone monuments, even those not destroyed during the war. The magnitude of the degradation and the rapidity of its development, even on recently restored monuments, needed a broad study to be carried out to investigate the causes of degradation and, if possible, to suggest ways for its limitation. It was hoped that the results of this research would lead to the immediate improvement of the current standards of restoration. This was also the period when many monuments and buildings were extensively cleaned. However, the real benefit of this type of «cleaning» was also questioned, especially because of the abrasion of the often protective calcite crust. Also questioned was the use of replacement stones that were more resistant than the original building stones.

It was in this context that studies were carried out on the cathedral of Strasbourg which, in contrast to the majority of French monuments, is constructed in a sandstone lacking in carbonate minerals. Very rapidly it could be established that this sandstone underwent the same alterations as other stones of different chemical and mineralogical compositions. If the erosion caused by frost was well advanced, given the richness of clay in this sandstone, the mechanisms for development of other forms of alteration could lead to the presence of salts and, in particular, to gypsum. All of this had not previously been established due to the difficulty of determining the original constituents of these minerals in the sandstones.

The work undertaken on the cathedral of Strasbourg for the last fifteen years has enabled a more precise distinction of the nature of all major forms of degradation observed. It has also favoured the use, in restoration, of sandstone more resistant to alteration than the original stones. This way of restoration has not been reviewed for number of years, but the initial results or outcomes for the stones used during this period, will be the basis for another research.

2 Progressive Understanding of the Alteration of Stones in Monuments

The alteration of stones in monuments was considered in the past as a normal phenomenon naturally inherent in building stones. This is why architects, wishing to ensure the durability of their constructions, were forced to select stones well adapted to their intended use. The criteria for selection were empirical, resulting in the majority of the cases from an analysis of the mechanical behaviour of the stones in their natural environment. The position of the stones on the buildings was also an important factor: for example, stone used for basements or for millstones would be of a type with large and vacuolar pores. The result was a limited rise of water from ground by capillarity. The superstructures of the buildings were, by contrast, built of more ordinary stone, whereas the less permeable and often more compact stones were used for structures which had to withstand streams of water, such as the cornices. The choice of stones also depended on their density, which was considered as a sign of strength, and mixtures of «strong» and «weak» stones, especially in damp areas of the building, were avoided because the hard stones increase the dampness (and possible alteration) of the neighbouring weaker stones.

From the Middle Ages to the 19th century, stone buildings or monuments were constructed on the bases of the above characteristics and of Vitruvius' recommendations about the placement of the stones. All of this was aimed at avoiding degradation of stones by air, water, salts or frost. The theories of Vitruvius were equally taken into account in the use of «fat» (high Ca content) lime mortars in controlling the mixtures of hydraulic limes (burned clay content). The alteration of stones and mortars has long been of great concern to builders who have continually searched for the ideal stone having good mechanical properties; in other words a high resistance to alterations and a grain fine enough to allow a high degree of sculptural detail. These preoccupations, which appeared in the writings of earlier architects like Philibert De L'Orme (16th century) and André Felibien (1681) also concerned Colbert for his buildings in Paris. However, the alteration of stones, particularly in the most exposed points of buildings, was regarded as unavoidable. The search for the ideal stone continued together with the quest for methods or products protecting or increasing the longevity of the stones, especially when highly sculpted. This included a range of gums, waxes and varnishes with variable and often improvised compositions.

At the beginning of the 19th century, problems caused by the alteration of stones on monuments were well known. At the same time, analysis showed the presence of salts in the alteration process, and by testing large numbers of rock and mortar samples, a clear analogy could be drawn between the degradation provoked by the presence of salts and ice. Products for improving the stone's resistance were first proposed by Kuhlmann (1841) who recommended the use of K silicate to transform «the weaker limestones into siliceous limestones». A number of applications were made, in particular on the cathedral of Paris, and Viollet Le Duc reported that the treated stones dried more quickly and that the product was not an obstacle to the evaporation of moisture. Other products such as Na silicate, known as «water glass», and some fluo-silicates were also advocated. In a number of countries, studies were carried out to determine the causes of alteration and also to investigate the claims of the numerous products being proposed for stones and mortars in all situations (Salisbury 1970). These products were used until the beginning of the 20th century and applied in all situations irrespective of the nature of the stone. The considerable disorder which has later appeared, has consequently thrown into discredit most of these processes of protection.

The second half of the 19th century corresponds to the use of siliceous cements often with high alkaline contents which enabled fabrication of impermeable mortars. The inter-calcination of this kind of joint also caused a number of disorders in the stones. Moreover, the properties of the cements allowed applications that would not otherwise have been possible with lime mortars. The pebble-dashed mortar used on the lower parts of monuments, and even used for «protecting» some sculptures, had led to stone degradation that reached unprecedented depths. The popular use of industrial products during the early 20th century, like bricks, cement and concrete which reduced the need for stones has, nevertheless, been employed without taking into account the actual physical properties or the conditions which led to degradation of these products.

During the first half of the 20th century, most studies considered alteration of shapes, as well as the conditions and factors controlling the development of such alteration. Based on careful observation, Kieslinger (1932) distinguished chemical and physical alteration, and analysed the role of moisture in the alteration of stones. However, without determining the actual mechanisms responsible for water solution transfer in stone, he highlighted the influence of the atmospheric conditions, especially smoke which leads to sulphate (gypsum) formation. He also studied the influence of evaporation in controlling the position of salt crystallization. Moreover, based on observation, he showed that surfaces washed by rain were usually unaffected by salt crystallization but rather by dissolution enhanced by rain acidity due to smoke. At the same time he pointed out the harmful consequences when stone pores became blocked by salt crystallization or by the application of superficial protective coatings.

Kieslinger established, in 1932, the principal concepts of stone alteration in monuments, which were used until 1970. In fact, most of the studies of Kieslinger were used afterwards to make precise the parameters of stone alteration that he pointed out earlier. Increasing importance was given to the influence of the atmospheric contribution due to industrial pollution which explains the high gypsum concentration and the increase of black crusts on monuments (Kaiser 1929; Camerman, 1948). During this period, all types of degradation, except those due to frost, were defined only by morphological, chemical and mineralogical criteria. The fact that carbonate and siliceous rocks, regardless of their chemical and mineralogical compositions, could be affected by these different degradations, increased interest in the influence of external parameters. As a consequence, the internal parameters governing the fluid transfer and the salt concentration have not been taken equally into account.

Despite the studies on the porous network of stones (Laurie and Milme 1926), the influence of physical properties on stone degradation has been emphasized only in the 1970s when some authors transposed their results on concrete to stone. In France, Mamillan (1966) was a precursor by emphasizing the importance of physical parameters like porosity, permeability and capillary transfer, on the alterability of stones in monuments. However, the influence of these parameters was only evaluated from in situ alteration tests, without any analysis of the values of these parameters nor any quantification of their role in solution transfer. Consequently, the chemical composition of the altered zone is still often used to describe the development of the alteration, whereas the relationships between capillary transfer and evaporation kinetics are not taken into account (Amoroso and Fassina 1983).

3 The First Studies on Sandstones in France

In this context, some research work was undertaken on the sandstones of the Strasbourg cathedral (Millot et al. 1967). Based on microscope observation, the alteration types were classified in terms of morphology and mechanisms. The alteration depending on frost and water channelling was distinguished from that resulting from chemical modifications like salt concentration (Plate I, 2–3), the former aspect having been especially studied on sandstones.

The first results showed high gypsum concentration associated with the degradation of sandstones of very low calcite content. The crystallization of this mineral, previously attributed to the reaction of sulphate ions in rainwater with the calcareous stone, emphasized the problem of the origin of the Ca ions. As the traces of calcite accumulated by repeated drainage of the stones could have been the origin of Ca, sandstones completely devoid of calcite were used for restoration. Nevertheless, the first results showed clearly that the development of gypsum crusts is mainly governed by external contributions.

A classification was established on the basis of a subdivision of the alteration processes in terms of the nature and position of the salts (Jeannette 1992). Efflorescence which is a relatively harmless salt-coating on the stone surface, has been distinguished from gypsum crusts slightly encrusted in the sandstone and leading to superficial disaggregation. The progress of water in the stones on monuments can be evaluated, considering the presence or lack of gypsum, or the depth of its crystallization (Jeannette and Hammecker 1992).

Despite these observations on solution migration in stone, only the atmospheric chemical contribution has been taken into account. Nevertheless, experimental studies have shown that the interaction of rainwater with sandstones allows the formation of gypsum (Fritz and Jeannette 1981).

Plate I.1. The dissolution of calcareous sandstones due to the leaching of rainwater prevents the development of gypsum crusts (the collegiate church St Thiébault of Thann, NE France). **2** A sandstone on a balustrade of the cathedral of Strasbourg (NE France) placed perpendicularly to stratification planes is fractured by frost. Under the cornice, the stone protected from rain is covered by a black gypsum crust. **3** Under the Romanesque vault of the cathedral of Strasbourg, the thick gypsum crusts develop only in areas sheltered from leaching. **4** In given exposure conditions, crusts develop on the surface of fine sandstones similar to those of the cathedral of Strasbourg. When the crusts diverge, the stone underneath appears untouched. **5** Section in a gypsum crust on sandstone (*white bar* 0.7 mm): in the *lower part*, the gypsum is only present in traces in the sandstone, the *darker area* above corresponds to ash and soot deposits strongly sealed in gypsum crystals. The external dark layer which consists of soot and ash, is very breakable because it is poorly cemented by gypsum. **6** Contour scaling on calcareous sandstones showing gypsum concentration at the base of the scales; they are located on surfaces very exposed to rain on the church St Antoine de l'Isère (E France)

4 Alteration Morphologies and Parameters Governing Their Development

Under natural exposure conditions, rocks are also affected by superficial modifications, leading to a loss of cohesion and matter. In altered zones, several transformations can be observed: chemical and mineralogical modifications and especially an increase in porosity and changes in the pore structures. All these variations can equally be observed in stones on buildings where abnormal exposures, such as overhangs, balconies or vertical stratification planes, enhance the decay of the stone. These extremely severe conditions, which are unknown in nature, are further increased on a monument by the heterogeneity of the surrounding stone and the presence of mortar joints with different chemical and physical properties. The common decay processes of stone are consequently faster in monuments than in nature.

Among all chemical mechanisms of alteration, the most likely are those related to salt crystallization leading to crust formation, contour scaling or disaggregation. These decay morphologies are used to emphasize the various factors having an influence in the alteration of stone in monuments. The black crusts represent the most commonly occurring alteration morphology. They develop on any kind of stone which is located in humid sheltered zones where leaching by rainwater cannot occur. Their presence is very frequent on the surfaces beneath the cornices or in the hollow parts of sculptures (Plate I, 2–3). The structure of these crusts can be observed in thin sections cut perpendicularly to the surface of sandstone forming the Strasbourg cathedral (Plate I, 5). On the external part of this section, a black and incoherent layer can be observed. It is formed by ashes, sand grains and various elements like mortar (Jeannette 1981). These constituents are agglomerated by a loose coating of gypsum crystals and sometimes by calcite. The latter results from solid deposits, whereas the gypsum crystallizes from solution. The entire layer, very crumbly in its external part, reaches sometimes a thickness of 1 cm. Beneath the surface, ash particles are scarcer but the fabric of gypsum crystals is denser. The layers of ashes sealed by gypsum constitute the present-day crust. When this crust remains on the stone (Plate I, 4), the surface underneath is undamaged despite the presence of traces of gypsum. On the other hand, when the transition between crust and stone is progressive, the stone underneath is impregnated with gypsum and it is affected by a disaggregation several centimetres deep.

Occurring on sandstones, granites and limestones, the gypsum crusts form regardless of the calcite content in stones. Their composition and thickness depend on their situation in the building. They are thicker and develop faster in urban areas where the contributions of the gypsum and of the ash are more important than in rural regions (Camerman 1948; Amoroso and Fassina 1983). The development of this kind of alteration shows that the atmospheric contribution plays a leading part in this process compared to the chemical composition of the stone. The structure of the gypsum crusts, their position in sheltered areas and the accumulation of ashes depend on the progress of the solutions generating them. Water moves in the stones by capillarity from the surface leached by the rain where nothing can accumulate, towards the sheltered zones. In these humid zones, ash adheres at the surface and it is agglomerated by gypsum crystallizing from concentrate solutions due to evaporation. Beside the solution composition, the capillary transfers and the evaporating conditions seem to be determinant factors in crust development.

If the same kinds of stone are placed in different conditions of exposure, they might be affected by different degradation morphologies. It is common to observe the same type of stones covered by a crust of black gypsum on sheltered areas of the monuments, whereas they are affected by superficial dissolution (Plate I, 1) or by contour scaling in exposed zones (Plate I, 6). In contour scaling, gypsum impregnates the entire scale, but its highest concentration is found behind the scale. After a direct rain imbibition, the solutions move through the stone by capillarity and evaporate during dry periods (Zehnder 1982). For both the superficial crust formation and the contour scaling, the chemical composition of the stone and the alteration product are similar. The exposure conditions which are the unique variable, influence the capillary transfer processes and the location of the main areas of evaporation and gypsum crystallization. Nevertheless, in some cases, outwardly similar stones placed in the same exposure conditions are affected by different decay morphologies corresponding to distinct imbibition–drying conditions. These differences can only be attributed to changes in the pore structures.

5 Pore Structures

The porous network is often the unique common factor between rocks as different as siliceous sandstones, micritic limestones or granites. However, the porous networks which are characterized by their extension and their geometry, depending on the organization of their minerals, differ in each kind of rock. The shapes and dimensions of the larger pores located in an intergranular position, are related to the grains or the intraclasts. On the other hand, the microporosity is distributed in clusters located in the centre of altered minerals, or in the argillaceous or carbonate matrix of sedimentary rocks. The pore structures observed microscopically on thin sections previously injected with coloured resins (Zinszner and Meynot 1982), can theoretically be transposed from one rock type to another.

In detrital rocks like the sandstones used on the cathedral of Strasbourg, the porosity consists in intergranular macropores and microporous clusters located in altered minerals and clays. The macropores form the main part of the porous volume which reaches 21% in this rock with edges showing irregular shapes (Plate II, 2). The interconnections between the macropores consist of necks located either in the grain joints or in argillaceous clusters. These necks ensure the connectivity of the porous network and control the solution transfer through the rocks. In the sandstones consisting of smaller grains, the intergranular pore space is partially occupied by clay clusters (Plate II, 2). The proportion of wide pores decreases in favour of the narrow pores, whereas the total porous volume remains similar. The

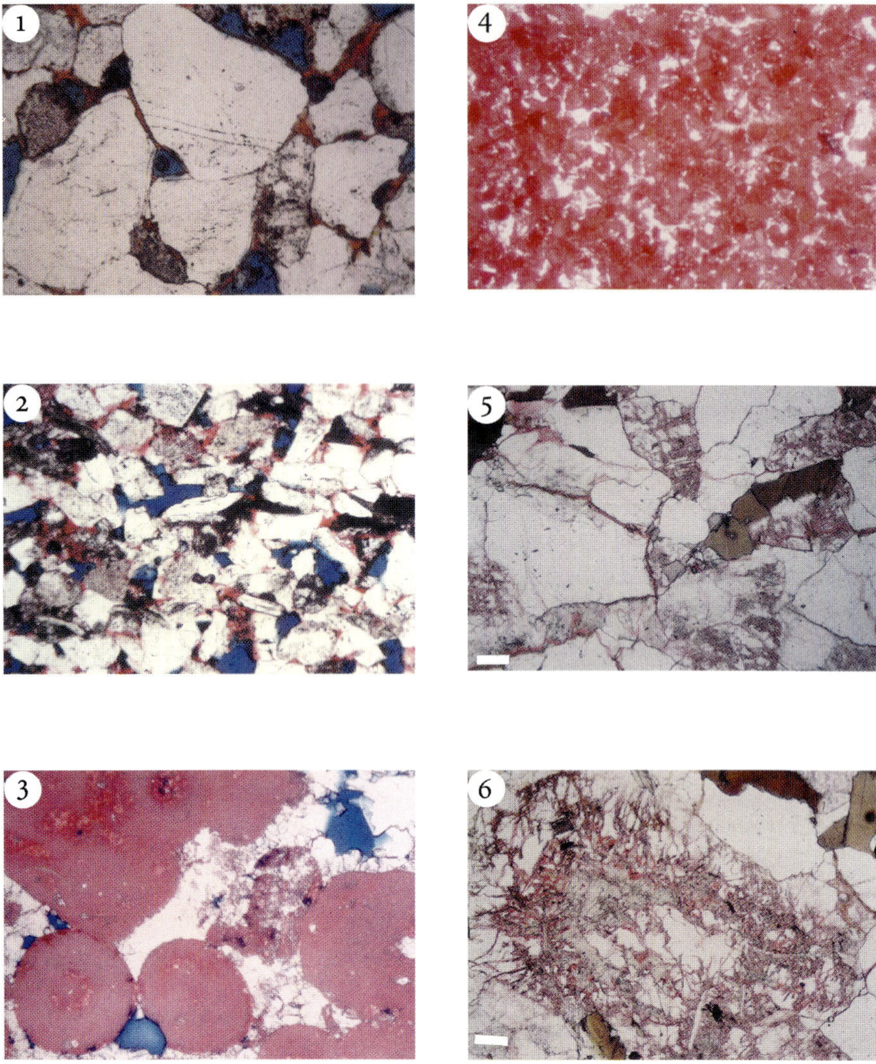

PlateII.1. Coarse sandstone in which the wide blue pores form the trapped porosity, the free porosity being essentially formed by narrow intergranular pores (*white bar* 0.2 mm)**2** A clayey sandstone in which the free porosity coloured in red is formed by microporous clusters associated to clays. The free porosity forms a homogeneous network, while the trapped porosity coloured in blue, is located in the intergranular voids (*white bar* 0.2 mm)**3** An oolitic limestone with trapped porosity located in the sparitic cement clearly differentiated from free porosity formed of micropores in the ooliths (*white bar* 0.2 mm)**4** A micrite with the microporosity located in the intraclasts forming a homogeneous porous network (*white bar* 0.2 mm)**5** Regular cracks in the granite forming a well interconnected lattice joining the microporous clusters in the altered plagioclases (*white bar* 0.2 mm)**6** An altered plagioclase with a core transformed into calcite and sericite. This core is porous where calcite is dissolved. Around the *central part*, in the microporous kaolinized aureole, the cracks are underlined by Fe oxi-hydroxides (*white bar* 0.8 mm)

uniform distribution of clay particles in a sandstone tends to homogenize the porous network.

The complexity of the porous network is amplified by the role of the different pores in the water-transfer processes. During a capillary imbibition, the pores with trapped air are not directly accessible to water. This so-called trapped porosity (Dullien 1979 ; Mertz 1991) is mainly formed by the widest pores coloured by blue resin, whereas the smaller pores normally accessible by capillary imbibition form the free porosity, coloured with red resin (Plate II, 1–2). The importance of the trapped porosity explains the low saturation rate in some rocks having a high total porosity. The kinetic rate of capillary imbibition in rocks depends on the radii of the interconnections controlling the incoming flow of water. Nevertheless, with equivalent interconnection radii, this kinetic rate of capillary imbibition is lower in heterogeneous free porosity, because the imbibition is slowed down by the widening of the pores which causes the capillary pressure, vector of the transfer, to drop (Hammecker et al. 1993). The influence of pore structures on transfer processes, explains the reason why rocks with similar total porosity values may have different kinetic rates for capillary imbibition. Thus, in sandstones with many wide pores, the heterogeneity of the porous network leads to lower imbibition kinetics than in finer sandstones where the free porous network is more regular (Plate II, 1).

Sedimentary carbonate rocks often have a contrasted porous network consisting in the juxtaposition of microporous clusters and intergranular macropores. In oolitic limestones with sparitic cement, the intergranular macropores are often trapped pores, whereas the free porosity is represented by the regular intra-oolitic micropores (Plate II, 3). In micritic limestones where the intraclasts which are reworked micritic mud, are sealed with fine sparite, the porosity is mainly constituted by micropores (Plate II, 4). These fine and regular pores are located in the intraclasts and form the free porosity. The degree of saturation and the kinetic rates of capillary imbibition are extremely high in these micrites.

Granites are the most important crystalline rocks used in buildings. They usually have a very low or even a lack of porosity, and are not affected by alteration. However, on some types of altered granite in the monuments, analysis reveals porosity values sometimes as high as in sedimentary rocks. The porous granites often show an ochre coloration attesting to the presence of Fe oxi-hydroxide. Porosity in granites occurs in two very different shapes. The most frequent porous network in granite consists of a lattice of micro-cracks spread over the rock (Plate II, 5). The uniform width of the cracks (5–30 mm) yields a regular fluid circulation, hence no trapped porosity is observed. In addition, some facies have microporous areas located in the plagioclases, where two alteration forms can be observed. In the centre of the plagioclases, the most basic part is superimposed by calcite and sericite. Usually this relatively compact part is not porous unless calcite has been dissolved. Other cases show kaolinite epigenesis on oligoclase crystals, which are spread by micro-cracks following twin and cleavage plans. These two types of plagioclase alteration, as well as the development of a regular fracturing, are not related to the weathering processes of granite in a monument. However, they result from plagioclase destabilization during the formation of the granite or are the consequence of hydrothermal circulation. These porous granites are yellowish coloured in the quarries, because of the migration of Fe oxi-hydroxides due to the transfer of meteoric water. Hence this colour is an indication of porosity.

Granites altered in a building usually show the characteristics of granites already altered in the quarry (Plate II, 5). The micro-cracks in the lattice interconnect the microporous plagioclase crystals affected by the two alteration forms described previously (Plate II, 6). The association of two microporous networks consisting of regular micro-cracks and clusters dispersed in the rock samples, guarantees a relatively high porosity as well as homogeneous and rapid capillary imbibition conditions.

By governing the processes of solution transfer through the rocks, the pore structures play a determinant role on the form and the nature of the alterations associated with salt concentrations. In fact, the alterations depend on the location of precipitation with respect to the surface of the stone. If kinetic rates of the capillary transfer are high enough to maintain an excess of water with respect to the loss due to evaporation, the precipitation of salt will take place at the surface of the rock without generating any damage. On the other hand, salt precipitation can take place in the stones and develop different types of degradation.

6 Water Balance and Stone Weathering

Depending on their location on buildings, stones can either be submitted to a constant capillary supply or placed in alternative imbibition and drying conditions. The first case corresponds to bridge pillars or basements of buildings, constantly in contact with the water table. Thus, the stones submitted to a constant flow, act like wicks. The second case which is more frequent, corresponds to surfaces supplied with water only during periods of rain.

In both situations the transfer of solutions through the stone depends on the kinetic rate of the capillary imbibition. These kinetic rates are similar to those describing the imbibition in cylindrical tubes. They are quantified by the Washburn law where the height L of the meniscus over the free water table is related to time t. A constant called B which is characteristic of the fluid, and the geometry of the tube relates the height L to square root of time:

$$L = B \times \sqrt{(t)}$$

where B is proportional to the radius of the tube and the surface tension of the liquid, and inversely proportional to liquid viscosity. This coefficient B is established experimentally in the stone. However, in these complex and irregular porous networks the influence of the geometry of the medium is not expressed as a function of a single real radius, but it is justified by the pore structure. The size of the interconnections controlling the flow, and the dimensions and the distribution of the wider pores, must be taken into account.

Fig. 1. Weight loss per unit of evaporating surface versus time in a sandstone and a limestone previously entirely saturated and dried at 33% relative humidity

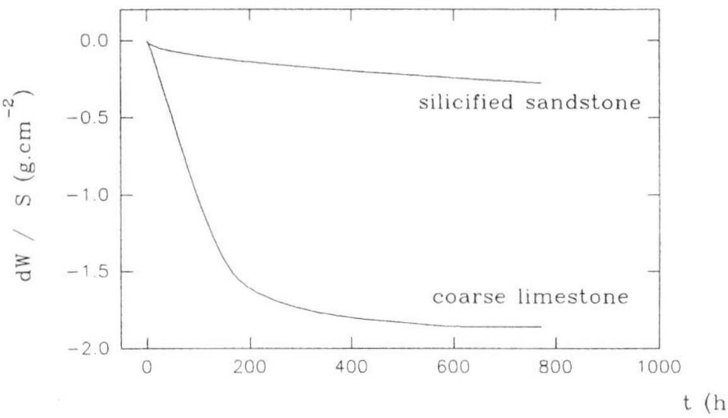

When the stone is continuously supplied with water by capillary imbibition, two different cases can be considered depending on the nature of the rock and the distance from the water table. In the first case, the capillary transfer is rapid and sufficient to maintain a constant moistering of the surface of the stone (Hammecker and Jeannette 1988). Hence, evaporation takes place on the surface and leads to the formation of efflorescences. This can often be observed in the lower part of walls near the floor: most of the stones used in the buildings have capillary imbibition rates (*B*) which are high enough to maintain, over short distances, a water supply greater than the loss by evaporation. The water balance representing the difference between the capillary supply and the loss by evaporation is positive. In the second case, for stones situated further from the ground or with lower capillary rates (*B*), the capillary supply is less important and the water balance is negative. In this case, evaporation takes place inside the rock.

When stones are alternately wetted and dried, the water found in the stones after a rainy period migrates towards the surface during the dry period. Both imbibition and drying are, thus, controlled by the capillary transfer properties of the stones. If the surface of the stones cannot be wetted from the internal supply of water, evaporation takes place inside the stones. This feature can be observed experimentally, when saturated rocks are dried at controlled temperature and relative humidity. For high capillary rates (Fig. 1, coarse limestone), moistening of the evaporating surface is possible as long as an hydraulic continuity is maintained for the water in the pores. When the water content drops under a limit called the critical hydric saturation, the hydraulic continuity is broken and water supply to the surface is no longer possible. Hence, the remaining small quantity of water evaporates inside the porous network of the stone, and is drawn to the surface by diffusion of water vapour. For stones saturated with water but with capillary rates lower than the evaporation rate (Fig. 1, silicified sandstone), the water balance is negative and the actual evaporation also takes place inside the stone.

The position of evaporation which can be surficial or internal, depends on the water balance D (Fig. 2). As the evaporation flow (Q_{ev}) is constant for stable external conditions (relative humidity, temperature, wind, etc.), the water balance is controlled by the flow of capillary transfer (Q_{cap}) which decreases with time. The water balance D also decreases with time, and remains positive until reaching equilibrium between capillary flow and evaporation flow. While the water balance

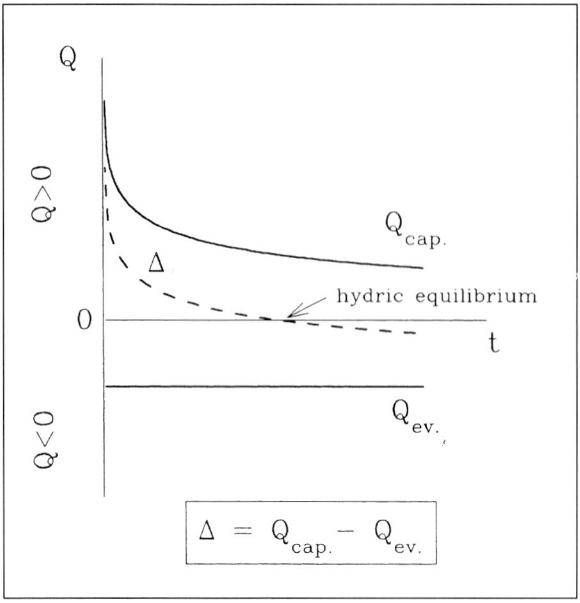

Fig. 2. Evolution versus time of Q_{cap} (the capillary imbibition flow in a rock), Q_{ev} (the superficial evaporation flow) and D (the resulting water balance)

is positive, evaporation takes place at the surface of the stone where salts can grow without damaging the stone. This condition corresponds to the case of stones which are located near a water table continuously supplied with water, or to the first stage of drying of stones submitted alternately to cycles of wetting and drying. Beyond the hydric equilibrium point, the water balance is negative because the capillary flow is lower than the evaporative flow. Consequently, the evaporation front progressively moves towards the inner part of the stone. The associated salt crystallization leads to severe degradation inside the stone.

7 Conclusion

Based on the index of forms and factors of degradation which are associated with salt concentration, the petrophysical approach specifies the importance of the mechanisms involved in the development of such degradation. Despite the fact that the formation of salts on buildings depends on the materials and the atmospheric contributions, it is mainly governed by the transfer and concentration of the solutions. In particular, the capillary transfer properties, which control the imbibition and the supply of water in the evaporating areas, play a determinant role in the form and position of the alterations. The different alteration developments in various stones can be explained by taking into account the analysis of the porous network of stones with the sizes and the positions of the pores, and the relations between pore thresholds and wider pore openings. Studying the pore structures also allows the simulation by numerical models and the prediction of the development of alteration in the stones (Hammecker 1993).

Despite the current use of chemical and mineralogical modifications for characterizing different types of degradation, the importance of the alterations that they introduce in stone is better illustrated by changes in pore structure. Therefore, the quantification of alteration in stones consists of an evaluation of the modifications of the pore structures.

Identification of the main parameters responsible for stone alteration in monuments led to the identification of methods and products aimed at protecting the stones or correcting their degradation. In both cases, the proposed methods modify the porous network, whether by a reduction of the wettability in order to restrain the capillary transfer (Snethlage 1983; Fischer 1993), or by the introduction of new cements sealing porous voids (Hammecker et al. 1992).

As a consequence of many studies made already on alteration and conservation of monumental stones, two main research points are, in particular, being investigated: the protection of stones by pore structure modifications and the importance of the atmospheric contribution of solid particles in the processes of stone degradation. Still poorly understood is the dual influence of these soot and ash deposits (1) on the chemical composition of solutions flowing through the stones; and (2) in their physical contribution in forming superficial coatings. Both of these influences modify the porosity, wettability and permeability of the stones.

Acknowledgements. I would like to thank Dr. Claude Hammecker for his help during completion of this contribution, as well as Mrs. Sarah Jane Hammecker for improvement of the English presentation.

References

Amoroso GG, Fassina V (1983) Stone decay and conservation. Materials Science Monographs 11. Elsevier, Amsterdam, 453 pp

Camerman C (1948) Sur l'altération des pierres de taille par les fumées. Ann ITBTP 14:147 pp

De L'Orme P (1568) Architecture. Pierre Mardaga, éditeur, 1981

Dullien FAL (1979) Porous media. Fluid transport and pore structure. Academic Press, New York, 396 pp

Felibien A (1681) Mémoire pour servir à l'histoire des Maisons Royales. J Baur Libraire, Paris, 1874, 106 pp

Fischer C (1993) Importance des propriétés pétrophysiques et des interactions physico-chimiques dans l'hydrofugation des roches par des silicones. Thèse, Univ Strasbourg, 191 pp

Fritz B, Jeannette D (1981) Pétrographie et contrôle géochimique expérimental de transformations superficielles de grès de monuments. Sci Géol Bull (Strasb) 34:193–208

Hammecker C, Jeannette D (1988) Rôle des propriétés physiques dans l'altération des roches carbonatées: exemple de la façade ouest de Notre-Dame La Grande de Poitiers (France). VIe Congr Int sur l'Altération de la Pierre, Torùn (Poland) 1988, pp 266–275

Hammecker C, Esbert RM, Jeannette D (1992) Geometry modifications of porous network in carbonate rocks by ethyl silicate treatment. Proc VIIth Int Congr on Deterioration and conservation of stone, Lisboa (Portugal) 1992, pp 1053–1062

Hammecker C, Mertz JD, Fischer C, Jeannette D (1993) A geometrical model for numerical simulation of capillary imbibition in sedimentary rocks. Transport Porous Media 12:125–141

Hammecker C (1993) Importance des transferts d'eau dans les altérations des pierres en oeuvre Thèse, Univ Strasbourg, 254 pp

Jeannette D (1981) Modifications superficielles de grès en oeuvre dans les monuments alsaciens. Sci Géol Bull (Strasb) 34:37–46

Jeannette D (1992) Les altérations des pierres des monuments. In: Philippon J, Jeannette D, Lefèvre R (eds) La conservation de la pierre monumentale en France. Presses du CNRS, Paris, pp 51–72

Jeannette D, Hammecker C (1992) Facteurs et mécanismes des altérations. In: La conservation de la pierre monumentale en France. Presses du CNRS, Paris, pp 73–81

Kaiser E (1929) Über eine Grundfrage der natürlichen Verwitterung und die chemische Verwitterung der Bausteine im Vergleich mit der in der freien Natur. Chemie der Erde IV, Jena, 291 pp

Kieslinger E (1932) Zerstörungen an Steinbauten. Verlag F Deutige, Leipzig

Kuhlmann F (1841) Mémoire sur la chaux hydraulique, les ciments et les pierres artificielles, suivi de considérations chimiques sur la formation des calcaires siliceux, et en général des espèces minérales formées par voie humide. CR Acad Sci Fr XII:850–855

Laurie AP, Milme J (1926) The evaporation of water and salt solution from surfaces of stone, brick and mortar. Proc R Soc Edinb 47:52–87

Mamillan M (1966) L'altération et préservation des pierres dans les monuments historiques. Int Conf Monument and sites, Bruxelles, pp 65–98

Mertz JD (1991) Structures de porosité et propriétés de transport dans les grès. Sci Géol Mém (Strasb) 90:149 pp

Millot G, Cogné J, Jeannette D, Besnus Y (1967) La maladie des grès de la cathédrale de Strasbourg. Bull Serv Carte Géol Als Lorr (Strasb) 20:131–157

Salisbury EJ (1970) Improvement in preserving and hardening stone and brick. US Patent 102, May 1970, p 869

Snethlage R (1983) Steinkonservierung. Bayerisches Landesamt für Denkmalpflege, München, Arbeitsheft 22, 203 pp

Zehnder K (1982) Verwitterung von Molassensandsteinen an Bauwerken und in Naturaufschlüssen, Matériaux pour la Géologie de la Suisse – Série Géotechnique, Lieferung 61. Commission Géotechnique Suisse, Kümmerly & Frey AG, Geographischer Verlag, Bern, 130 pp

Zinszner B, Meynot C (1982) Visualisation des propriétés capillaires des roches réservoirs. Rev Inst Fr Pétr 37:337–361

10 Continental Silicifications: A Review

MÉDARD THIRY

1 Introduction

The hypothesis of climatic silicification has been known for almost 200 years, as Coupe (1804) documented such a mechanism. To explain the silicification of the Fontainebleau Sands, he cited the Libyan dune fields suggesting that each winter Si went into solution in the soils and «thickened» by desiccation during the summer. Later, Gosselet (1888) created the expression «atmospheric metamorphism» to explain the Eocene quartzites of the Paris Basin. But the study of the wide silicified stretches in South Africa and Australia really imposed the idea of a surface bound silicification to the geologists working in these areas. Kalkowsky (1901) and Stortz (1928) in the Kalahari desert, Lamplugh (1907) in Rhodesia and Woolnough (1927) in Australia considered that these silicifications are recent, or that they are even developing today, and they linked them to the relatively arid climates of these regions.

In France, the detailed petrographic inventory of siliceous sedimentary rocks by Cayeux (1929) showed that the composition of the original material has an influence on the habit of Si. Chalcedony develops in limestones, opal dominates in clayey materials and quartz-grain overgrowth is the rule in «clean» sandstones without primary cement. This model was outlined by a French school that had North Africa and the Sahara Desert as a «playground».

Georges Millot approached this problem from a geochemical and crystallochemical perspective. He synthesized observations gathered from studies of the Saharan silicifications and recognized the influence of two main factors (Elouard and Millot 1959; Millot et al. 1959; Millot 1964, 1970).

«The main question is the morphology of the landscape. Quartzification occurs readily at the surface, whereas in basins and within the groundwater system more disordered forms of Si, i.e. chalcedony and opal, develop preferentially. The influence of parent rocks is the second-most important factor. Generally, carbonates silicify preferably into chalcedony and clays into opal. The nature of the bedrock subjected to silicifying solutions determines the type of silicification.»

Georges Millot interpreted these facts in term of solution chemistry and crystal growth. First, he endeavoured to kill the «myth» of colloidal Si in natural waters. Then, he showed that «silicifications, flints, cherts, etc. result from the growth of crystals fed by natural solutions containing monomeric Si in solution» (Millot 1960, 1961, 1970; Millot et al. 1963). He emphasized the fact that Si precipitation and dissolution are similar to polymerization and depolymerization reactions. The different types of Si result from a more or less ordered crystal growth with greater disorder caused by the incorporation of impurities into the crystal lattice. The degree of crystal ordering and the composition of the mother solutions control the type of the produced silicification facies:

1. Quartzification results from regular growth of crystals fed by clear, diluted solutions with low concentrations of impurities. Detrital quartz grains with clean surfaces act as seeds.
2. Chalcedony and microcrystalline quartz result from thwarted quartz development. The mother solutions contain higher concentrations of impurities and of other cations, or variable nucleating materials (clay minerals and various crystalline surfaces). When numerous defects occur in the tetrahedral stacking sequences (cations and misfittings in the nucleation networks), the crystals remain small.
3. Opal-rich silicifications form from solutions containing even more other cations, but also more Si. There is a proliferation of crystal nuclei, the tetrahedral chain being disrupted by other cations. The crystalline network does not develop, producing amorphous material.

Si minerals often display a sequence going from disordered to the more ordered forms in silicified materials (Thiry and Millot 1987). This sequence is determined mainly by the concentration of Si and other cations in the mother solutions. It also depends partly on the solubility of the Si phases in the host material and on the rate of percolation of solutions through the system. Rapid percolation favours dissolution, but in zones with slow percolation, equilibrium is attained and no transformation of the existing phases is visible. In zones of intermediate porosity, the disordered Si phases recrystallize progressively into quartz. To achieve this, solutions must be enriched faster in Si that in any other element, and reach quartz saturation before any other silicate (Parron et al. 1976). The effect of the rate of percolation on dissolution and recrystallization explains the vertical distribution of different Si polymorphs observed in numerous silcrete duricrusts.

New interest in surficial silicifications and revised concepts have appeared during the two last decades. These recent studies are based on a better understanding of the geochemical phenomena of the hydrosphere and also on new analytical techniques. In particular, interpretation of the micro-morphological structures of soils was essential for the understanding of pedogenic silcretes. A somewhat «quantitative» approach in terms of mass balance calculations also helped to consider groundwater silicifications. In the following sections, facts will be presented as examples to discuss the mechanisms and unanswered questions will be addressed.

2 The Geological Data and Their Interpretation

Most of the surficial silicifications described in ancient sedimentary deposits, as well as in present-day landscapes, can be classified into three main types.

1. Pedogenic silicifications displaying specific structures characteristic of soils, which obliterated completely the structures of the parental material. Si accumulates in a relative fashion, while other cations are removed from profiles.
2. Groundwater silicifications retaining largely the primary structures of the host material and are always characterized by absolute Si accumulation, Si deposits in the pores, or epigenetic replacements.
3. It is often difficult to document at what stage of the sedimentary history silicification connected with evaporites developed. Mineralogy of the Si phases allows distinction from other types.

2.1 Pedogenic Silicifications

Pedogenic silicifications develop near the surface, probably within the soil environment. They display typical soil structures related to infiltration and downward percolation of water (such as differentiated horizons and illuviation structures). The silicified materials exhibit a variety of forms, depending to some extent on the character of the rock or sediment precursors, but also on the processes involved in the silicification process. Two main varieties of materials can be distinguished: (1) silcretes mainly formed of secondary microquartz without clay minerals and Fe oxides, but enriched in Ti; and (2) "hardpans" or "duripan"-like silicifications formed mainly of opal with the original clay minerals and Fe oxides still present. Descriptions of these silicification types come mainly from Eocene occurrences in the Paris Basin. Specific physical characteristics and mechanisms of formation are also derived from observations on similar materials from Australia and central America.

2.1.1 Quartzose Silcretes: Relative Silica Accumulation

Silicified pans and lenses cover the Eocene detrital kaolinitic deposits of the southern part of the Paris Basin. These features are very hard, tightly cemented quartzites that break up in variable sized blocks with puffy, tear-like shapes, coated with deposits of yellow-white opaline Si masking the internal structures (Thiry 1981; Rasplus 1982).

2.1.1.1 Profile Structure

Silcretes have a typically columnar structure with a characteristic laminated capping (Fig. 1). They are about 2 m thick and display several distinct horizons with systematic micromorphological and mineralogical structures (Thiry 1981). The lower granular horizon contains granules occurring throughout a yellow, sandy clay. These granules consist of microcrystalline quartz and opal. Transmission-electron-

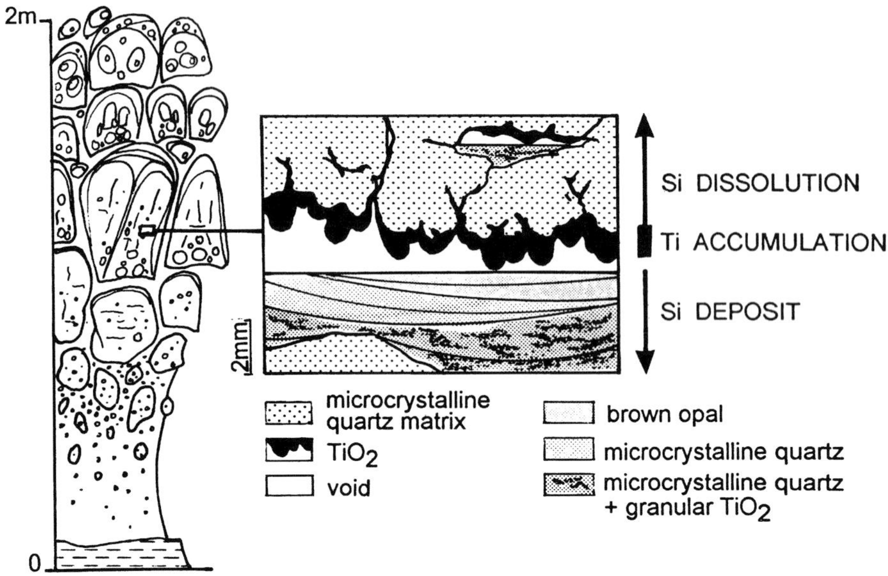

Fig. 1. Macro- and micro-morphological structures of the pedogenic silcretes. Note the columnar structures of the profiles capped by wide illuviations. The siliceous matrix is the location of successive silica dissolution and deposition leading to relative accumulation of the crystalline quartz

microscope and electron-diffraction studies of the clay matrix show that hexagonal-shaped kaolinites have corrosion embayments and are coated and welded together by a Si gel. A columnar horizon results from the silicification process. The matrix is formed of quartz grains with irregular overgrowth apophyses which grade into Ti-enriched microcrystalline quartz. Opal cutans develop at the walls of the vertical fractures, at the floors of the horizontal joints and cap the top of the columns. The horizontal joints display specific features of successive reorganisations. Illuviation laminae of fine silica are accumulated along the base. The youngest laminae, at the top of the illuviations, consist always of opal with low Ti content. The first laminae, at the base of the illuviations commonly have a nodular structure and are formed of microcrystalline quartz with higher Ti content.

The described sequence indicates a progressive recrystallization of the opal. Recrystallization is accompanied by loss of Si which induces the destruction of the primary structures and relative Ti accumulation. Mammillary deposits of Ti, similar in form to stalactites, cover the ceilings of the joints. The deposits penetrate between the coarse quartz grains, replacing the microcrystalline matrix. In some places, the ceilings have collapsed and fragments of the stalactites, together with skeleton grains, have accumulated on the floors of underlying voids. At the joints of the ceilings, preferential dissolution of the microcrystalline quartz and preservation of the coarse quartz grains can be seen.

A more massive horizon with complex nodular and pseudo-breccia fabric, is found at the top of the profile. It is devoid of opal and the microcrystalline quartz matrix is partly dissolved. Abundant irregular voids around pseudo-nodules allow exposure and account for their displacement. Ti accumulates in rims around the nodules, euhedral quartz develops in the voids and some quartz grains have overgrowths.

2.1.1.2 Structure Interpretation

At the base of the profiles, amorphous and cryptocrystalline forms of Si dominate. Upwards with the development of the columnar structures, microcrystalline quartz is the main Si form. The base and ceiling of the voids, however, work in different fashions. Opal forms at the base of the voids where circulation is slow and waters stagnate, while the microcrystalline matrix dissolves at the ceiling of the voids where the solutions seep and pass. Quartz crystals develop at the top of the profiles. The mineral sequence is:

$$\text{opal} \rightarrow \text{microcrystalline quartz} \rightarrow \text{euhedral quartz.}$$

In this sequence, every mineralogic and petrographic Si mineral derives from the previous generation by on site dissolution and recrystallization indicating re-equilibrium of the mineral with its environment. Recrystallizations are irreversible and favour development of quartz during successive dissolutions and recrystallizations (Thiry and Millot 1987).

Profiles clearly emphasize a migration of Si from top to bottom and an inheritance of micromorphological features in the uppermost horizons, which developed in the lowermost horizons. This inheritance shows that the silicification profiles sink progressively into the landscape, just like ferricrete and calcrete duricrusts (Millot et al. 1976) or like a weathering front. The greatest amount of Si comes from dissolution at the top of the profile, followed by a sequence of precipitation and re-dissolution from top to bottom. This close link between leached and confined environments does not imply, however, a strict synchronism of degradation and construction. The two systems work in an alternating fashion, whereas periods of loss and accumulation follow each other sequentially.

2.1.1.3 Palaeogeographical Framework

Silcretes of the Paris Basin are mainly residual and crop out in such a widely scattered fashion that does not allow reconstruction of the palaeolandscapes in which they developed. In Australia, on the other hand, silcretes appear to have developed in contrasted landscapes and not on plateaus and peneplains. They indurate glacis at the edges of the plateaus (Thiry et al. 1991; Milnes and Thiry 1992) or around palaeoreliefs (Simon-Coinçon et al. 1996). At the edges of the scarps, silicifications are restricted to thin skins with high Ti contents (Hutton et al. 1972).

At the glacis scarps, they are more extensive and almost only formed of thick toppings on fallen rocks, whereas they form a regular horizon with columnar structure in the plains. Thicker silcretes in the transition zones between the glacis and the plains, where water discharge is still important but flow rates slow down, reflect the contribution of lateral waterflows along glacis by developing pedogenic silicifications.

2.1.1.4 *Tepetates: Beginning Silcretization?*

Soil profiles that show many similarities with silcretes have been described recently in Mexico. The resistant horizons (=tepetates) develop within thick halloysite profiles over volcanic materials, under subhumid to subarid climates with a well-defined dry season. The profiles are about 2 m thick and display two to three superimposed prismatic horizons separated by indurated laminar horizons (Fig. 2; Campos and Dubroeucq 1990). Destruction of the clayey matrix occurs within the prisms, whereas clay minerals accumulate in the laminar horizons together with Si (Dubroeucq 1992). These laminar horizons are plastic and gelatinous at depth, but harden at the outcrop to form duripans. The waters which flow through the profile, have high Si contents. During the period of high flow, the Si/Al ratio is low, whereas during the period of low water flow, the Si/Al ratio is high (Campos 1992). This is probably related to different solubility rates of Si and Al. Flow measurements would be required to compute a mass balance.

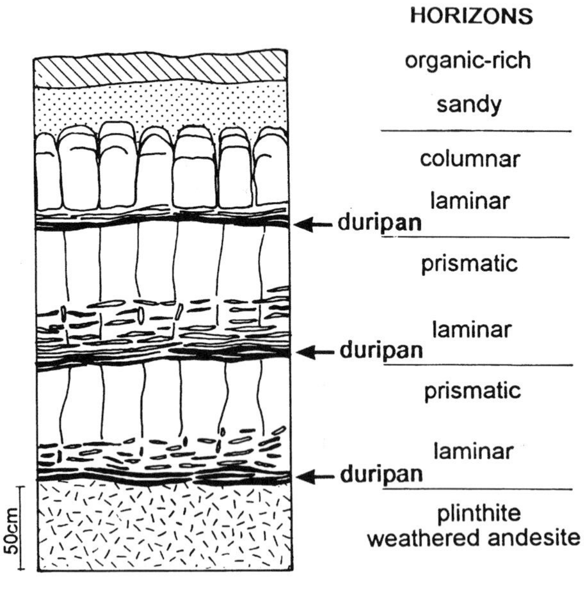

Fig. 2. Structure of the indurated prismatic horizons (=tepetates) developing at the top of halloysite-rich weathering profiles from Mexico. Silica accumulates in the laminar horizons and hardens at the outcrop. (According to Campos and Dubroeucq 1990)

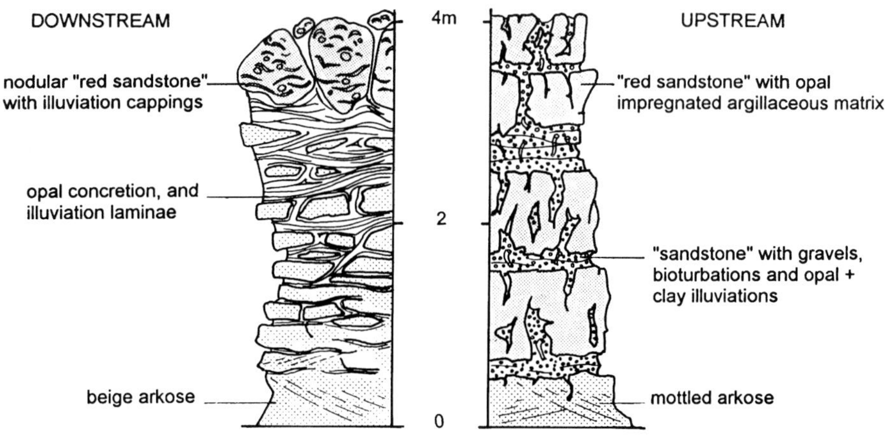

Fig. 3. Opaliferous pedogenic silicifications of hardpan type. Silicifications occur in catenas displaying distinctive structures that are a function of position in the palaeolandscape: siliceous illuviations dominate upstream, due to percolations, while silica concretions develop downstream in wide planar structures due to lateral flows

2.1.2 Hardpans: Absolute Silica Accumulation

At the edge of the Paris Basin in the northern Massif Central, red sandstones (siderolithic facies) form around the basement palaeorelief and cross laterally into the Tertiary sediments. In fact, these red sandstones are ferruginous palaeosols with typical pedogenic features. They form a contrasting catena from the highs of the palaeolandscape towards the Eocene sedimentary basin. Silicification was imprinted on the ferruginous palaeosols and produced peculiar structures as a function of their position in the palaeolandscape (Thiry and Turland 1985).

2.1.2.1 Structures of the Profiles

Red sandstones are visible upstream with a rough columnar structure forming superimposed horizons (Fig. 3). The matrix of the red sandstones is made of hardened sandy clay with angular quartz grains floating in a red argillaceous and porous matrix. Vertical and horizontal fissures are filled with an ochre sandstone composed of loosely cemented millimetric gravels (Thiry et al. 1983b). This filling displays numerous burrows and red sandstone nodules (from 1–4 cm in diameter), capped by thin laminar illuviations. Cappings can include several nodules and may join together to form large undulating laminae.

Downstream, the profiles are characterized by a strong development of braided subhorizontal fractures that produce a pronounced but irregular layering. Thick deposits (up to 3–4 cm) of thin Si laminae develop in the horizontal fractures. Nodules of red-ochre sandstone wrapped in undulating illuviation laminae, form the uppermost horizon of the profiles. In the basin, jasper and chert lenses with opal nodules cemented by complex concretionary structures are observed in association with carbonate material.

2.1.2.2 Micromorphology and Chemistry

Upstream, beige to brown cutans occur in all open pores of the sandstone and are composed of a mixture of 50% opal-CT and 50% kaolinite. Quartz (either microcrystalline or fibrous) does not appear. In the matrix, opal is intimately associated with non-expandable clay minerals. The granular filling of the joints is made up entirely of gravels or pseudo-ooliths of argillaceous and siliceous materials arranged in concentric layers around skeleton grains. The gravels are cemented by a mixture of opal and kaolinite similar to that of the sandstone matrix.

In the downstream profiles, thick Si-laminae deposits show a succession of mammillary opal concentrations and of illuviated granular material (quartz grains, clay chips, silicified debris, gravels, etc.). The concretionary opal is virtually devoid of Al but contains relatively high K. The K/Al ratio is much higher than in the clay minerals of the profiles.

Cherts and jaspers from the basin are composed of opal-CT, chalcedony and microcrystalline quartz, but not of detrital grains. They contain translucent centimetric nodules with concentric and radial cracks, typical of gel dehydration. The relationships between the opal and quartzose zones clearly indicate recrystallization of primary opal. Opal is devoid of Al but still has a significant K content. Small barite crystals develop in some nodules.

2.1.2.3 Structure Interpretation

Silicifications are recorded in the structures of ferrallitic palaeosols. Topography and position in the palaeolandscape control the nature and the distribution of the Si phases. Illuviations due to water infiltration and percolation are dominant upstream. Downstream, horizontal and concretionary structures develop in relation mainly to lateral drainage close to the regional base level. The succession of concretions and illuviations indicates that periods of percolation and of waterlogging alternated, probably controlled by the level of the lake. In the basin, no sign of illuviation is visible, as the environment is permanently water saturated.

Silicification processes appear to be differentiated by the morphology. Illuviations of Si are mixed upstream with clay minerals, while downstream pure Si is precipitated from oversaturated solutions. Si gels formed in the basin, in the presence of sulphates.

2.1.2.4 Hardpans and Duripans in Modern Landscapes

The red sandstones of the northern Massif Central are similar to the hardened soils called red-brown hardpans in Australia. The hardpans consist in mixtures of clay minerals, Fe oxides and opal (Bettenay and Churchward 1974; Chartres 1985; Milnes et al. 1991). Duripans described in the arid areas of the United States display similar characteristics (Flach et al. 1974; Chadwick et al. 1989).

These formations tend to occur in subdued landscapes and immediately underneath contemporary soils. They are closely related to contemporary landscapes in terms of formational environments and some appear to be still developing. Bettenay and Churchward (1974) believe that the hardpans develop in subdued drainage areas, which are periodically flooded with 250–300 mm annual rainfall. The American formations are also related to alluvial and colluvial deposits in regions where the rainfall did not exceed 300 mm during Pleistocene (Harden et al. 1991).

2.1.3 Mechanisms of Pedogenic Silicifications

The origin of silica has not been well established in the red-brown duricrusts. Dissolution of the Si phases is not observed (neither quartz nor clay minerals) and, therefore, Si appears to have been added to the profiles. This could have been accomplished by leaching of either regional highs, or siliceous oozes deposited at the top of the profiles during periods of flooding or lake high stands. Harden et al. (1991) hypothesized an aeolian supply of volcanic materials in modern soils.

Whatever its source, Si must have been removed and then solution concentrated to ensure deposition of Si gel and opal. Contrasting wet and dry seasons or periods seem to be the most likely conditions for this to occur. The geochemistry of the system appears relatively simple. There is no dissolution of quartz, nor Fe oxides, nor even clay minerals. The environment was restricted and of neutral pH, and formation of Si gels was probably favoured by the presence of brines which lowered Si solubility (Marshall 1980).

In the quartzose silcretes, secondary Si comes from the profile itself. There is much evidence for Si dissolution in the profiles and destruction of clay minerals at their bases with release of Si as gel and opal. The formation and destruction of organic complexes in the profiles, which are able to cause quartz dissolution even in oversaturated environments, should also be considered as possible Si sources (Morris and Fletcher 1987; Bennett et al. 1988).

The mineral sequence found in the quartzose silcretes with quartz at the top, microcrystalline quartz in the middle and opal at the base, probably results from a progressive concentration of Si and other cations in the downward-moving solutions. At the top of the profile, rainwater may interact with the quartzose sediment, during the wet season, to produce a solution containing up to 6 ppm SiO_2. At the base of the profile, the solution may be concentrated two to four times through evaporation, during dry periods. This would be sufficient to allow silica precipitation. Apparent opposition between dissolution and precipitation results from cyclic stages, due to alternating dry and wet seasonal or climatic periods. The concentration of other elements may have a direct effect on the size of the precipitated quartz crystals as the greater the content of impurities, the smaller the crystals (Millot 1960, 1970).

The incongruent dissolution of kaolinite at the base of the profile is more difficult to explain. It implies an acidic environment in which Al is more soluble that Si and is leached from the sediment while Si is retained as opal. Two mechanisms can be proposed: (1) in low hydrated plasmic or microporous systems, the pH can decrease subsequently to an intense dissociation of H_2O into H^+ and OH^- (Fripiat et al. 1960; Bourrié and Pédro 1979); and (2) in hydromorphic environments two reactions may occur, either a release of an electron during Fe oxidation to produce a H^+, or retention of another H^+ during formation of hydroxides (Brinckmann 1970; Fripiat 1971; Brinckmann et al. 1973; Chaussidon and Pédro 1979; Espiau and Pédro 1983). Progressive «autolysis» of the clay minerals is induced (Pédro and Delmas 1979) with migration of Al and retention of Si in disordered Si phases. Both cases require alternating dry periods during which the acidic environment develops and Al is released, and wet periods during which Al, possibly complexed with organic components, is removed from the profile.

2.2 Groundwater Silicifications

In contrast to the pedogenic silicifications, groundwater silcretes develop at depth. Their main characteristics are the superposition of silicified lenses and the preservation of the host-rock structures (stratifications, bioturbations, etc.). There are two main varieties of groundwater silicifications: those occurring in sands and developing into massive sandstone facies by cementation of the pores, and those created by epigenetic replacement of limestones and claystones and, therefore, having more irregular and discontinuous shapes.

2.2.1 Sandstone Silicification: the Fontainebleau Sandstones

2.2.1.1 Silcrete Distribution and Deep Weathering

In the Fontainebleau Sands (Oligocene, Paris Basin), flat-lying quartzite lenses (1–8 m thick) occur at different levels within the formation (Thiry et al. 1988). The quartzites are very pure and consist of detrital quartz grains with well-developed subeuhedral overgrowths. Drill-hole data show that wherever quartzite lenses crop out at the edges of the plateaus or in the valleys, they pinch out rapidly under the limestone-capped plateaus. Several drill holes near the edges of the plateaus, reveal that the quartzite lenses do not extend more than a few hundred metres beneath the plateaus. The quartzites are, thus, restricted to the outcropping zones of the Fontainebleau Sands. This strong link between quartzite distribution and present-day morphological shape suggests that the quartzites developed relatively recently near the outcropping zones of the Fontainebleau Sands.

In addition, bleaching and intense Si leaching occur in the upper unit of the Fontainebleau Sands. At the edge of the plateaus, the sands of the sections are slightly ferruginous and contain unweathered round flint pebbles at the base. At the top of the sections, however, the sand is white and very pure (more than 99.5% SiO_2), and the flint pebbles are deeply corroded. In addition, the uppermost quartzite lens has a warped shape with numerous concavities and a friable, porous outer fringe. In this outer fringe, the quartz overgrowths are corroded, suggesting dissolution of

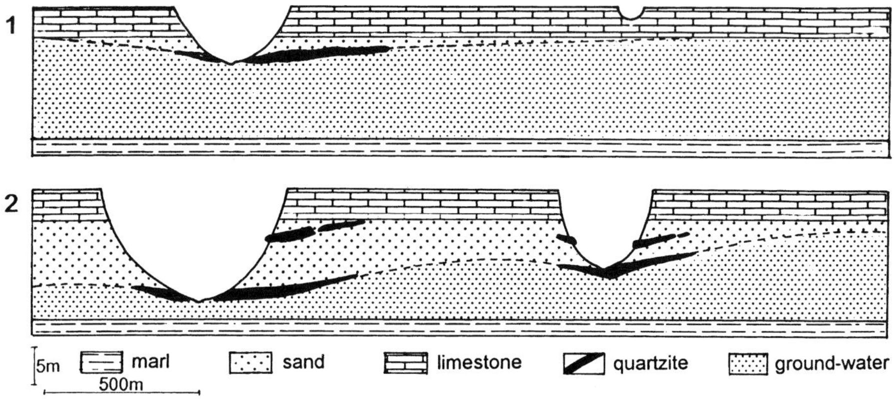

Fig. 4. 1–2 Development of silicifications in groundwater discharge zones. 1 Silicifications develop first at the base of the incision behind spring lines, after dissection of the limestone cover. 2 Resumption of erosion leads to formation of a quartzite pan in the lower level, former silcretes at the top of the sand undergo dissolution

the upper quartzite pan. White sand and quartzite have been altered by percolating water after silicification (Thiry et al. 1984). The leaching is less intense at the base of the sand unit.

2.2.1.2 *Silicification by Groundwater Outflow*

The layout of quartzite in horizontal lenses suggests a relationship with water table levels within the sand formation. The hydrological data on the sand formation confirm this relationship with the water table. The areas in the Fontainebleau Sands where a major quartzite development is visible, are also those where the groundwater discharges under a high hydraulic gradient. The silicification process seems to be associated with the groundwater outflow and developed behind springs or spring lines, in the modern incised landscape. It developed at the water table, whereas the sands above the water table were leached. Landform evolution directly controls the development of the superimposed quartzite lenses and of the leached profile, as shown in Fig. 4. The key processes are as follows:

1. Dissection of the landscape exposes the sand units in thalwegs. An initial silicification may develop at the top of the outcropping sand formation behind the spring lines at the base of the valleys.
2. The drainage network continues to cut down the landscape. Each erosion step leds to a concomitant drop of the water table and eventually to the development of a lower level of silicification. In the mean time, former silicification levels were stranded at the top of the sand formation, in the vadose zone where they were subjected to dissolution at their edges.

2.2.1.3 Discussion

Si originates in groundwaters contained in the sand formation. The Si content in groundwaters ranges from 10–15 ppm, which is about twice the saturation value of quartz (6 ppm) at surface temperature and pressure. This raises two questions: why is the groundwater beneath the plateaus oversaturated with respect to quartz and why does quartz precipitation only occur near the outcropping zones? It might be suggested that quartz precipitation is inhibited within the groundwater zone, but not in the silicification zones. Only speculative interpretations can be offered about the effect of organic compounds, pH, ion activity, etc., and how these factors may influence the surface properties of the quartz, inhibiting or facilitating overgrowths. It is likely that very small changes in the physico-chemical conditions are sufficient to cause precipitation of a part of the Si in solution.

Hydraulic and geochemical calculations show that 30000 years represent a sufficient time to cement a quartzite lens from Si available in the water of an average spring. The consequence is that cementation by quartz can take place under common surface conditions from tap waters and in a short geological time span. This is in total contrast to previous interpretations of number of silicifications.

2.2.2 Claystone Silicification: the Australian Tertiary Regolith

Silicification in central Australia are associated with deep, bleached profiles of Tertiary regoliths. Bleached profiles, up to 50 m deep, have developed widely in Cretaceous shale and siltstone formations, as well as in Jurassic and Palaeozoic sediments and Precambrian granites. Opal and gypsum are usually present in these profiles together with kaolinite, alunite and jarosite. About 90% of the precious opal of the world is mined from these bleached formations. Despite the presence of kaolinite, the profiles are not leached, as they retain soluble elements such as SO_4, Ca and K (Thiry and Milnes 1992). The presence of alunite indicates highly acidic environments. Groundwater silcrete layers are almost systematically confined in these profiles.

2.2.2.1 Description

The section of the Stuart Creek Opal Field, near the northern tip of Lake Torrens in southern Australia, shows a typical silicified facies (Thiry and Milnes 1990). The Cretaceous marine shales at the base of the section are eroded and overlain by Tertiary fluviatile sandstones (Fig. 5). The section shows various groundwater silcretes consisting of opalite and glassy sandstone at depth, and a nodular to columnar pedogenic silcrete at the top of the section. Silicification is largely confined to an erosion channel. It extends to a significant depth in the underlying Cretaceous shales at the edge of the channel.

The upper quartzite pan is a quartzose. Silicification proceeds by micro-crystalline quartz replacement of the primary clayey matrix and by quartz overgrowths

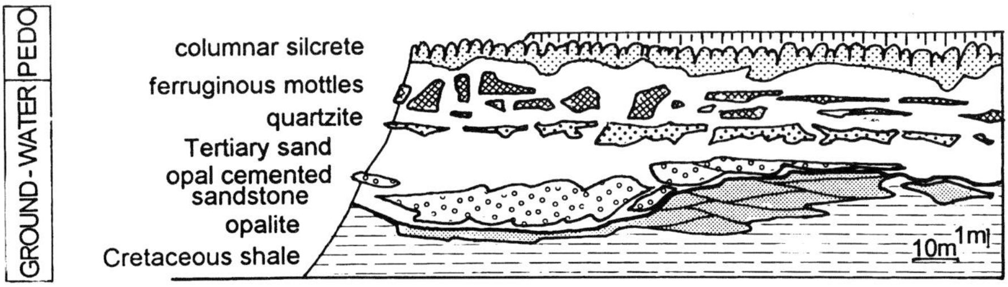

Fig. 5. Schematic sketch of field relationships of silcretes in the Stuart Creek Opal Field (South Australia). Two groundwater silicification pans are superimposed. The lower pan is largely confined to an erosion channel filled with coarse-sand which hosts a groundwater drainage. Significant silica penetration occurs in the Cretaceous shales only at the edge of the channel

in the initially «clean» facies without clay matrix. The degree of silicification is highly variable, changing, on a centimetric scale, from a friable into a tight quartzite facies without residual porosity. Silicifications displaying very contoured shapes are obviously weathered, and show deep dissolution features of the overgrowths and original quartz grains. The lower opal cemented sandstone contains about 10–20% of opal-CT. Two types of cement coexist, one resulting from epigenetic replacement of the primary clay minerals and the other from Si deposits in the former pores. In the clay matrix, the primary flaky morphology, which is typical for smectite, can still be recognized under scanning electron microscope, even though it is composed of about 95% of SiO_2 and only 1% Al_2O_3 remains. The void deposits show relatively complete sequences from opal to quartz with transitional varieties of fibrous Si.

The opalites of the Cretaceous shales are composed of 50 to 70% of opal-CT; they are true opalites. Clay minerals are no longer present, even though the primary sediment contained more than 50% clay. All sedimentary structures are perfectly preserved, including morphologies of the primary clay minerals. The voids (joints and gypsum crystal moulds) display a complex succession of Si deposits:

brown opal $\rightarrow$ limpid opal $\rightarrow$ lussatite $\rightarrow$ chalcedony
or quartzine or lutécite $\rightarrow$ quartz.

2.2.2.2 Interpretation

Detailed mineralogical studies show that the opal-CT of the opalite layers is composed of high temperature tridymite. It contains appreciable Al and its hexagonal rings of SiO_4 tetrahedrons are slightly twisted as seen in clay mineral structures. This similarity of structure suggests that the opal particles are derived from clay minerals by removing Al, while retaining the SiO_2 network (Rayot et al. 1992). Clay can potentially change into opal-CT without destroying the entire original structure, through a solid state alteration. The Si for these opalites could, therefore, come directly from the original clay minerals, by leaching of the octahedral cations.

The acidic environment and the bleaching of the rocks appear at an early stage of weathering. This acidic environment caused leaching of Fe and Al. Clay mineral structures collapsed. The Si network was reorganized into opal-CT or was removed to feed silicification. This process, therefore, provides an important source of highly soluble Si to sustain the opal deposits and it eliminates the necessity to concentrate Si-bearing solutions by evaporation.

Development of the deep bleached rock profiles and of the acidic environment is yet not fully understood. In the case of the Cretaceous shales, the acidic conditions appear to have been created by oxidation of primary organic matter and pyrite. Geochemical modelling shows that by combining the effects of sulphate-rich brines and pyrite oxidation it is possible to dissolve feldspars and 2:1 clay minerals, and to form kaolinite, alunite, gypsum during a single weathering process, while accumulating highly soluble Si phases (different varieties of opal; Rayot 1994). Development of high acidity implies that the released protons do not migrate as they form, and that, therefore, water circulation must have been restricted, the water table being at a low stand. However, subsequent development of the groundwater silcretes with mobilization of Si released by the acidic conditions, implies elevation of the water table.

The phenomenon is, in fact, much more extensive and general, and it exists in other arid regions of the world. Some Tertiary palaeoweathering profiles and silcretes of the Iberian peninsula, which contain alunite, have probably been created by similar processes (Blanco and Cantano 1983; Meyer and Pena dos Reis 1985; Blanco 1991).

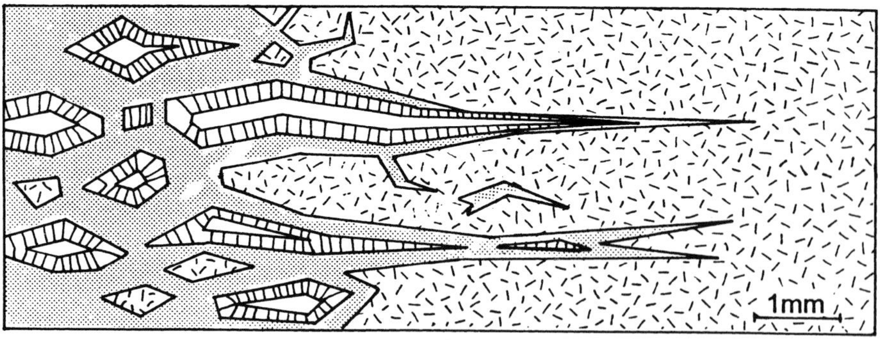

Fig. 6. Silicification intensity is related to porosity in limestones. Epigenetic replacement occurs only at the edges of the voids and joints, areas directly dependent on drainage. When water flow decreases, as at the termination of a fissure, silicification intensity decreases and dies out

2.2.3 Limestone Silicification: the Limestone Plateaus of the Paris Basin

Many limestone formations contain irregular masses of silicification, an example of which being the lacustrine limestones that form the plateaus of the Paris Basin. These silicifications have been described since the last century (Cuvier and Brongniart 1808; Cayeux 1929; Ménillet 1988). This Si was generally thought to be chemically precipitated during the dry periods, at times of limestone deposition in confined environments. A petrographic and geochemical approach has been made in order to understand these silicifications.

2.2.3.1 Description on the Silicified Limestones

A section of the Calcaire de Champigny, which is an Upper Eocene formation, has been studied in detail by Ribet (1990). The limestone is about 20 m thick, massive, without bedding, very pure and has a nodular and brecciated structure with numerous traces of early reworking features (mud cracks, microkarst, etc.). The silicifications develop inside a layer about 5–8 m thick and are most often completely discordant with respect to the host sedimentary structures, even if the silicified bodies are extending horizontally.

Two types of silicification exist together: quartz and chalcedony deposits in voids, and epigenetic replacement of the limestone matrix with preservation of the primary sedimentary structures. The Si epigenetic replacements are composed of microcrystalline quartz, small amoeboid quartz, or large euhedral quartz (Ribet and Thiry 1990). The nodular structure of the primary limestone is easily distinguished in the Si zones by the variable grain size of the microcrystalline quartz found among the nodules or between the nodules and the matrix.

Thin sections show a systematic link between silicification and zones of high porosity. Porosity is either preserved or partly cemented by Si deposits and consists of pores between nodules or micronodules or more irregular microkarst-like voids. When porosity is reduced laterally, for example towards the end of a fissure, the amount of Si deposited in the void and the extent of the replacement at the edge of this void decreases in the same fashion (Fig. 6). Limestone epigenetic replacement by silica occurs only on the edges of the voids and pores, over a distance that generally does not exceed 0.5–1 mm.

2.2.3.2 Interpretation and Mechanisms

Because the silicified limestones are pure without any sandy or clayey layers, Si had to be imported from other formations to create the silicification. This can only be done by substantial water flow, which would explain the observed relationship between porosity, and intensity and extent of the silicification. Porosity and flow determine the degree of silicification. Silicification decreases with reduction of the fissures and of the water flow. Silicification in these limestones is related to groundwaters, in a similar fashion to those described in the sandy formations.

Substantial groundwater out flows are only possible after uplift and downcutting of the limestone formation. In the example of the Paris Basin, such conditions only existed during the Pliocene and the Quaternary periods and, therefore, the implication is that silicification had to occur relatively fast. A mathematical model of the Si deposition and epigenetic replacement predicts that 10 000 to 100 000 years are necessary to create such silicifications (Ribet 1990).

Limestone replacement by silica with preservation of the structures implies that silica deposition and calcite dissolution have to occur simultaneously, and that actually Si deposition induces calcite dissolution. To make the mechanism work, Si influx and calcite release must be able to diffuse between the pores and the replacement sites. The diffusion potential, therefore, restricts the extent of the epigenetic replacement.

Groundwater flow was previously proposed by Conrad (1969) to explain intense silicification of the Moroccan Hamada limestones along the main fluvial axis and nearby groundwater discharge zones. Khalaf (1988) and Arakel et al. (1989) also related some silicifications in modern calcretes from the Persian Gulf and inland Australia to groundwater flow.

2.2.4 Variability of the Groundwater Silicifications

The three case studies of groundwater silicification presented here are very different from each other: they have been chosen to illustrate several types of silicification mechanisms. They also show the extreme variability and heterogeneity of groundwater silicifications. This variability is the common feature and it provides diverse and instructional cross sections of natural examples:

1. Spatial discontinuity of the silicifications. Their geometric distribution is dependent on groundwater flow, on granularity and porosity of the host rocks, and also on amount of landscape downcutting. In most cases, there are sharp boundaries between the silicified zones and the host rocks. The geometries are the most complex in the limestones, even showing development of karst networks.
2. Facies variability as a function of the nature of the host rock, where primary structures are preserved. There is also a variability in the degree of cementation, from friable sands to tight sandstones with a lustrous fracture appearance.
3. Petrographic variability and heterogeneity. All petrographic forms of Si can be present in a single thin section. This variability results from interaction between the host rocks and the solution which changes in time and space during the silicification process.
4. Lastly, the variability of the environments where silicifications developed, includes environments as diluted and neutral as the groundwaters where potable water is drawn, up to very concentrated environments with extreme acidic pH. There is no diagnostic climatic signature, as with the pedogenic silicifications, because the continental silicifications are dependent on groundwater Si and, therefore, beyond control of landscape evolutionary processes.

2.3 Silicifications Associated with Evaporites

Silicifications associated with evaporites are complicated and difficult to understand because of the known relationship between sulphate silicification and distinctive quartz forms such as quartzine (length slow chalcedony), lutecite and «flame"-like or «cubic» quartz (Munier-Chalmas 1890; Cayeux 1929; Folk and Pittman 1971; Arbey 1980). Confusion occurred consequently because these distinctive quartz forms have often been regarded as results of evaporitic environments. To contribute towards the understanding of these silicifications, silica behaviour in modern evaporitic environments will first be examined and then the question of the sulphate replacements will be addressed.

2.3.1 The Modern Evaporitic Environments

2.3.1.1 Salt Lake Deposits

Si-gel formation by direct inorganic precipitation at the time of sedimentation was initially reported from the seasonal Coorong lakes (South Australia) by Peterson and von der Borch (1965). The dominant precipitates in the lakes are Mg-calcite, dolomite and magnesite. The Si-bearing carbonates are only a few hundred years old by ^{14}C dating. The pH of the lake water commonly rises to 10.2 during photosynthesis periods and decreases to 8.2 towards maximum concentration when the lake dries up to a muddy surface. Beneath the surface of the sediment, the interstitial solution of the organic-rich sediments has a pH as low as 6.5. Near the boundary between these two pH domains, gelatinous-hydrated silica precipitates. Detrital quartz is corroded in the lake sediments, indicating that it has probably provided much of the source Si for the Si-gel precipitation. An environment where quartz dissolutions coexist spatially with amorphous Si deposition exists, although the mechanisms are not synchronous.

Direct silica precipitation in salt lakes has only been shown exceptionally. Laminated chert layers were, however, often described within alkaline evaporites (Eugster 1969; Surdam et al. 1972; Maglione and Servant 1973; Sheppard and Gude 1974). These cherts develop from hydrated Na silicates with a high Si/Na ratio, mainly magadiite and kenyatite. Alteration of Na-silicates took place by removal of Na, either by percolating meteoric water or by rising groundwater. Quartz usually forms the chert layers, but the presence of crystallized hydrated silicate as silhydrite ($3SiO_2, H_2O$), has been reported (Sheppard and Gude 1974).

2.3.1.2 Brine Geochemistry

Brine/mineral equilibrium studies and evaporation models provide important information about Si behaviour in these environments. Two geochemical environments can be distinguished: an alkaline (carbonate-rich) one and a neutral (chlorine and sulphate-rich) one, the silica behaviour being different in each.

In the Chott El Jerid (southern Tunisia), considerable amounts of gypsum and halite precipitate and the Si concentration decreases as salinity increases (Gueddari et al. 1982). The pH is neutral to slightly acidic. The phase diagrams show that the Si content is controlled by development of Mg-rich clay minerals and, therefore, is always below the point of amorphous Si saturation.

Lake Natron (East African Rift) is alkaline with Na carbonate deposits. According to dissociation of H_4SiO_4 silicic acid at high pH, the Si content can reach 1800 ppm in the most alkaline brines (Fritz et al. 1987). Speciation calculations and models show that Si concentrates in the brines until saturation of the Na-silicates, which is probably a kinetic function. As activity of the neutral silica molecules (H_4SiO_4) reaches saturation, amorphous silica is able to precipitate. In such brines, a rapid decrease in the pH, either by mixture or dilution, will reduce the dissociation of H_4SiO_4 and lead to a high excess of Si in the solution.

In inter-dunal depressions of the northern part of Lake Chad, the two geochemical types develop from waters with identical initial chemistry (Maglione 1976; Droubi et al. 1975). In euxinic ponds, as sulphates change to sulphides, an alkaline geochemical trend appears with carbonate development. In these brines, the pH and the Si contents rise with the concentration of salts. In oxidizing environments, the brines follow a neutral geochemical trend characterized by Ca, Mg and Na sulphate deposits, in which the Si content decreases with the concentration of salts.

2.3.2 Sulphate Replacements in Geological Formations

The epigenetic replacement of various sulphate minerals like gypsum, anhydrite and barite by Si, has been described in many evaporite deposits. The Carboniferous formations of the United States, the Triassic littoral facies of Europe and North Africa containing Pb–Ba–F ore deposits, the Purbekian of the United Kingdom and the Tertiary gypsum formations of France and Spain, are among the formations most frequently referred to. However, only a few studies provide tangible arguments defining the time or the environment of silicification. Most often the only reliable argument is the reworking of specific facies in the overlying deposits (Durand and Meyer 1882).

The Pb, Ba and F ore deposits of the Triassic arkoses are a good example of barite silicification. Numerous authors refer to these replacements of barite by Si and show that they have been reworked in successive arkosic deposits (Lefavrais-Raymond et al. 1965; Bernard and Samama 1970; Rouchy 1974; Schmitt 1976). A typical example of these deposits comes from Morocco (Thiry and Schmitt 1977). Arkose-type rocks fill palaeovalleys cut into the granitic basement at the edge of a confined littoral lagoon (Fig. 7). Silicification is very intense and the silica is composed of microcrystalline quartz towards the streamhead with euhedral quartz and overgrowths downstream towards the lagoon. Barite is replaced by flame-like microcrystalline quartz, quartzine (length slow chalcedony) and by overgrowths of detrital quartz grains. These typical silicified facies are reworked in the successive sedimentary sequences and, therefore, are undoubtedly early. The silicification

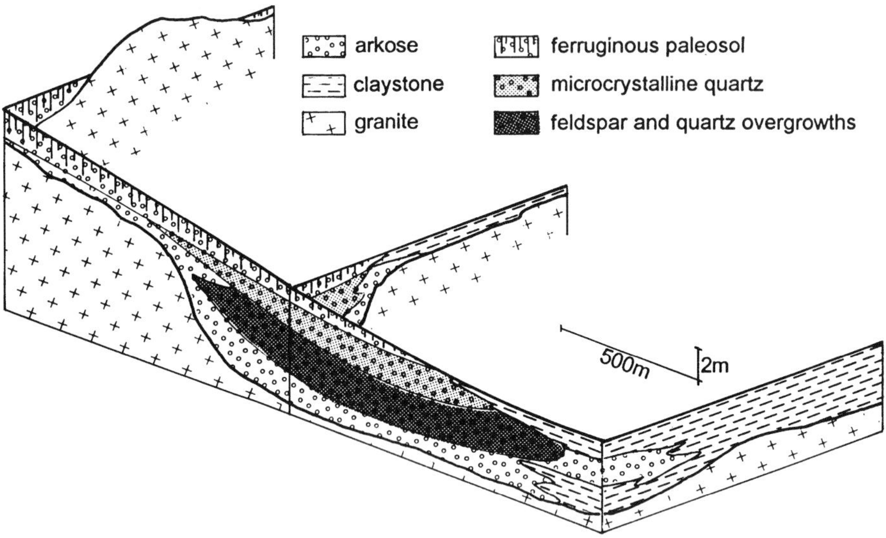

Fig. 7. Silicification of arkose-filled channels at the edges of littoral lagoons in the Triassic deposits of Morocco. Continental groundwater silica precipitates at the contact with sulphate-rich brines of the lagoon. Microcrystalline quartz develops in the upper part of the arkose, where clay minerals, iron oxides and sulphate minerals are abundant. Quartz overgrowths develop in the lower part, where the arkose is «cleaner», without abundant matrix

process is restricted to coarse facies filling palaeovalleys and channels, and it is related to groundwater discharge into the sulphate-rich lagoon.

In most of the silicified evaporites, there is as yet no evidence documenting the time frame of silicification. Silica deposition could occur throughout burial diagenesis and even after surface exposure of the sediments. Oxygen isotope analyses of silicifications in the Carboniferous formations of North America indicate that the chemical composition of the solutions changed between the start and the end of the silicification process (Milliken 1979). The first silicification which was composed of microcrystalline quartz, came from sea water; the fibrous quartz varieties being formed in waters of intermediate composition. The subeuhedral quartz in the centre of geodes is formed in meteoric waters. Milliken (1979)believes that the mechanism can still operate in present-day environments.

2.3.3 Discussion

In modern evaporitic environments, silicifications occur primarily through changes in pH values and salt concentration. These changes are both spatial and temporal, involving mixing of fresh continental waters with brines, or a succession of periods of desiccation and subsequent dilution by rainwater. The mixing can be achieved either by surficial or groundwater at the border of the evaporitic depressions, or by artesian circulations in the joints and cracks beneath evaporitic deposits (Guillou and Ndiaye 1988). Si deposition is particularly favoured when evaporites are penetrated by fresh water. By the way, it has to be noticed that evaporitic levels deposited during a regressive phase are naturally intruded by fresh water, thus explaining the frequency of the phenomenon.

These silicification processes are mainly formed by epigenetic replacement, that means by diagenetic replacement following deposition. Even though some Si mineralogical forms are diagnostic of sulphate-rich environments, they do not necessarily indicate that they were formed in saline environments. As an example, gypsum can be replaced by groundwater silicification, at least at the replacement front in the presence of high sulphate concentrations, but not all silicifications develop in saline environments. Interpretation of these epigenetic replacements and of their palaeoenvironments must, therefore, be made with care.

3 Mechanisms at Work

The variety of facies, deposition processes and mineralogical compositions of surficial silicifications results from combined and, sometimes, opposite effects of the solution geochemistry, and of crystal nucleation and growth. A few simple mechanisms can be used to explain many facts. However, oversimplification must be avoided. Properly constrained models are built through careful observation and patient collection of supporting evidence.

3.1 The Geochemical Rules

3.1.1 Silica Solubility

It seems paradoxical that the solubility of the most abundant element found at the surface of the earth (after oxygen) can be a problem. The existence of colloidal silica has been discounted by the work of chemists since the 1950s (Millot 1964, 1970). Various studies have shown that the presence of Al and Mg cations hardly affect silica solubility (Jephcott and Johnston 1950; Hast 1956; Siffert 1962; Iler 1979; Delmas et al. 1982). Since then, geologists and geochemists have argued in terms of the solubility of individual Si forms in relation to each other, irrespective to the influence of other cations, organic compounds or possible complexes. The 1980s brought a renewal of these concepts.

3.1.1.1 Solubility in Brines

The studies of Marshall (1980) and Marshall and Warakomski (1980) showed the importance of salt in solution for silica solubility. With increasing concentration of salt, the solubility of amorphous silica consistently decreases. Amorphous silica solubility is lowered as much as ten times in a solution saturated with $MgCl_2$ or $CaCl_2$ (5% of its solubility in water) and by 2.5 times in a solution saturated in NaCl (40% of its solubility in water). Sulphates play a similar role. These experimental data explain the behaviour of Si in evaporitic environments. As Si-loaded fresh waters mix with brines, they become greatly oversaturated and, therefore, are able to feed silicification processes. This explains why evaporitic deposits appear so favourable for silicification.

3.1.1.2 Complexation: Dissolution in Saturated Solution

The role of organic material in relation to silica solubility has been addressed for some time, but has often been refuted due to a lack of precise analytical data. A decrease in the amount of dissolved Si in the presence of sulphate-reducing bacteria has been reported by Birnbaum and Wireman (1985). The exact role of the bacteria remains obscure: activation of the polymerization or nucleation on the walls of the cells?

Furthermore, strong evidence that organic acids increase the dissolution rate of quartz was reported by Bennett and Siegel (1987). They presented evidence for an increase in the solubility of quartz in groundwater, caused by dissolved organic compounds produced by the biodegradation of petroleum. Quartz grains in contact with the contaminated water displayed active chemical etching, whereas these of the uncontaminated zones were not altered. Apparently the organic compounds accelerated dissolution of quartz in water with neutral pH. Subsequent experiments of quartz dissolution in dilute aqueous solutions with organic acids was investigated using the batch dissolution method. These experiments showed that there is a significant organic–silica interaction (Bennett et al. 1988). In these experiments, citric, oxalic and salicylic acids increased the solubility and dissolution rate of quartz in water with the greatest effect found at near-neutral pH. Quartz dissolution rate was ten times higher in the presence of citrate than in pure water and was accompanied by a 100% increase in apparent quartz solubility at 25 °C. Formation of complexes at the quartz grain surfaces may favour Si release and then cause a decrease in Si activity in the solution (Bennett 1991).

Other experiments demonstrated that Fe^{2+} in solution also has an influence on quartz solubility (Morris and Fletcher 1987). In the presence of Fe^{2+}, quartz is not dissolved. Si goes into solution only after oxidation of the solution. The concentration of Si in solution is in direct correlation with the contact time of the ferrous solution with quartz and can exceed the solubility of quartz in pure water. Formation of Fe^{2+}/Si complexes at the quartz grain surfaces may inhibit dissolution; their decomposition during solution oxidation may release their Si. In this way, quartz solubility could be increased ten times and approach the solubility of amorphous silica.

These experiments show that ordinary organic acids within concentration grades found in nature, or common oxidizing–reducing reactions, are able to appreciably increase quartz solubility and dissolution rate. Such complexing mechanisms can provide significant amounts of Si in some soils and are likely to explain quartz grain corrosion even in environments where silicifications develop. These areas are quite new for geologists and geochemists and numerous developments can be expected in the future.

3.1.2 Mineral Sequences

The successive silica deposits in voids display a systematic sequence from «amorphous» or cryptocrystalline to microcrystalline forms (the different varieties of opal), towards better «crystallized» forms (generally the fibrous quartz varieties), throughout large euhedral quartz crystals in the centre of the voids (Cayeux 1929; Chowns and Elkins 1974; Milliken 1979; Arbey 1980; Thiry and Milnes 1991). These sequences are generally interpreted as resulting from changes in the composition of the parent solutions. First deposits would have resulted from highly oversaturated solutions containing other cations and impurities, while later ones would develop in «fresh», diluted and only slightly oversaturated waters (Millot 1964; Milliken 1979; Thiry and Millot 1987).

The compositional changes in the solutions result from progressive silicification of the host rocks. Initially, the circulating solutions are in contact with the minerals and the pore water of the host rocks; they pick up various cations from the latter and microcrystalline forms develop. As silicification progresses, the solutions are insulated from host rocks by the newly formed silica rims, and the influence of other cations decreases, allowing development of larger quartz crystals.

This explains why no recurrence or fluctuation is observed in these mineral sequences. The last crystallization products primarily reflect the nature of the parent solutions which, therefore, are always dilute and contain only low concentrations of cations other than Si.

3.2 The Crystal Growth Rules

3.2.1 Crystal Nucleation and Growth

Quartz crystals never precipitate directly from solution. Quartz crystals always develop on rock surfaces, as in geodes. There is no quartz nuclei development in oversaturated solutions; disordered polymers form, then gels (Iler 1979). Quartz crystals and other silicates make up the crystal germs.

3.2.1.1 Quartz Precipitation

Mechanisms of quartz precipitation have been described using a kinetic approach (Garcia Hernandez 1981; Delmas et al. 1982). Cristobalite dissolution experiments through percolation show that quartz crystallization requires a germination support, cristobalite in this case. If solutions are dilute and «clean», quartz effectively develops. If the growing surface is disturbed, germination is not possible. The disturbance of the surface can be a function of either oversaturation, or to «poisoning» of the surface by adsorbed cations that cover the growing sites. If quartz germination is blocked, Si concentration can rise until formation of amorphous silica occurs.

The notion of contamination of the growing surfaces by adsorbed cations may be an explanation for the general quartz oversaturation of groundwaters (Hem

1985). Quartz precipitation can occur as soon as the cations are released from the surface of the grains which can then be overgrown or become germination sites. This is probably one of the reasons for development of groundwater silicification in selected zones, while the source groundwaters are kept in an oversaturation state.

3.2.1.2 *Petrographic Facies*

Heterogeneous nucleation and germination mechanisms also explain the variety of facies that comprise the superficial silicifications. In silicified limestones, the primary structures can be recognized by the crystalline–granular texture of the quartz, which goes from very tiny microcrystalline quartz, to amoebic quartz with crystals several tens of microns in diameter. The replaced supports influence, therefore, quartz crystallization. In some cases, a relationship between the crystalline–granular texture of the primary carbonate and of the replacing quartz appears, whereas in other cases, such a relationship does not exist. The difference in crystalline texture appears, therefore, to be ascribed to the presence of minor compounds (trace elements, organic matter, etc.). There was probably some interference linked to crystal germination and surface properties of the grains at the time of silicification.

The modelling of incorporation of other cations into quartz crystals also brings a fresh perspective. Development of twisted and non-twisted chalcedony grains has been modelled using the pre-requisite that the crystal lattice distortion is due to Al substitutions (Wang and Merino 1990). Growth of specific quartz varieties in the presence of sulphate-rich solutions must probably be considered in similar terms including crystal lattice distortion by substitution, incorporation of other ions, or selective blocking of some growing sites.

3.2.2 Recrystallization

One of the difficulties in the interpretation of the diverse silica varieties comes from the fact that they often result from transformation and recrystallization, and that the nature of the primary silica deposit is thus obliterated. Sometimes, the opal concretions around pores change laterally to fibrous quartz or appear as if they were punched by this latter material (Thiry and Millot 1987; Thiry and Milnes 1991). In these samples, scanning electron microscope observations show that the fibrous quartz consists of micro-spherolites about 0.1 μm in diameter, welded together to form laths or crude sheets. Opal recrystallized into fibrous quartz. The varieties that display this micro-spherolitic structure are mostly the quartz crystals with positive extension, as well as the large lutecite crystals found in sulphate-rich environments (Arbey 1980; Arnold and Guillou 1983).

These quartz varieties obviously result from recrystallization of micro-spherolitic opal deposits. Recrystallization occurs by successive dissolution–precipitation steps with elimination of the cation impurities. This is most likely guided by the composition of the primary opal, but also by the texture and the microporosity that control solution percolation and composition. Recrystallization into quartz is favoured when an equilibrium is established between opal dissolution, quartz nucleation and quartz growth rates (Delmas et al. 1982; Williams et al. 1985). Other parameters that also constrain the recrystallization process, include preferential dissolution of the smallest particles, infilling of the hollows and growth of the largest particles (Ostwald ripening; Iler 1979) and variation of Si activity in small pores (Dandurand et al. 1982).

3.2.3 Epigenetic Replacements

Silicification of limestone proceeds mainly by epigenetic replacement of the carbonate phase. This occurs by silica replacement of carbonate, molecule by molecule, with preservation of the primary structures without forming voids. This is frequently seen, not only in surficial silicifications, but also in burial diagenesis (cf. flints). Other minerals such as calcite, Fe oxides and sulphates can also form epigenetic replacements.

A mechanism for epigenetic replacements has been proposed by Maliva and Siever (1987). The concept appears particularly applicable to many observed occurrences. Carbonate dissolution and silica precipitation occur simultaneously. Away from the replacement front, calcite is stable and the pore solutions are not undersaturated. If calcite dissolves at the replacement front, it is because silica growth leads to calcite dissolution. The calcite dissolution reaction is caused literally by silica crystallization. As the growth of the silica phase applies a pressure across the silica–carbonate contact, the solubility of the carbonate is increased. This local force causes an increase in calcite solubility on the pressured surface. In this way, a mineral can dissolve even if the environment is oversaturated relative to this mineral.

Si supply and Ca removal are accomplished by diffusion in thin solution films between the sicate–carbonate phases (Weyl 1959; Tada et al. 1987). The restriction of the solution phase to thin films would allow ghost features to be preserved in the replacement by Si. Si supply controls the progress of the replacement. As epigenetic replacement progresses, there is an increase of the distance between the Si source and the replacement front. In the mean time, the diffusion gradient, the diffusion rate and the kinetic factors for replacement decrease until cancelling out. Modelling of the epigenetic replacement by crystallization pressure has been done by Dewers and Ortoleva (1990) who calculated that it takes 17000 years to form a centimetric siliceous concretion.

This mechanism adequately explains the facts observed in the silicified limestones. It is also possible that it works, completely or partly, in clay mineral replacements seen in the groundwater silicification from central Australia and at the base of pedogenic quartzose silcretes.

4 Conclusions

Facies and diverse occurrences of continental silicifications are due to different mechanisms that intervene and interfere with each other. Understanding these silicifications requires that the recorded messages be adequately deciphered and that respective roles due to climate, landscape, geochemistry and crystal growth are discerned.

Morpho-climatic factors are of prime importance in the evolution of continental surfaces. They are well illustrated in continental silicifications, leading to three major silicification types:

1. Pedogenic silicifications develop under climates with alternating humid periods during which silica goes into solution and dry periods during which solutions are concentrated by evaporation. Si accumulation and removal of the other cations from weathering profiles require water flows through the landscape. Pedogenic silicifications are symptomatic of glacis and piedmont landscapes.
2. Groundwater silicifications develop at depth, in environments insolated from direct climatic influence. Removal of mobilized elements and Si supply are achieved through the movement of groundwaters. Superposed groundwater silcretes are the consequence of lowering of the water table after landscape dissection. These silicifications are diagnostic of dissected landscapes.
3. Evaporite silicifications are the result of the mixtures of brines and «fresh» water containing Si. They are dependent on brines in which silica solubility is greatly reduced. They are diagnostic of restricted environments and arid climates.

Solutions oversaturated in Si are required for development of silicifications. Recent work shows that there are several geochemical ways to achieve this oversaturation:

1. Solution concentration by evaporation is the option which was frequently used to explain continental silicifications, the origin being, therefore, interpreted as climate controlled.
2. Oversaturation can be reached by dissolution of silicates or some Si varieties, as long as precipitation of the more stable Si varieties is prevented or occurs at a slower rate than dissolution.
3. Complex formation that, in some way, hides the released Si, could be a mechanism to explain why some minerals dissolve far beyond their solubility, additional research being required to understand these processes.
4. Lowered silica solubility in brines certainly intervenes in silicifications associated with evaporites. This can be achieved by solution mixing either on a regional scale, or on a microenvironment scale, at the sulphate-replacement front.

If cations such as Al and Mg are abundant, clay minerals will preferentially form and scavenge the available Si. Silicification will not develop. This explains the frequent development of pedogenic silicifications in kaolinitized landscapes, which have been leached for their K, Na, Ca and Mg cations, and also of groundwater silicifications in leached and bleached sandstones or in pure limestone formations without clays.

The different silica forms reveal solution compositions and their development controls the Si content of the solutions. Numerous interrelated mechanisms operate in this case. The reaction kinetics are fundamental to the process, but they are often not well understood:

1. Oversaturated mother solutions, cation- or impurity-enriched, lead to the development of «disordered» mineralogical forms. Multiple crystal defects restrict the size of the crystal structures. Diluted and «clean» mother solutions lead to less disordered crystal structures with larger crystals.
2. If growth surfaces are neutralized by adsorbed cations or have particular electrical properties, crystals do not grow and the Si content in the solution can rise and be maintained at high oversaturation levels.
3. Rate of silica precipitation is in direct proportion to the surfaces available for growth. If nucleation is rapid, then the growth surface increases rapidly and Si is rapidly fixed. With identical Si influx, clean and fine-grained sandstones offer larger growth surfaces and are silicified more quickly then coarser sandstones.

Acknowledgement. I wish to acknowledge several individuals who helped me in my research on silicifications. First at Strasbourg, Georges Millot, who brought discipline to my reasoning at the time of my thesis, and Norbert Trauth, for the arguments and contradictions discussed at the outcrops. At Fontainebleau, discussions with Régine Simon-Coinçon, Jean-Michel Schmitt and Georges Grandin assisted the emergence and maturation of the ideas; each of them will recognize his own contribution. In Australia, Malcom Wright and Anthony Milnes of CSIRO, who guided me to the hundreds of square kilometres of silcrete outcrops and provided me with their knowledge during long discussions around camp fires. Lastly, Phoebe Hauff, who knows many of the described outcrops, in the Paris Basin and in Australia, tried to edit my poor English.

References

Arakel AV, Jacobson G, Salehi M, Hill CM (1989) Silicification of the calcrete in paleodrainage basins of Australian arid zone. Aust J Earth Sci 36:73–89

Arbey F (1980) Les formes de la silice et l'identification des évaporites dans les formations silicifiées. Bull Centre Rech Explor Prod Elf-Aquitaine 4:309–365

Arnold M, Guillou JJ (1983) Croissance naturelle de paracristaux de quartz dans une saumure sulphatée calcique à basse température. Bull Minér 106:417–442

Bennett PC (1991) Quartz dissolution in organic-rich aqueous systems. Geochim Cosmochim Acta 55:1782–1797

Bennett P, Siegel DI (1987) Increased solubility of quartz in water due to complexing by organic compounds. Nature 326:684–686

Bennett PC, Melcer ME, Siegel DI, Hassett JP (1988) The dissolution of quartz in dilute aqueous solutions of organic acids at 25 °C. Geochim Cosmochim Acta 52:1521–1530

Bernard A, Samama JC (1970) A propos du gisement de Largentière (Ardèche). Essai méthodologique sur la prospection des "Red Beds" plombo-zincifères. Sci Terre (Nancy) 15:207–264

Bettenay E, Churchward HM (1974) Morphology and stratigraphy of the Wiluna Hardpan in arid Western Australia. J Geol Soc Aust 21:73–80

Birnbaum SJ, Wireman JW (1985) Sulphate-reducing bacteria and silica solubility: a possible mechanism for evaporite diagenesis and silica precipitation in banded iron formations. Can J Earth Sci 22:1904–1909

Blanco JA (1991) Cuarta parada: Los procesos de silicificacion asociados al Paleogeno basal del borde SW de la Cuenca del Duero: II sobre los sedimentos Paleocenos. In: Blanco JA, Molina E, Martin-Serrano A (eds) Alteraciones y paleoalteraciones en la morphologia del Oeste Peninsular. Inst Tecn Geominero Esp, Monogr 6:239–249

Blanco JA, Canteno M (1983) Silicification contemporaine à la sédimentation dans l'unité basale du Paléogène du Bassin du Duero (Espagne). Sci Géol Mém (Strasb) 72:7–18

Bourrié G, Pédro G (1979) La notion de pF, sa signification physico-chimique et ses implications pédogénétiques. I. Signification physico-chimique. Relation entre pF et activité de l'eau. Sci Sol 4:313–322

Brinckmann R (1970) Ferrolysis, a hydromorphic soil forming process. Geoderma 3:199–206

Brinckmann R, Jongmans AG, Mioedema R, Maaskant P (1973) Clay decomposition in seasonally wet, acid soils: micromorphological, chemical and mineralogical evidence from individual argilans. Geoderma 10:259–270

Campos A (1992) Les «Tepetates» de Xalapa, Veracruz (Mexique): une induration pédologique dans les sols d'origine volcanique. 1. Structure et transformation de la couverture pédologique. Cah ORSTOM, Sér Pédol 26:227–234

Campos A, Dubroeucq D (1990) Formacion de tepetates en suelos provientes de las alteraciones de materiales volcanicos. Terra 8:137–147

Cayeux L (1929) Les roches sédimentaires de France. Roches siliceuses. Mém Carte Géol Fr, Paris, 774 pp

Chadwick OA, Hendricks DM, Nettleton WD (1989) Silicification of Holocene soils in northern Monitor Valley, Nevada. Soil Sci Am 53:158–164

Chartres CJ (1985) A preliminary investigation of hardpan horizons in North-West New South Wales. Aust J Soil Res 23:325–337

Chaussidon J, Pédro G (1979) Rôle de l'état hydrique du système poreux sur l'évolution du milieu. Réalité de l'altération dans les systèmes à faible teneur en eau. Sci Sol 2:223–237

Chowns TM, Elkins JE (1974) The origin of quartz geodes and cauliflower cherts through the silicification of anhydrite nodules. J Sediment Petrol 44:885–903

Conrad G (1969) L'évolution continentale post-hercynienne du Sahara algérien (Saoura, Erg Chech-Tanezrouft, Ahnet-Mouydir). Centre Rech Zones Arides, Géologie, 10, CNRS, Paris, 527 pp

Coupé JM (1804) Sur la formation des pierres meulières. J Phys Chim Hist Nat 61:175

Cuvier G, Brongniart A (1808) Sur la géographie minéralogique des environs de Paris. Mém Acad Sci Inst Fr 11:278 pp

Dandurand JL, Mizele J, Schott J, Bourgeat F, Valles V, Tardy Y (1982) Premiers résultats sur la solubilité de la silice amorphe dans les pores de petite taille. Variation du coefficient d'activité de la silice en fonction de l'activité de l'eau. Sci Géol Bull (Strasb) 35:71–79

Delmas AB, Garcia-Hernandez JE, Pédro G (1982) Discussion sur les conditions et les mécanismes de formation du quartz à 25 °C en milieu ouvert. Analyse réactionnelle par voie cinétique. Sci Géol Bull (Strasb) 35:81–91

Dewers T, Ortoleva P (1990) Force of crystallization during the growth of siliceous concretions. Geology 18:204–207

Droubi A, Cheverry C, Fritz B, Tardy Y (1975) Géochimie des eaux et des sels dans les sols des polders du lac Tchad: application d'un modèle thermodynamique de simulation de l'évaporation. Chem Geol 17:165–177

Dubroeucq D (1992) Les «Tepetates» de Xalapa, Veracruz (Mexique): une induration pédologique dans les sols d'origine volcanique. 2 Microscopie et minéralogie des indurations. Cah ORSTOM, Sér Pédol 26:235–242

Durand M, Meyer R (1982) Silicifications (silcrètes) et évaporites dans la zone-limite violette du Trias inférieur lorrain: comparaison avec le Bundsandstein de Provence et le Permien des Vosges. Sci Géol Bull (Strasb) 35:17–39

Elouard P, Millot G (1959) Observations sur les silicifications du Lutétien en Mauritanie et dans la vallée du Sénégal. Bull Serv Carte Géol Als Lorr (Strasb) 12:15–21

Espiau P, Pédro G (1983) Etude du phénomène de ferrolyse par voie expérimentale. Production d'acidité d'échange et mise en évidence du rôle catalytique des minéraux argileux. Sci Sol 3:173–184

Eugster HP (1969) Inorganic bedded cherts from the Magadi Area, Kenya. Contr Miner Petrol 22:1–31

Flach KW, Nettleton WD, Nelson RE (1974) The micromorphology of silica cemented soil horizons in western North America. In: Rutherford GK (ed) Soil microscopy. Limestone Press, Kingston, Ontario, pp 715–729

Folk RL, Pittman JS (1971) Length-slow chalcedony: a new testament for vanished evaporites. J Sediment Petrol 41:1045–1058

Fripiat JJ (1971) Interaction eau-argile. Bull Gr Fr Argiles 23:1–8

Fripiat JJ, Chaussidon J, Touillaux R (1960) Study of dehydration of montmorillonite and vermiculite by infrared spectroscopy. J Phys Chem 64:1234–1241

Fritz B, Zins-Pawlas MP, Gueddari M (1987) Geochemistry of silica-rich brines from Lake Natron (Tanzania). Sci Géol Bull (Strasb) 40:97–101

Garcia-Hernandez JE (1981) Interprétation de la géochimie d'altération de la silice à basse température (25 °C). INRA, Versailles, 213 pp

Gosselet J (1888) L'Ardenne. Mém Carte Géol Fr, Paris, 881 pp

Gueddari M, Fritz B, Tardy Y (1982) Géochimie de la silice dans les milieux évaporitiques sulphato-chlorurés. Etude des saumures du Chott El Jerid en Tunisie. Sci Géol Bull (Strasb) 35:41–54

Guillou JJ, Ndiaye PM (1988) Distribution des filons de surface à quartz fibreux et pseudomorphoses siliceuses, liés aux évaporites. Cas des filons sous sebkha. CR Acad Sci Paris II, 306:141–144

Harden JW, Taylor EM, Reheis MC, McFadden LD (1991) Calcic, gypsic and siliceous soil chronosequence in arid and semiarid environments. In: Nettleton WD (ed) Occurrence, characteristics, and genesis of carbonate, gypsum, and silica accumulations in soils. Soil Sci Soc Am, Spec Publ 26:1–16

Hast N (1956) A reaction between silica and some magnesium compounds at room temperature and at 37 °C. Ark Kemi 9:343–360

Hem JD (1985) Study and interpretation of chemical characteristics of natural water, 3rd edn. US Geol Survey Water Supply Pap 2254, US Govt Printing Office, Washington, DC

Hutton JT, Twidale CR, Milnes AR, Rosser H (1972) Composition and genesis of silcretes and silcrete skins from the Beda Valley, southern Arcoona Plateau, south Australia. J Geol Soc Aust 19:31–39

Iler RK (1979) The chemistry of silica: solubility, polymerization, colloid and surface properties and biochemistry. Wiley, New York, 866 pp

Jephcott CM, Johnston JH (1950) Solubility of silica and alumina. Arch Ind Hyg Occup Med 1:323–340

Kalkowsky E (1901) Die Verkieselung der Gesteine in der nördlichen Kalahari. Abh Naturwiss Ges Isis, Dresden:55–107

Khalaf FI (1988) Petrography and diagenesis of silcrete from Kuwait. Arabian Gulf. J Sediment Petrol 58:1014–1022

Lamplugh GW (1907) The geology of the Zambezi Basin around the Batoka Gorge (Rhodesia). Geol Soc Lond Q J 63:162–216

Lefavrais-Raymond A, Lhégu J, Renaud L, Scolari G (1965) Contribution à l'étude géologique et métallogénique du Nivernais septentrional (Région de Chitry-les-Mines, Nièvre). Bull BRGM 2:1–22

Maglione G (1976) Géochimie des évaporites et silicates néoformés en milieu continental confiné. Les dépressions interdunaires du Tchad Afrique. Trav Doc ORSTOM 50:335 pp

Maglione G, Servant M (1973) Signification des silicates de sodium et des cherts néoformés dans les variations hydrologiques et climatiques holocènes du bassin tchadien. CR Acad Sci Paris 277, D:1721–1724

Maliva RG, Siever R (1987) Mechanism and controls of silicification of fossils in limestones. J Geol 96:387–398

Marshall WL (1980) Amorphous silica solubilities. I. Behaviour in aqueous sodium nitrate solutions; 25–300 °C, 0–6 molal. Geochim Cosmochim Acta 44:907–913

Marshall WL, Warakomski JJ (1980) Amorphous silica solubilities. II. Effects of aqueous salt solutions at 25 °C. Geochim Cosmochim Acta 44:915–924

Ménillet F (1988) Les accidents siliceux des calcaires continentaux à lacustres du Tertiaire du bassin de Paris. Bull Inf Géol Bass Paris 25:4:57–70

Meyer R, Pena dos Reis RB (1985) Paleosols and alunite silcretes in continental Cenozoic of western Portugal. J Sediment Petrol 55:76–85

Milliken KL (1979) The silicified evaporite syndrome. Two aspects of silicification history of former evaporite nodules from southern Kentucky and northern Tennessee. J Sediment Petrol 49:245–256

Millot G (1960) Silice, silex, silicification et croissance des cristaux. Bull Serv Carte Géol Als Lorr (Strasb) 13:129–146

Millot G (1961) Silicification et néoformations argileuses: problème de genèse. Coll Int CNRS 105:167–173

Millot G (1964) Géologie des argiles. Masson, Paris, 499 pp

Millot G (1970) Geology of clays: weathering, sedimentology, geochemistry. Springer, Berlin Heidelberg New York, 429 pp

Millot G, Radier H, Muller-Feuga R, Defossez M, Wey R (1959) Sur la géochimie de la silice et les silicifications sahariennes. Bull Serv Carte Géol Als Lorr (Strasb) 12:3–14

Millot G, Lucas J, Wey R (1963) Researches on evolution of clay minerals and argillaceous and siliceous neoformations. In: Proc 10th Natl Conf on Clay and Clay Minerals, Austin, Texas, 14–18 Oct 1961. Clays Clay Miner Monogr 12:399–412

Millot G, Bocquier G, Paquet H (1976) Géochimie et paysages tropicaux. Recherche 65:236–244

Milnes AR, Thiry M (1992) Silcretes. In: Martini IP, Chesworth W (eds) Weathering, soils and paleosols. Dev Earth Surf Proc 2:349–377

Milnes AR, Wright MJ, Thiry M (1991) Silica accumulations in saprolites and soils in South Australia. In: Nettleton WD (ed) Occurrence, characteristics, and genesis of carbonate, gypsum, and silica accumulations in soils. Soil Sci Soc Am, Spec Publ 26:121–149

Morris RC, Fletcher AB (1987) Increased solubility of quartz following ferrous-ferric iron reactions. Nature 330:558–561

Munier-Chalmas I (1890) Sur les formations gypseuses du Bassin de Paris. II. Sur les dépôts siliceux qui ont remplacé le gypse. CR Acad Sci Fr 110:663–666

Parron C, Nahon D, Fritz B, Paquet H, Millot G (1976) Désilicification et quartzification par altération des grès albiens du Gard: modèles géochimiques de la genèse des dalles quartzitiques et silcrètes. Sci Géol Bull (Strasb) 29:273–284

Pédro G, Delmas AB (1979) Regards actuels sur les phénomènes d'altération hydrolytique. Leur nature, leur diversité et leur place au cours de l'évolution géochimique superficielle. Cah ORSTOM, Ser Pédol 18:217–234

Peterson MNA, von der Borch CC (1965) Chert: modern inorganic deposition in a carbonate-precipitating locality. Science 149:1501–1503

Rasplus L (1982) Contribution à l'étude géologique des formations continentales détritiques tertiaires du Sud-Ouest du Bassin de Paris. Sci Géol Mém (Strasb) 66:227 pp

Rayot V (1994) Altérations du centre de l'Australie: rôle des solutions salines dans la genèse des silcrètes et des profils blanchis. ENSMP Mém Sci Terre, Paris 22:142 pp

Rayot V, Self P, Thiry M (1992) Transition of clay minerals to opal-CT during surficial silicification. In: Schmitt JM, Gall Q (eds) Mineralogical and geochemical records of paleoweathering. ENSMP Mém Sci Terre, Paris 18:47–59

Ribet I (1990) Silicifications continentales de calcaires. Observation, interprétation, modélisation. Rapport d'option, Ecole des Mines, Paris, 77 pp

Ribet I, Thiry M (1990) Quartz growth in limestone: example from water-table silicification in the Paris Basin. Chem Geol 84:316–319

Rouchy JM (1974) Etude géologique et métallogénique de la Haute-Vallée de l'Orb (Hérault). Relations socle-couverture. Problème des silicifications et des minéralisations barytiques. Bull Mus Hist Nat 3, 214:1–93

Schmitt JM (1976) Sédimentation, paléoaltération, géochimie et minéralisation en plomb de la série triasique de Zeida (Haute Moulouya, Maroc). Thèse Ecole des Mines, Paris, 97 pp

Sheppard RA, Gude AJ (1974) Chert derived from magadiite in a lacustrine deposit near Rome, Malheur County, Oregon. J Res US Geol Surv 2:625–630

Siffert B (1962) Quelques réactions de la silice en solution: la formation des argiles. Mém Serv Carte Géol Als Lorr (Strasb) 21:86 pp

Simon-Coinçon R, Milnes AR, Thiry M, Wright MJ (1996) Evolution of landscapes in northern South Australia in relation to the distribution and formation of silcretes. J Geol Soc Lond 153:467–480

Stortz M (1928) Die sekundären authigenen Kieselsaure in ihrer petrogenetischgeologischen Bedeutung. Monogr Geol Paleontol II:481 pp

Surdam RC, Eugster HP, Mariner RH (1972) Madadi-type chert in Jurassic and Eocene to Pleistocene rocks, Wyoming. Bull Geol Soc Am 83:2261–2266

Tada R, Maliva RG, Siever R (1987) A new mechanism for pressure solution in porous quartzose sandstone. Geochim Cosmochim Acta 51:2295–2301

Thiry M (1981) Sédimentation continentale et altérations associées: calcitisations, ferruginisations et silicifications. Les Argiles Plastiques du Sparnacien du Bassin de Paris. Sci Géol Mém (Strasb) 64:173 pp

Thiry M, Millot G (1987) Mineralogical forms of silica and their sequence of formation in silcretes. J Sediment Petrol 57:343–352

Thiry M, Milnes AR (1991) Pedogenic and groundwater silcretes at Stuart Creek Opal Field, South Australia. J Sediment Petrol 61:111–127

Thiry M, Milnes AR (1992) Bleachings related to continental landscapes. Role of groundwater circulation and geochemistry. Cuarta Reunion Argentina Sedimentologia, La Plata, Actas 3:201–208

Thiry M, Schmitt JM (1977) Minéralisation en plomb par évolution pédogénétiques d'une série arkosique du Trias (Zeida, Haute Moulouya, Maroc). Bull BRGM, Sect II, 2:113–133

Thiry M, Turland M (1985) Paléotoposéquences de sols ferrugineux et de cuirassements siliceux dans le Sidérolithique du nord du Massif central (bassin de Montluçon-Domérat). Géol Fr 2:175–192

Thiry M, Delaunay A, Dewolf Y, Dupuis C, Ménillet F, Pellerin J, Rasplus L (1983a) Les périodes de silicification au Cénozoïque dans le Bassin de Paris. Bull Soc Géol Fr 25:31–34

Thiry M, Schmitt JM, Trauth N, Cojean R, Turland M (1983b) Formations rouges "sidérolithiques" et silicifications sur la bordure nord du Massif Central. Rev Géol Dyn Géogr Phys 24:381–395

Thiry M, Panziera JP, Schmitt JM (1984) Silicification et désilicification des grès et des sables de Fontainebleau. Evolutions morphologiques des grès dans les sables et à l'affleurement. Bull Inf Géol Bass Paris 21, 2:23–32

Thiry M, Bertrand Ayrault M, Grisoni JC (1988) Ground-water silicification and leaching in sands: Example of the Fontainebleau Sand (Oligocene) in the Paris Basin. Geol Soc Am Bull 100:1283–1290

Thiry M, Simon-Coinçon R, Milnes AR (1991) Marcos Morfologicos de desarrollo de silcretas. In: Blanco JA, Molina E, Martin-Serrano A (eds) Alteraciones y paleoalteraciones en la morphologia del Oeste Peninsular. Inst Tecn Geominero Esp, Monogr 6:161- 183

Wang Y, Merino E (1990) Self-organizational origin of agates: banding, fibre twisting, composition, and dynamic crystallization model. Geochim Cosmochim Acta 54:1627- 1638

Weyl PK (1959) Pressure solution and the force of crystallization. A phenomenological theory. J Geophys Res 64:2001–2025

Williams LA, Parks GA, Crerar DA (1985) Silica diagenesis, II. Solubility controls. J Sediment Petrol 55:301–311

Woolnough WG (1927) Presidential address. Part I. The chemical criteria of peneplanation. Part II. The duricrust of Australia. J Proc R Soc New South Wales, Sidney 61:1–53

11 Clay Minerals, Paleoweathering, Paleolandscapes and Climatic Sequences: The Paleogene Continental Deposits in France

Marie-Madeleine Blanc-Valleron and Médard Thiry

1 Introduction

Georges Millot established the basis of the geochemistry of continental landscapes. After his thesis on clay mineral distribution in sedimentary series (Millot, 1949), he tried to understand the mechanisms producing weathered landscapes. He worked in collaboration with geologists, geographers and geochemists, attempting to reconstruct the different components of the landscape. He was interested in every type of weathering and investigated the regolith evolution on different rocks under the same climate and then across climatic zones, from the equator to temperate areas.

Georges Millot focused his interest on geodynamic aspects and geomechanical mechanisms rather than on an inventory of materials. He established the assessment of lateritic weathering by isovolumetric mass balance calculations (Millot and Bonifas 1955), focused on the dynamics of particles and ions throughout landscapes, showed how the upstream feeds the downstream and then, because of deposits forming barriers to following drifts, how the downstream invades the upstream (Bocquier et al. 1970; Millot et al. 1976). He analysed the so-called geochemical culbuto mechanism of successive dissolutions and reprecipitations, and showed how geochemistry shapes landscapes (Boulet et al. 1977; Nahon and Millot 1977; Ruellan et al. 1977).

Experience of present-day weathering is essential to understand the genesis of clay minerals on continents, and consequently to interpret the paragenesis of ancient series and reconstruct paleolandscapes, paleoclimates, and paleoclimatic successions. The contribution of the scientific work supervised by Georges Millot is critical for understanding of continental successions. In such successions, kaolinite and the different varieties of smectites and palygorskite can now be interpreted, as well as calcretes, silicifications, ferruginous and bauxitic crusts, in both geochemical and climatic terms, and in terms of landscape development.

Continental formations of the Early Tertiary in France and the neighboring countries provide various examples of clay mineral associations. Their succession in time and space illustrates the interaction between climate, topography, erosion, source rock and vegetation cover, and so allows us to apply some of Georges Millot's lessons.

2 Landscapes and Regoliths at the End of the Cretaceous

The major regression at the end of the Cretaceous left behind a wide peneplain made of sands, limestones and chalks. Notching of these deposits followed lowering of the base level and promoted their weathering. In France, a widespread regolith with extensive karstic phenomena developed under warm and humid climatic conditions.

At the border of the Aquitaine Basin, in the Périgord–Quercy area, karstic sinkhole fillings testify that karst topography developed as deep as 500 m below the Tertiary paleosurface. Some paleosols reach thicknesses of as much as 100 m (Astruc 1988). Later, an erosional surface intersected the karstic series. Thus, one can find from the Aquitaine Basin to the Massif Central: (1) superficial carbonate-free horizons of the paleokarst with many paleosol features; (2) some sinkholes filled with collapsed sediments, partly weathered; and (3) a deep zone, sometimes plugged by slightly weathered sands, where phosphate accumulations took place (Simon-Coinçon and Astruc 1991).

A thick regolith developed also in the eastern part of the Paris Basin where weathering affected the sandy-clayey formations of Early Cretaceous, Liassic and Triassic ages, while elsewhere the flint-bearing Cretaceous Chalk was weathered and dissolved to over 200 m deep (Klein 1970). This resulted in a thick regolith of flint-bearing clays blanketing the whole area. The upper horizons of the profiles were kaolinitic, whereas the lower consisted of smectite and kaolinite/smectite mixed layers (Thiry et al. 1977). At the base, down to the "Chalk" the flint layers were deeply altered and coated by smectites without layer disturbances (Fig. 1).

This paleoweathering profile of the Cretaceous Chalk shows a noteworthy geochemical system. The upper horizons are carbonate free, the environment being acidic with removal of the Fe oxides, flints and clay minerals. The lower horizons are calcitic: at the contact with the Chalk, downward moving solutions were rapidly saturated with respect to carbonates. In contrast to the upper horizons. Si dissolution is slower, although flints are still weathered at depth and release Si. Combined with Al and transported towards these horizons, Si induces neoformation of smectite.

Comparison of the paleosols and paleolandscapes of Western Europe, which formed after the Cretaceous regression and before the beginning of the Tertiary, with the present landscape is difficult. The present continental configuration does not provide an example of such a large carbonate platform subjected to humid tropical climates. The kaolinite/smectite mixed-layer minerals are typical of these regolith weatherings developed in a leached environment buffered by omnipresent carbonates. The associated kaolinites are always disordered.

The Hercynian crystalline basement was also coated by thick kaolinitic paleosols formed during Cretaceous time. Kaolinite was already reworked into Late Cretaceous transgressive sands which form the Upper Cretaceous transgressive units (Louail 1984; Platel 1989) and constitute the first Tertiary deposits on the Massif Central (Lapparent 1930; Simon-Coinçon 1989). Residual bauxitic duricrusts corresponding to the bauxite deposits interbedded in Cretaceous formations of

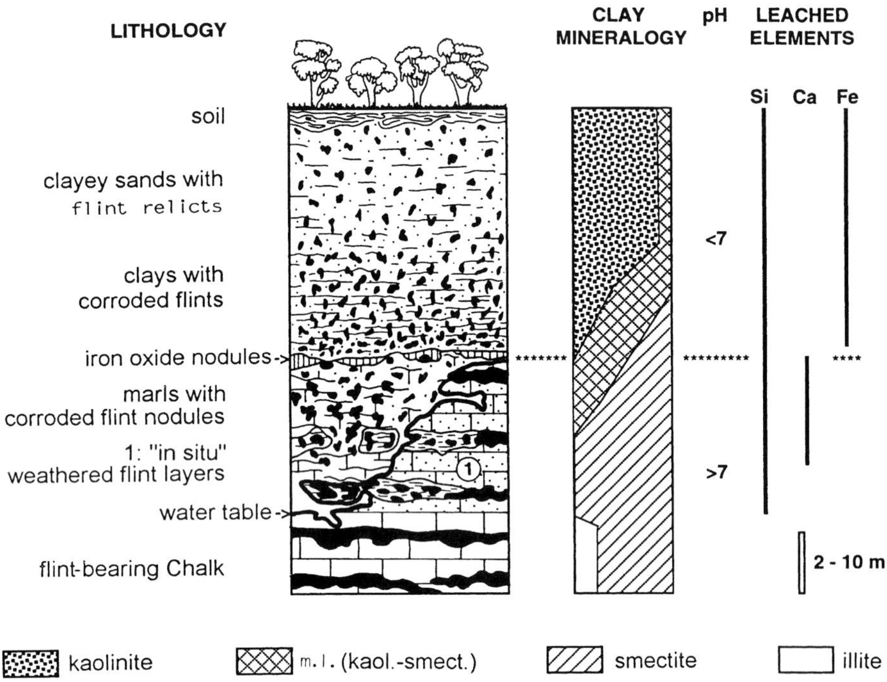

Fig. 1. Sketch of the paleoweathering profiles developed in the Paris Basin on flint-bearing "Chalk" during the Late Cretaceous. The weathering induced the development of a thick mantle with corroded flint. Mineralogy is characterized by the abundance of kaolinite/smectite mixed layers. Kaolinite is linked to the upper sandy horizon, whereas smectite is related to the deeper part of the profile where the flint layers undergo dissolution in a still carbonate-rich environment

southern France and Spain (Guendon et al. 1983; Combes 1990), were also preserved regionally. Kaolinite related to such basement weathering is always better ordered than kaolinite related to weathering of the carbonate platforms.

At the beginning of the Tertiary, the clay minerals were varied in the landscapes of France and neighboring countries. On the one hand, wide plains with a limestone substratum were covered by a thick weathering blanket characterized by disordered kaolinite and kaolinite/smectite mixed layers. On the other hand, plateaus corresponding to the weathered crystalline basements were covered by kaolinitic paleosols, *pro parte* inherited from former periods, mainly made up of well-ordered kaolinite. These thick alterite blankets were the source of material that fed the so-called siderolithic discharge at the beginning of the Tertiary.

3 The Siderolithic Discharge

The first Alpine tectonic movements were perceptible during the early Tertiary, and the warm and humid climate of the Cretaceous time became drier (Frakes et al.

1992). The instability of the relief, together with the climatic change, induced rhexistasy (Erhart 1956): the alterites were eroded, inducing the onset of the most important detrital discharge of the whole Tertiary. This «siderolithic» discharge was not synchronous everywhere, but protracted, at least from Paleocene to Middle Eocene. It represents always the first detrital sequence in the Tertiary basins and is made up of sandy, clayey, kaolinitic and Fe-rich formations. This discharge gave the kaolinitic deposits mined for ceramics and refractories in the Hesse (Germany) and Charentes regions and in the Paris Basin (France). The "Argiles Plastiques" Formation of the Paris Basin is a good example to illustrate the mechanisms involved in the development of these deposits. It shows the interaction of the environment, the sedimentology, the inheritance of the hinterland and the paleoweatherings in the facies differentiation.

3.1 Formation of the "Argiles Plastiques"

At the southern edge of the Paris Basin, the formation of the "Argiles Plastiques" of «Sparnacian» age (Early Eocene) overlies the Cretaceous Chalk and forms the first continental deposit. Two distinct deposits can be recognized (Fig. 2): an early one in the west, consisting of mottled clays and nodular carbonates, and a later one in the East, formed of kaolinitic clays and sands (Thiry 1981).

3.1.1 The Mottled Clays

In the western part of the basin, the Formation of the "Argiles Plastiques" consists of an oxidized facies with yellow and red mottling. It contains several pedogenic calcrete horizons and is composed of aluminous smectites (montmorillonite-beidellite) and kaolinite/smectite mixed layers with traces of kaolinite.

These mottled clays were deposited in a flood plain between two ridges. Weathering developed between flooding periods and resulted in the formation of calcretes and more or less hydromorphic vertisols, pseudogley and gley soils. During repeated weathering, the detrital material was progressively altered and the clays progressively purified by geochemical processes. This flood plain extended NW towards southern Britain where the Reading Beds Formation displays similar paleosol features (Buurman 1980).

Such an environment can be compared to the flood plain of the southern part of the Lake Chad Basin, where sedimentary and pedologic processes interfere with each other, developing smectite-rich vertisols with carbonate nodules (Bocquier 1972; Gac 1980). The paleoclimate was also comparable to that of the southern Lake Chad area with 800 mm rainfall during 5–6 months during the year and a mean temperature of 28 °C.

3.1.2 The Kaolinitic Clays

In the eastern part of the basin, the Argiles Plastiques were deposited in euxinic environments. The formation contains finely dispersed organic matter with lignite

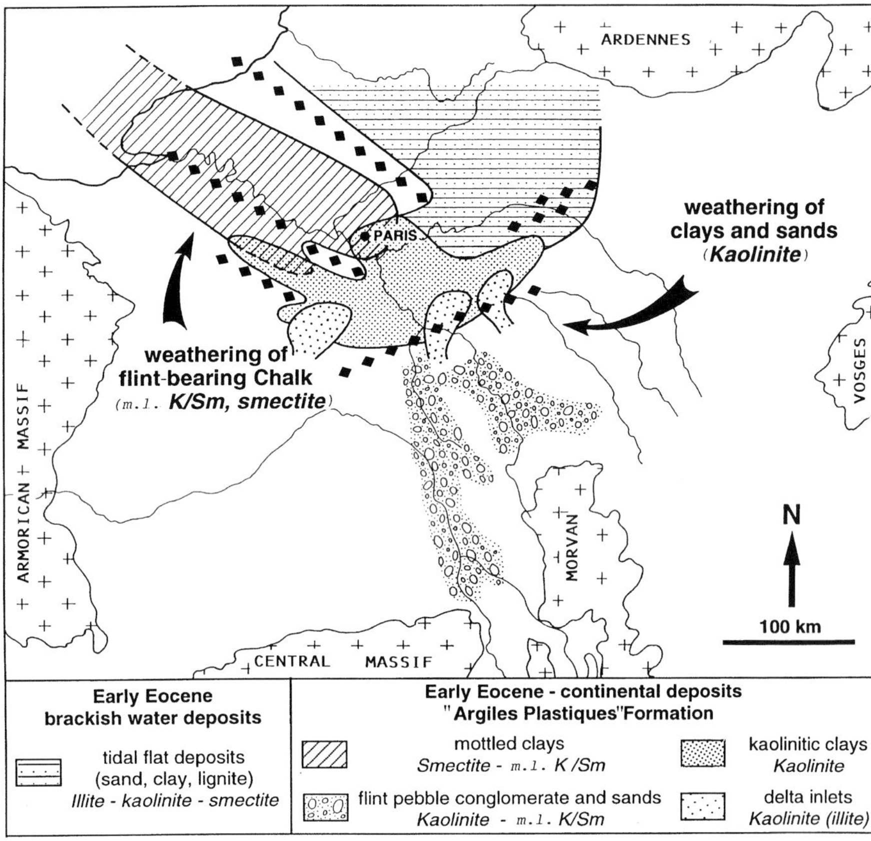

Fig. 2. Early Eocene paleogeography of the Paris Basin after the "siderolithic" detrital discharge. Two distinct environments can be recognized: mottled clays with many paleosol horizons accumulated in an alluvial plain; kaolinitic clays settled in a lacustrine environment at the front of prograding delta inlets. The deposits areas are framed by anticlinal axis of Hercynian and Variscian trends

lenses. Above a basal colluvium of reworked chalk, the Argiles Plastiques Formation is made up of a quartz-free kaolinite unit containing only small amounts of pyrite and dispersed organic matter. It is overlain by lenses of silty kaolinitic clays evolving to more or less clayey sands with interbedded lignite lenses towards the top.

Kaolinite and traces of illite mica form the main part of the deposit, feldspars being absent. Small amounts of smectite and kaolinite/smectite mixed layers are restricted to the colluvium at the basis of the series. Traces of illite mica appear in the sandy layers towards the top of the formation. The kaolinite crystallinity is low in the basal unit of pure clay and increases in the upper sands. Regionally, kaolinite crystallinity decreases from the border towards the center of the basin (Thiry 1982).

The euxinic character of the deposits indicates sedimentation under a permanent and widespread body of water. The bedding of the formation and the granulometric distribution of kaolinite indicate that the basin filled on a deltaic front with progressive sorting of the minerals. Sedimentation rates of the kaolinite particles are related to their morphology and surface properties. The kaolinite of better crystallinity consists of larger crystals settling first near the border of the basin; the most disordered being finer grained and deposited farther into the basin.

3.1.3 Origin of the Clays

The distribution of smectite and kaolinite in the basin is related to erosion of the paleosols developed on various parent rocks around the basin, under hot and warm climate conditions at the end of the Cretaceous. In the western area where the hinterland corresponds to flint-bearing chalk, the weathering environment was always buffered by the occurrence of Ca carbonate and Si enriched by the flint weathering. It, therefore, favored precipitation of smectite and kaolinite/smectite mixed layers. In the eastern part of the basin, argillaceous and sandy sediments of Lower Cretaceous, Liassic and Triassic age were extensively exposed, providing more acidic weathering environments in which kaolinite developed.

Clay minerals did not evolve much during sedimentation. In the western area, weathering developed between successive flooding periods, favoring smectite and calcrete development, but the bulk composition of the clay mineral assemblage did not change. In the East, inherited kaolinites were not modified in the lacustrine environment where they were deposited continuously.

3.2 Extension of the «Siderolithic» Facies

In the Aquitaine Basin, the detrital Eocene deposits are thick and well preserved. Towards the Massif Central, the deposits are thinner than 10–20 m and are generally discontinuous. They are often restricted to paleokarstic sinkholes, are usually oxidized, and show well-preserved weathering structures (Archanjo 1982; Trauth et al. 1985; Simon-Coinçon and Astruc 1991). In the whole basin (Charentes), the deposits are more than 100 m thick and comprise several upward coarsening negative deltaic sequences. The dominant mineral components are quartz and kaolinite of variable crystallinity (Kulbicki 1956; Dubreuilh et al. 1984; Dubreuilh 1989). Two different types of kaolinite deposits are present: well-ordered kaolinite with some muscovite (5–10%) and traces of feldspars, and disordered kaolinite, similar to those of the "Argiles Plastiques" Formation of the Paris Basin, containing anatase (1–3%) but devoid of micas and feldspars. These kaolinites correspond to distinct sedimentary episodes and reflect a change in the source areas (Dubreuilh et al. 1984). The first deposit with ordered kaolinite and mica was supplied by reworking of paleoweathered profiles developed on the granitic and gneissic basement of the Massif Central. The second episode with disordered kaolinite and anatase appears to have been fed from paleoprofiles set on sedimentary rocks, especially Jurassic and Cretaceous limestones from the northern Aquitaine platform.

The siderolithic facies exists at the base of most Tertiary basins in France. In the Provence, several basins which were initiated by early Alpine movements, show kaolinite-rich red sandy-clayey series (Triat and Trauth 1972; Roulin 1985). In the Massif Central, red formations were found in Limagnes (Lapparent 1930; Deschamps 1973; Larqué 1981) and residual deposits and paleoweathering profiles rest on the basement (Krauth and Vatan 1938; Simon-Coinçon 1989). Similar formations are known in Brittany (Milon 1930, Estéoule-Choux 1983). These paleoweathering profiles which developed on the basement, fed the detrital discharge in Languedoc (Freytet 1971), Vendée and Poitou (Steinberg 1969). In the Jura mountains, the "siderolithic" formations are known to exist as paleokarst fillings (Fleury 1909; Vernet 1969). They allowed active mining of Fe ores (Bohnerz) and associated products such as refractory clays (kaolinite) and white sands used for Fe casting.

The "siderolithic" facies is widespread in western Europe. In Spain, deep kaolinitic regolith occurs on the basement of the Central Hercynian Range and feeds the detrital discharge at the base of the Tertiary basins, here of Paleocene age (Blanco et al. 1982; Blanco and Cantano 1983; Martin-Serrano 1991). In Germany, the Bohnerz facies is present at the borders of the Rhinegraben (Schalch 1922; Wittmann 1955) and on the wide Jurassic limestone platform of Bavaria (Körber and Zech 1984; Borger 1992). Further north in the Hesse Graben, an important thickness of kaolinite accumulated (Bühmann 1974) associated with lignite seams (Steckman 1952; Brosius and Gramann 1958).

3.3 Paleolandscapes

Soil erosion and correlative sedimentary discharge of the Lower Eocene shaped a wide glacis covered with alluvium. This pediment intersected Cretaceous and Jurassic formations around the Tertiary basins and it even extended onto the crystalline basement. In the basinal areas, sediments were deposited both in aggraded flood plains where pedogenic and geochemical processes transformed inherited clay minerals and in lacustrine or lagoonal environments where mechanical processes were dominant and inherited clay minerals underwent little or no alteration.

The lowlands formed humid landscapes clogged by clay accumulation, where marshes, gallery forests and mangrove coast lines developed (Gruas-Cavagneto 1968; Gruas-Cavagneto et al. 1980; Roche 1988). Warm climates favored high organic production in these humid areas. Lignites associated almost systematically with the kaolinite deposits of the Lower Eocene in the Paris Basin, in Charentes, and in the Hesse Graben (where they are still mined), developed in these humid landscapes. Pollen characteristics of drier episodes are also present, they probably reflect best the climatic conditions prevailing during that period (Gruas-Cavagneto 1968). The glacis which is made by the "siderolithic" discharge and is covered by a mantle of kaolinitic sand and clays, underwent different weathering processes and transformations under drier climates during the Middle and the Upper Eocene.

4 The Indurated Landscapes

The evolution towards a drier climate occurred during the Middle Eocene time (Schuler 1990). The wide piedmonts built up around sedimentary basins by the "siderolithic" discharge, underwent a long continental evolution, resulting in the formation of calcretes and silcretes in the landscapes. In lowland areas, lakes with carbonate sedimentation were widely outspread.

4.1 Pedogenetic Silcretes

The upper part of the kaolinitic formations is often capped by silcretes. Their micro-morphological organization and the successive silica dissolutions and precipitations observed in these profiles, indicate a pedogenetic origin (Thiry 1981). These silcretes form a true siliceous induration, more or less continuous, which reinforces the top of the siderolithic detrital formations.

In the Paris Basin, the silcretes cap the kaolinitic "Argiles Plastiques" and move upwards to the borders of the peripheral crystalline massifs (Fig. 3). They are differentiated on a regional scale. At the periphery of the glacis, they are discontinuous and superimposed on ferrallitic paleoweathering profiles. Towards the basin, the drainage axis and the lowlands are preferentially affected. Near the basin, they form an almost continuous induration (Thiry et al. 1983; Thiry and Turland 1985). Pedogenetic silcretes are known in the same lithostratigraphic position in the Aquitaine Basin (Kulbicki 1956; Archanjo 1982), in Brittany (Estéoule-Choux 1967), in Provence (Roulin 1985), in Spain (Blanco and Cantano 1983; Blanco 1991) and as far north as the Rheinish Massif (Lange 1912; Teichmuller 1958).

From a geochemical point of view, the development of these silicifications shows clearly the shift from leached environments to more restricted ones. The mineral sequence in silcretes, with quartz at the top of the profile and opal at the bottom, corresponds to a progressive concentration of the percolating solutions. The successive precipitations and dissolutions of silica in the profiles clearly suggest alternate dry and humid periods. However, the climate was not the only factor at work. The glacis shaped by the siderolithic discharge favored silcrete by setting up an open environment. On the glacis, slow water outflow allowed export of the

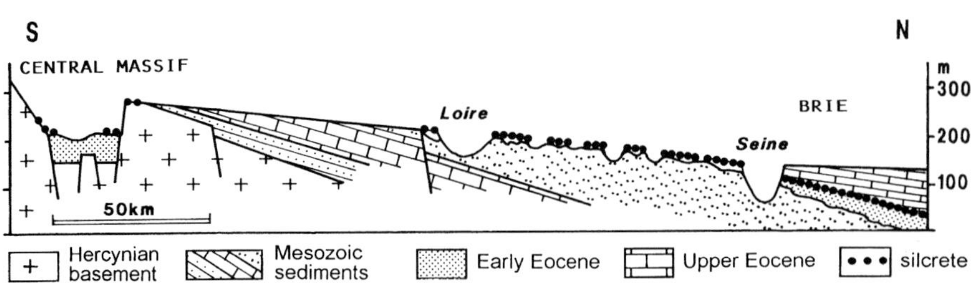

Fig. 3. Schematic section across the southern border of the Paris Basin. Pedogenetic silcretes are sitting on the Eocene palaeosurface. This was covered by a veneer of Lower Eocene flint-pebble conglomerates, sands and kaolinitic clays

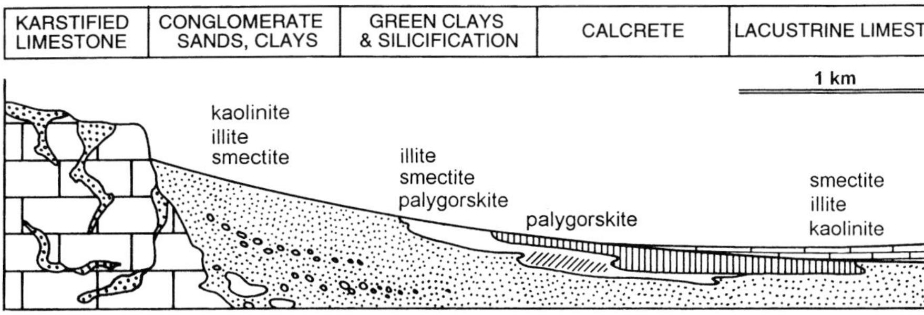

Fig. 4. Schematic section across the Eocene formations in the basins of southeastern France. Calcretes with palygorskite developed first in the detrital formations surrounding the lakes and then progressively invaded the whole alluvium pediment

most soluble cations and fixed the moderately soluble cations, such as Si in the landscape. The leached character of the "siderolithic" host material is a further condition for silcrete development. If alkaline or alkaline-earth cations, especially Mg, are present, Si combines with them to form clay minerals. That combination happened in the lowest parts of the landscape, together with the development of calcretes.

4.2 Calcretes

Calcretes are especially well developed in southeastern France. Three factors favored their expansion: a tectonic framework individualizing basins and creating contrasted morphologies, a hinterland mostly made up of limestone and a relatively dry climate.

The typical series begin with the deposition of clastics with inherited clay minerals, mainly kaolinite, smectite and illite. They are overlain by green clays which are more or less silicified, and contain illite and/or smectite and/or palygorskite. These sandy and clayey formations are overlaid by a Middle Eocene limestone where palygorskite can become the dominant clay mineral (Fig. 4; Triat and Trauth 1972; Trauth 1977; Truc 1975; Valleron 1981; Valleron et al. 1983; Valleron-Blanc et al. 1985; Roulin 1985). The structures of the first limestone layers with nodules, columns, «floating» quartz grains, glaebules, etc., indicate calcrete horizons. On the other hand, the lack of vadose features and the transition to palustrine or lacustrine facies indicate that calcitization developed in a groundwater environment. Shortly after calcitization, raising of the water table led to the setting of lakes and to the deposit of fossiliferous lacustrine limestone with common clay mineral assemblages, devoid of palygorskite. The silicifications are likely to be related to several phases. An early phase corresponds to a progressive alteration of the green clays into opalite beneath the calcrete unit. A later phase induced silicification of the calcrete facies.

The transition from green clay to calcrete shows clearly a transformation from smectite or illite into palygorskite. This transformation is shown by the growing of palygorskite laths on smectite plates (Trauth 1977). Beyond this modification, chemical analysis of the clay fraction suggests following relationship:

$$\text{Al-Fe beidellite} \rightarrow \text{Al-Mg smectite} \rightarrow \text{palygorskite.}$$

In addition, chemical analyses of single particles show that the octahedral chemical composition of smectites and palygorskite overlap significantly (Paquet et al. 1987).

From a genetic point of view, this transformation is thought to be an alteration of smectite to palygorskite or crystallization of palygorskite from smectite. Palygorskite is linked to the calcrete facies, whereas smectites are exclusively linked to primary clastic sediments. The opalites developed by leaching of the octahedral cations of the clay minerals. They appear as end members of a sequence of increasingly siliceous minerals:

$$\text{beidellite} \rightarrow \text{palygorskite} \rightarrow \text{opal.}$$

A similar succession of silcrete, calcrete and lacustrine limestone is known in the Aquitaine Basin (Trauth et al. 1985). There, calcrete developed in superimposed layers within the "siderolithic" facies, trapped on the edge of paleoreliefs and in paleokarst sinkholes (Fig. 5). Ferruginous paleosols developed and were silicified in a stable landscape during a period of tectonic quietness. Calcrete development came with a resumption of detrital input, heralding an interruption of the drainage network and the development of lakes. Calcitization was accompanied by the alteration of the inherited kaolinite into smectite and palygorskite (Archanjo 1982; Trauth et al. 1985).

The same kinds of facies are known in most of the French Eocene basins. In Alsace, the «perlitique» facies of the Bouxwiller limestone described by J. de Lapparent (1934) is related to similar calcrete facies with successive phases of calcitiza-

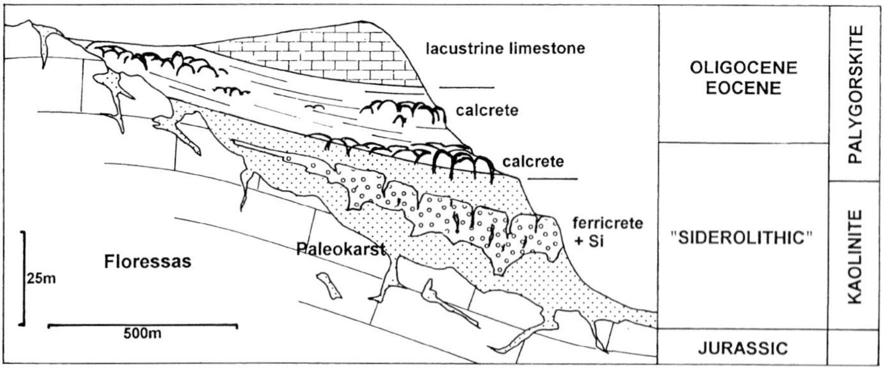

Fig. 5. Schematic section across the Eocene formations deposited above the karstified limestones of the northern Aquitaine platform. Development of calcretes with palygorskite in the kaolinite pediment followed development of silicified palaeosols and heralds restricted lacustrine areas

tion and palygorskite formation. In Touraine, calcitization at the top of the "siderolithic" clastic discharge accompanies petrographic and mineralogical modifications, especially in the clay fraction, including disappearance of kaolinite, increase of illite content and appearance of sepiolite towards the top of the limestone formation. (Saugrin 1982; Valleron et al. 1983). In the Paris Basin, calcretes with palygorskite invaded obliquely the clayey sands at the top of the kaolinitic "Argiles Plastiques" Formation, and almost completely «digested» the silcrete indurations (Thiry 1981, 1989; Thiry and Simon-Coinçon 1996).

4.3 Encrusted Paleolandscapes

The often observed succession of calcitization following silicification corresponds less to a temporal succession of two types of weathering than to their relative positions in the paleolandscapes. Silcretes correspond to relative accumulations of Si, with loss of all the other elements. They developed on glacis and piedmonts, in fairly open environments, from which the more soluble elements were discharged downstream. Calcretes which correspond to absolute accumulations, formed in the low areas of the landscape, at the border of lacustrine basins (Fig. 6). As lowlands were aggraded by sedimentary deposits and became flattened, calcretes went up the glacis and progressively invaded the silcretes.

Palygorskite development requires high Si and Mg concentrations and high pH values. Under specific geomorphologic circumstances, these conditions can be obtained in soils or sediments by rising groundwater subjected to severe evaporation. Studies conducted on present weathering profiles suggest semi-arid or periodically dry climatic conditions with rainfall of less than 400 mm/year (Paquet 1983; Singer 1984).

The frequent occurrence of calcretes with palygorskite in Middle Eocene formations can be interpreted as a stratigraphic marker, but this has not been proven yet. In fact, palygorskite-bearing limestones are diachronous. Palygorskite in marls and carbonates is reported in Provence–Languedoc as early as the Lower Eocene, maybe even in the Upper Cretaceous (Sittler 1965; Freytet and Plaziat 1982). Some of these limestones are obviously calcretes (Durand 1984). In Spain, palygorskite-rich paleoweathering profiles are believed to be of Paleocene age (Fernandez Macarro and Blanco 1991). Existence of a climate with a pronounced dry season allowed its formation as early as the end of the Cretaceous. The first palygorskite-bearing formations occur within basins where Pyreneo–Provencal tectonics had already begun. Their development is chiefly controlled by tectonics that disrupted the drainage network and induced the formation of restricted areas or basins.

5 Pre-Evaporitic Clay Minerals in Restricted Basins

Pre-evaporitic deposits developed in the Paris Basin and tectonic troughs, as early as Middle Eocene. They are precursors of the evaporite deposits that spread out during the Upper Eocene and Oligocene, on the platforms and in the grabens.

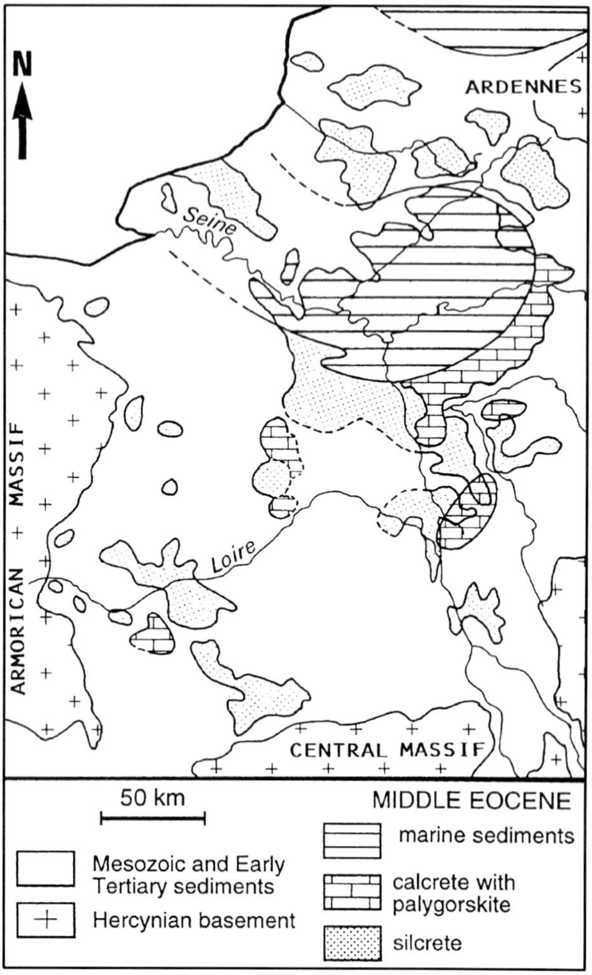

Fig. 6. Distribution of silcretes and calcretes in the Middle Eocene of the Paris Basin. Notice the development of silcretes around the southern border of the basin, up to the Hercynian basement. Calcretes developed extensively at the edge of the sedimentary basin and in local low areas

5.1 In Southeastern France

In southeastern France, the already mentioned Middle Eocene paleosols and duricrusts are topped by various facies among which are many relicts of chemical deposits. In the Mormoiron Basin, for instance, argillaceous deposits with Al–Fe smectites are overlain by a gypsum body flanked by two dolomitic episodes. As dolomites and sulfates were deposited, clay minerals became progressively enriched in Mg (Al–Mg smectites, sepiolite and stevensite; Triat and Trauth 1972; Trauth 1977). The chemical character of the clay mineral assemblages is especially well marked in basins with low accumulation rates, like in the Mormoiron Basin. On the other hand, in the grabens with thick sulfate and chloride deposits (as in the Camargue, Valence, the Bresse, etc.), the clay minerals are inherited and underwent no, or only few transformations during deposition (Moretto 1987; Dumas 1988).

That can be explained by the high burial rate of the sediments, which allowed preservation of the most soluble facies, but did not provide enough time to develop aggradation, or even neoformation, of the clay minerals.

5.2 In the Paris Basin

In the Paris Basin, pre-evaporitic and evaporitic deposits are especially well developed in the Middle Eocene and Oligocene. The formation of the «Masses du Gypse» comprising four gypsum layers interbedded with marls, reaches up to 35 m in thickness. It was deposited in the center of the Paris Basin during the «Marinesian» (late Middle Eocene) and the «Ludian» (Upper Eocene). During the Marinesian, fibrous clay minerals (i.e., sepiolite and palygorskite) developed in several clayey and marly formations overlying «Auversian» marine sands (Sautereau and Decarreau 1973; Decarreau et al. 1975). The lower gypsum bed ("Quatrième Masse du Gypse") rests on these marls and clays. They display clay mineral parageneses which develop into successive zones on an east-west basin axis where marine influences are noticeable (Fig. 7). Illite prevails in the central zone. Illite and palygorskite make up the clay mineral paragenesis of the intermediate zone. Illite, palygorskite, smectites and sepiolite are present in the peripheral zone.

Aggradation and transformation processes determine the geochemistry of the clayey sediments. Most of the chemical elements came from erosion and leaching of continental superficial formations. The Marinesian paleogeography induced the

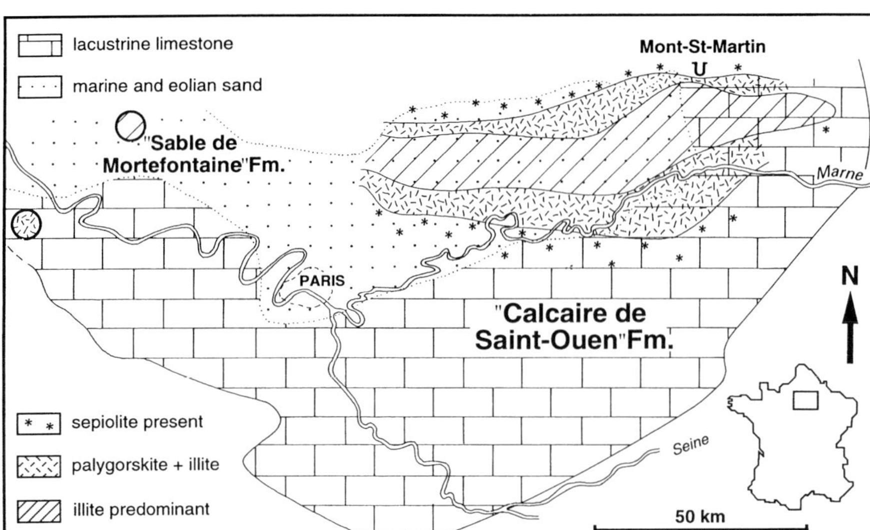

Fig. 7. Middle Eocene ("Marinesian") paleogeography of the Paris Basin. The clay parageneses are organized around a sedimentary axis. The most chemical deposits containing sepiolite, developed at the periphery of this axis where peak solution concentration is reached in local depressions protected from detrital influx

geochemical polarity. Smectite and kaolinite are part of the inherited material and are linked with coarser layers of detrital character. Illite came from aggradation of Fe-beidellite by incorporating K supplied by marine incursions. Palygorskite can result from transformation of inherited Fe-beidellite by a similar mechanism to that considered for its genesis within calcretes. Sepiolite is sporadically present in the more peripheral areas. It is authigenic by precipitation from solutions.

The study of a section of the peripheral area (Mont-Saint-Martin, Fig. 8) allows precise determination of the paleolandscapes and mechanisms. The substratum consists of "Auversian" eolian-sand dunes deposited soon after withdrawal of the sea. The "Marinesian" deposits are limited to interdune depressions, the pure clay lenses being composed of sepiolite and smectite. The sandy layers include smectite, illite/smectite mixed layers and traces of illite and kaolinite. The more regular clayey units, which overlapped the eolian dunes, contain palygorskite, smectite and illite. After a long sedimentary hiatus, clay minerals radically changed with deposition of pure illite in the green clay units of Oligocene age. The highest concentration in the solutions occurred in the interdune-clay pans and ponds, into which groundwater was discharging. Sepiolite formed in these areas, which were sheltered from detrital influxes under strong evaporation.

During the "Ludian" period, thick gypsum units were deposited in the center of the basin with magnesian marls and lacustrine limestones in the peripheral areas (Hébert 1860). The clay minerals formed three main mineral suites (Fig. 9; Trauth 1977): (1) palygorskite and Al–Mg smectites are the most common clays; (2) Al–Fe smectites are present on the southern margin of the basin; and (3) sepiolite and Mg smectites formed on the southwestern margin of the gypsum deposit. The distribution of these clay-mineral suites does not fit with the paleogeographic zones. There is an inconsistency between the bulk-rock mineralogy and the clay-mineral suite. Mg clays are not always associated with dolomite: sepiolite is frequently associated with calcite-rich units, whereas dolomite units usually contain Al–Mg clay minerals (smectites and palygorskite).

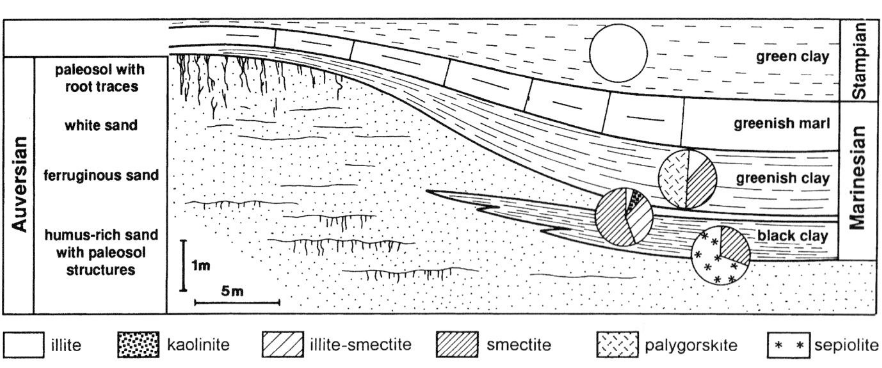

Fig. 8. Section of the "Marinesian" clay formations above the "Auversian" eolian sand (top of the Middle Eocene) at Mont-Saint-Martin (Marne). The Mg-clay minerals (smectite and sepiolite) developed in the interdunal depressions

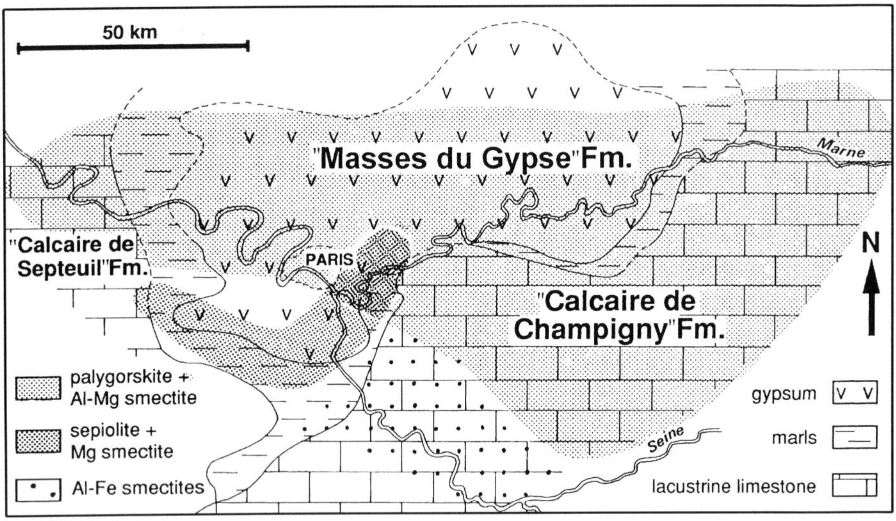

Fig. 9. Late Eocene ("Ludian") paleogeography of the Paris Basin. The distribution of the clay paragenesis matches the paleogeographic zones (after Trauth 1977). The sepiolite-rich zone can be related to the influx of Si-rich waters deriving from flint-bearing chalk that cropped out at the southwestern margin of the basin

Clay minerals are in competition with carbonates for incorporation of Mg, with Si controlling the process. When Si is absent, Mg is incorporated in carbonates and dolomite is formed. When Si is present, sepiolite and Mg smectites develop and scavange Mg: calcite precipitates together with Mg silicates. The distribution of sepiolite and Mg smectite along the southern margin of the basin probably has to be related to Si-rich water inflows, originating from flint-bearing chalk which formed the local bedrock.

5.3 Geochemical Mechanisms

The geochemistry of the solutions can be assessed from simulating evaporitic systems. Two different geochemical trends exist, depending on the oxic or anoxic character of the environment (Maglione 1976; Droubi et al. 1976; Gueddari et al. 1982; Fritz et al. 1987). In a reducing environment, sulfides form and brines are increasingly alkaline as they become concentrated, their Si content increasing with concentration. In an oxidized environment, brines become sulfate-rich and are of acidic or neutral pH. In these sulfate-rich brines, silica solubility decreases with concentration (Marshall and Warakomski 1980), which is one key to the development of Mg-rich clay minerals. In a sulfate-rich environment, silica solubility is low, allowing only transformation of inherited clay minerals (development of Al–Mg smectites and palygorskite). In an alkaline environment, solubility of silica increases with concentration and clay minerals do not develop while Mg is incorporated by carbonates. Alkaline waters, as well as fresh waters, transport Si. Mixing of such solutions with sulfate-rich solutions can induce high Si over-saturation. Around lagoons with sulfate-rich brines, continental fresh waters release the Si, which in turn scavanges Mg: sepiolite and Mg smectite precipitate together with calcite.

5.4 Paleolandscapes and Mechanisms

Pre-evaporitic deposits develop in very flat paleolandscapes corresponding to margino-littoral environments (paralic realm, Guelorget and Perthuisot 1983) with very shallow lakes or lagoons which were often dry. Surficial water flows and runoffs were limited and most of the water came from discharge of ground or marine waters in depressed areas, as in present systems (Logan 1987; Jacobson et al. 1988). However, the maximum of solution concentration did not occur in the largest bodies of water. Solution concentration was the most intense in the small restricted peripheral depressions (interdune areas, insolated channels, etc.), and in the capillary fringe directly above the emerging water tables. These environments can be compared to those described by Gac (1980) and Maglione (1976) in the interdune depressions to the north of Lake Chad where concentration of the brines by evaporation occurs until Na carbonate precipitates.

Silcretes and calcretes, previously developed around the sedimentary basins, left a deep imprint on the landscapes. They were later preserved from mechanical erosion by siliceous and carbonaceous indurations which protected the underlying sandy and clayey formations. Thus, chemical deposits developed without significant detrital influx. Landscapes can be compared with the present-day wide "Hamada" limestone plateaus in northern Africa and with the silcrete capped plateaus of inland Australia. Dunes may probably have existed, but they are only identified in some formations of the Paris Basin.

There is no evidence of paleoweathering correlative to evaporitic deposits. Weathering must have been very restricted under the dry climate of the Upper Eocene, corresponding mainly to sulfate accumulations, as scattered crystals in the soils or as true gypcretes, similar to present-day ones, in areas where precipitation is less than 250 mm/year (Jessup 1951; Pouget 1968; Wells et al. 1987; Eswaran and Gong Zi-Tong 1991). These surficial and thin accumulations could have been swept away during later marine transgressions, or dissolved under later more humid climates. Large amounts of sulfate, and possibly of chloride, have probably been stocked in the soils around the endoreic basins, up to the peripheral crystalline massifs. Dissolution and reworking of this pedogenic gypsum during the following more humid period might have, at least *pro parte*, contributed to the accumulation of the thick gypsum deposits in depressions.

6 Geochemical Sequence of Climatic Origin

Goldschmidt (1937, 1945) was the first to propose a geochemical sequence of sedimentary deposits from insolubles (sands and sandstones with their suite of heavy minerals) to evaporation residues. As a geochemist, he underlined the distribution of the trace elements in the different components of the sequence. Millot (1957, 1964) discussed it from a geological point of view, determining that the main factors were tectonism and climate.

The continental deposits of the French Paleogene form a noteworthy example of such a geochemical sequence. If we exclude marine sediments interbedded during transgressions and consider only the continental series, the evolution of continental environments can be deduced from mineralogical suites. The clay mineral parageneses of the Paleogene sediments in France illustrate well the diversity of the minerals developing in continental environments. Tectonism and climate are the driving forces of the system to which they provide energy. The geological framework and the geomorphologies are the controlling factors of the regolith and the sedimentary deposits because they provide the detrital material and the solutions.

The main stages of this sequence are controlled first by the climate. This latter has an influence on the vegetation cover, the nature of the materials formed, the erosion and the relief of the landscape. The original nature of the materials, the water chemistry and the relief are inter-dependent elements: they are part of the landscape geochemistry (Millot et al. 1976). The successive landscapes and environments result from geochemical differentiation. Clay minerals are in equilibrium with these environments and are often indicators of them.

6.1 Development of the Weathering Mantles

Only warm and humid climates are likely to induce the formation of large amounts of deeply weathered materials. Other climates generate less evolved and less typical materials, or in too small amounts to have a major impact on the detrital input of a sedimentary basin. Thick kaolinitic weathering profiles mark strongly continental series. They are the starting point of the observed evolutions and transformations.

A long period of tectonic stability (the biostasic period; Erhart 1956) is necessary for the development of a thick weathering mantle. Under warm and wet climates, the residual phase of the weathering (Al clays, Fe oxides, etc.) is stored under the protective forest cover. Soluble products are drained off the landscape. Weathering profiles deepen and persist as long as tectonic stability and humid climate continue. The nature of the bedrock imprints its mark on subsequently formed materials (ordered or disordered kaolinites, gibbsite, ...).

6.2 Landscape Disruption

Tectonic movements or climatic changes induce erosion and landscape disruption. Tectonic movements have a local impact, whereas climatic changes have a wider response and more global effects. Climate aridification leads to destruction of the forests and subsequent erosion of the weathering mantles. Peripheral basins then receive detrital products, recording rhexistasy periods (Erhart 1956). Such a bioclimatic break-up is responsible for the great "siderolithic" discharge known in Europe, Africa and America during the Paleocene to Middle Eocene period. Lowlands are filled with alluvial sediments which are progressively buried. Luxuriant vegetation develops in these humid zones and is buried by successive alluvial deposits. Lignite layers, frequently associated with siderolithic deposits, form during this phase in the humid parts of the landscapes.

Most of the clay minerals are inherited from former landscapes. High accumulation rates for these detrital deposits do not allow enough time for their transformation in the basin. Kaolinite, halloysite and aluminous smectites are characteristic records of this episode.

6.3 Landscape Transformation

Under drier climates, erosion of landscapes is limited, because rains are rare. As scarce vegetation favors sheet flood and formation of glacis, linear runoffs are reduced. Weathering conditions change radically: leaching and accumulation alternate as climatic periods or seasons occur in the same region. The landscape becomes geochemically diversified. Only the most soluble elements are exported. Depending on the landscape and on the geological framework, Si, Ca or Fe are successively dissolved and re-precipitated further downstream as seasons alternate, corresponding to what Georges Millot called «the geochemical culbuto.» There is the domain of vertisols «rising» upslope and of siliceous or calcareous duricrusts, which all promote flattening of the landscapes (Millot et al. 1976). These landscapes are very sensitive to tectonic movements. Actually, there is no pervasive erosion and any movement is able to break the drainage networks and to induce individualization of lacustrine basins.

In these environments, clay minerals undergo successive transformations. First, «grafting» of silica on the inherited kaolinitic minerals which are transformed into smectite, then progressive Mg fixation induces formation of Al–Mg smectites and finally of palygorskite.

6.4 Evaporitic Accumulations

When climates become very dry and arid, sedimentary deposits are restricted to endoreic domains and littoral lagoons. Everywhere else eolian erosion prevails when loose formations are exposed. Among the materials moved throughout the landscape by the wind, only the most soluble elements are taken away and accumulate in the superficial part of the soils. A large stock of sulfates and chlorides can, therefore, be trapped and then redissolved during more humid periods. Only after a transit in the groundwaters, will they precipitate in the basins. The chemistry of the bedrock controls the geochemistry of the groundwaters.

Away from lowlands where groundwaters flow out, clay minerals are not modified. Mg minerals develop in the depressions protected from detrital influx. Detrital Al–Fe clay minerals are transformed and aggraded when Si concentration is low. Purely Mg clay minerals, Mg smectites and sepiolite precipitate when the Si content of the groundwaters is high. These areas are the domain of the evaporitic clay minerals (Trauth 1977).

6.5 Sequence Polarity

The evolution of the climatic sequences recorded in continental successions always goes from humid to arid conditions. The reverse climatic evolution would lead to a rapid alteration and erosion of the weathering materials formed during former dry climates. This problem has to be considered when trying to do paleoclimatic reconstructions. Even without tectonic movement and erosion, the sequence is often truncated, either because the climatic evolution has not reached the most arid period, or because of recurrent more humid conditions which obliterate the imprint of former drier conditions.

7 Conclusion

The surfaces of the continents display a mosaic of extremely varied environments and almost every kind of clay mineral can be formed. Three factors, which can act in a complementary or in an opposite way, are of major importance for the understanding of the clay-mineral genesis: climate, topography and erosion. The bedrock and the vegetation cover also induce differences that can be valuable indicators in the reconstruction of landscapes.

1. Climate controls the concentration of solutions through rainfall and evaporation and the reaction rates through temperature.
2. Topography controls the run-off and hence solution concentrations. High relief limits the residence time of solutions and prevents increase in the concentration. Conversely, flat landscapes favor concentration of solutions through water stagnation.
3. Erosion controls the detrital influxes to basins and thus regulates the residence time of particles at the water-sediment interface. If detrital input is high, particles are rapidly buried and the depositional environment has only limited influence on them. On the other hand, if influx is restricted, the residence time of particles is increased and the environment can imprint the sediments.
4. Rocks of the landscape basement determine the nature of the detrital influx and the chemistry of the waters reaching the basin.
5. Vegetation cover regulates detrital influxes through bio-rhexistasy. It also interferes through organic reactions. This component of the environment is the most difficult to determine: most often organic matter is not preserved or is modified during diagenesis.

The evolution of clay parageneses in the continental Paleogene of France shows a succession of minerals starting with Al phases (kaolinite, kaolinite/smectite mixed layers and smectite), which are followed by Al–Mg minerals (smectite and palygorskite). This is a geochemical sequence beginning with leached environments, closing up progressively and finishing with a chemical precipitation of clay minerals in restricted evaporitic environments.

The sequential deposits result from complex interactions between several processes. Under somewhat similar hot climates with contrasted seasons, several types of weathering can develop. On the pediplain generated by the "siderolithic" detrital discharge, consisting only of Al and Si, clay minerals are destroyed and silicification develops. If natural dams are created (by tectonism or rising of base levels) carbonate accumulates in the "siderolithic" materials. Calcretes and vertisols develop, Mg is fixed, clay minerals are aggraded: Al–Mg smectites and palygorskite develop. Under more arid climates, «evaporitic» clay minerals appear, *pro parte* because landscapes have been formerly capped by silcretes and calcretes that preserved them from erosion and greatly reduced detrital influx. Yet, in areas where tectonism plays a major role (grabens), detrital influx dominates the clay spectrum and the climatic sequence of clay minerals is truncated.

Several factors are involved in the development of this sequence. The geochemical systems involved have self-regulation and self-development properties. If a chemical condition of the system is not expressed or, alternatively, becomes exacerbated, some components of the sequence are delayed or appear sooner during the climatic evolution. A diachronism thus arises between the sequences of different basins or within a basin, according to the paleogeographic situation.

This overview has to be considered when studying clay minerals of continental deposits. Interactions between climate, relief and geology of the landscapes are multiple: their results are recorded in the deposits. To interpret them, we need to differentiate the respective roles of these different factors.

Acknowledgements. This work has benefited from financial support of the CNRS (UA 723 «Physico-chimie des processus biosédimentaires»), of the MNHN (BQR «Evolution des climats et paléoenvironnements») and is a contribution to IGCP 317 «Paleoweathering Records and Paleosurfaces.» The authors thank personally Norbert Trauth for sharing his wide experience of paleoweatherings and evaporitic environments. Students or external researchers working with N. Trauth also contributed greatly to this study, in Strasbourg or during field trips: J. Archanjo, F. Ménillet, R. Simon-Coinçon, F. Sommer, G. Truc and G. Vilas-Boas especially. The authors also thank R. J. Gilkes and P. C. de Graciansky for critical reading and corrections which improved and strengthened the manuscript and C. Nigrini and P. Ditchfield for editing the English text.

References

Archanjo JD (1982) Le Sidérolithique du Quercy blanc (France). Altérations polyphasées sur roches sédimentaires. Essai de datation, Thèse, Univ Strasbourg, 148 pp

Astruc JG (1988) Le paléokarst quercynois au Paléogène; altérations et sédimentations associées. Doc BRGM 133, 135 pp

Astruc JG, Simon-Coinçon R (1992) Enregistrement de l'évolution climatique et tectonique par les paléokarsts (Exemple du Quercy et de ses marges). In: Salomon JN, Maire R (eds) Karst et évolutions climatiques. Presses Univ, Bordeaux, pp 497–508

Blanco JA (1991) Cuarta parada: Los procesos de silicificacion asociados al paleogeno basal del borde SW de la Cuenca de Duero: II sobre los sedimentos Paleocenos. In: Blanco JA, Molinar E, Martin-Serrano A (eds) Alteraciones y paleoalteraciones en la morphologia del Oeste Peninsular. Inst Tecn Geominero Esp, Monogr 6:239–249

Blanco JA, Cantano M (1983) Silicification contemporaine à la sédimentation dans l'unité basale du Paléogène du Bassin du Duero (Espagne). Sci Géol Mém (Strasb) 72: 718

Blanco JA, Corrochano A, Montigny R, Thuizat R (1982) Sur l'âge du début de la sédimentation dans le bassin tertiaire du Duero (Espagne). Attribution au Paléocène par datation isotopique des alunites de l'unité inférieure. CR Acad Sci Paris 295, D:259–262

Bocquier G (1972) Genèse et évolution de deux toposéquences de sols tropicaux du Tchad. Interprétation biogéodynamique. Mém ORSTOM, Paris, 62, 350 p

Bocquier G, Paquet H, Millot G (1970) Un nouveau type d'accumulation oblique dans les paysages géochimiques: l'invasion remontante de la montmorillonite. CR Acad Sci Paris 270, D:460–463

Borger H (1992) Paleotropical weathering on different rocks in Southern Germany. Z Geomorph NE, Berlin, Suppl-Band 91:95–108

Boulet R, Bocquier G, Millot G (1977) Géochimie de la surface et formes du relief. I. Déséquilibre pédoclimatique dans les couvertures pédologiques de l'Afrique tropicale de l'Ouest et son rôle dans l'aplanissement des reliefs. Sci Géol Bull (Strasb) 30:235–243

Brosius M, Gramann F (1958) Das ältere Tertiäre von Grossalmerode (Hessische Senke). Z Dtsch Geol Ges 111:543–558

Bühmann D (1974) Die Tonmineralzusammensetzung in den Sedimenten der Niederhessischen Senke als Indikator festländischer Verwitterung und brackisch-mariner Tonmineralneubildung. Diss, Univ Göttingen, 83 pp

Buurman P (1980) Paleosols in the Reading Beds (Paleocene) of Alum Bay, Isle of Wight, U.K. Sedimentology 27:593–606

Combes PJ (1990) Typologie, cadre géodynamique et genèse des bauxites françaises. Geodinamica Acta 4/2:91–109

Decarreau A, Sautereau JP, Steinberg M (1975) Genèse des minéraux argileux du Bartonien moyen du Bassin de Paris. Bull Soc Fr Minér Cristallogr 2–3:142–151

Deschamps M (1973) Etude géologique du Sidérolithique du Nord-Est, du centre du Massif central français et des terrains qui lui sont associés. Thèse Univ Paris IV, 1270 pp

Droubi A, Cheverry C, Fritz B, Tardy Y (1976) Géochimie des eaux et des sels dans les sols des polders du lac Tchad: application d'un modèle thermodynamique de simulation de l'évaporation. Chem Geol 17:165–177

Dubreuilh J (1989) Synthèse paléogéographique et structurale des dépôts fluviatiles tertiaires du nord du bassin d'Aquitaine. Passage aux formations palustres, lacustres et marines. Doc BRGM 172:461 pp

Dubreuilh P, Marchadour P, Thiry M (1984) Cadre géologique et minéralogie des argiles des Charentes, France. Clay Min 19:29–41

Dumas D (1988) Le Paléogène salifère du bassin de Valence (Sud-Est de la France): géométrie et sédimentologie des dépôts, synthèse de bassin. Thèse Univ Lyon, 293 pp

Durand JP (1984) Basse Provence: Paléocène et Eocène. In: Debrand-Passart (ed) Synthèse géologique du Sud-Est de la France. Doc BRGM 125:425–429

Erhart H (1956) La genèse des sols en tant que phénomène géologique, Masson édit, Paris, 83 pp

Estéoule-Choux J (1967) Contribution à l'étude des argiles du Massif armoricain. Argiles des altérations et argiles des bassins sédimentaires.» Thèse Univ Rennes, 319 pp
Estéoule-Choux J (1983) Kaolinitic weathering profils in Brittany. Genesis and economic importance. Geol Soc, Spec Publ 11:33–38
Eswaraen H, Gong Zi-Tong (1991) Properties, genesis, classification, and distribution of soils with gypsum. In: Nettleton WD (ed) Occurrence, characteristics, and genesis of carbonate, gypsum, and silica accumulations in soils. Soil Sci Soc Am, Spec Publ 26:89–119
Fernandez Macarro B, Blanco JA (1991) Septima parada: le depression de Talavan-Torrenjon el Rubio (caceres, Espana). In: Blanco JA, Molinar E, Martin-Serrano A (eds) Alteraciones y paleoalteraciones en la morfologia del Oeste Peninsular. Inst tecn Geominero Esp, Monogr 6:263–286
Fleury E (1909) Le Sidérolithique suisse. Contribution à la connaissance des phénomènes d'altération superficielle des sédiments. Mém Soc Fribourg Sci Nat 6:262 pp
Frakes LA, Francis JE, Syktus JJ (1992) Climate modes of the Phanerozoic: the history of the earth's climate over the past 600 million years. Cambridge Univ Press, Cambridge, 274 pp
Freytet P (1971) Les dépôts continentaux et marins du Crétacé supérieur et des couches de passage de l'Eocène en Languedoc. Bull BRGM 1, 4:1–54
Freytet P, Plaziat JC (1982) Continental carbonate sedimentation and pedogenesis – Late Cretaceous and Early Tertiary of southern France. Contrib Sediment 12:213 pp
Fritz B, Zins-Pawlas MP, Gueddari M (1987) Geochemistry of silica-rich brines from Lake Natron (Tanzania). Sci Géol Bull (Strasb) 40:97–101
Gac JY (1980) Géochimie du bassin du lac Tchad. Bilan de l'altération, de l'érosion et de la sédimentation. Trav Doc ORSTOM, Paris, 123, 251 pp
Goldschmidt VM (1937) The principles of distribution of chemical elements in minerals and rocks. J Chem Soc Lond 140:655–673
Goldschmidt VM (1945) Fondements géochimiques de la répartition des oligo-éléments. Soil Sci 60:1–7
Gruas-Cavagneto C (1968) Etude palynologique des divers gisements du Sparnacien du bassin de Paris. Mém Soc Géol Fr 110:1–144
Gruas-Cavagneto C, Laurain M, Meyer R (1980) Un sol de mangrove fossilisé dans les lignites du Soissonnais (Yprésien) à Verzenay (Marne). Géobios 13:795–801
Gueddari M, Fritz B, Tardy Y (1982) Géochimie de la silice dans les milieux évaporitiques sulfato-chlorurés. Etude des saumures du Chott El Jerid en Tunisie. Sci Géol Bull (Strasb) 35:41–54
Guelorget O, Perthuisot JP (1983) Le domaine paralique. Expressions géologiques, biologiques et économiques du confinement. Trav Lab Géol Ecole Norm Sup, Paris, 16, 136 pp
Guendon JL, Parron C, Triat JM (1983) Incidences des altérations crétacées sur la notion de sidérolithique dans le S.E. de la France. Bull Soc Geol Fr 1:41–50
Hébert E (1860) Note sur le travertin de Champigny et sur les couches entre lesquelles il est compris (coupe de Bruy-Marne). Bull Soc Géol Fr 17:800
Jacobson G, Arakel AV, Chen Y (1988) The central Australia groundwater discharge zone: evolution of associated calcrete and gypcrete deposits. Aust J Earth Sci 35:549–565
Jessup RW (1951) The soils, geology and vegetation of north-western South Australia. Trans R Soc Aust 74:92–105
Klein C (1970) La «surface de l'argile à silex.» Rev Géogr Phys Géol Dyn 12/3:185–220
Körber E, Zech W (1984) Zur Kenntnis tertiärer Verwitterungreste und Sedimente in der Oberpfalz und ihrer Ungebung. Relief, Boden, Paläoklima 3:67–150

Krauth F, Vatan A(1938) Sur l'origine des roches argileuses des environs de Confolens (Charente) attribuées au Sidérolithique. CR Acad Sci Paris 206:443–445

Kulbicki G(1956) Constitution et genèse des sédiments argileux sidérolithiques et lacustres du Nord de l'Aquitaine. Sci Terre Nancy 4:5–101

Lange O (1912) Ueber Silikatsteine für Martinöfen. Stahl Eisen 42:1731

Lapparent J de (1930) Comportement minéralogique et chimique des produits d'altération élaborés aux dépens des gneiss du Massif Central français avant l'établissement des dépôts sédimentaires de l'Oligocène. CR Acad Sci Paris 190:1062–1064

Lapparent J de (1934) D'une latérite, d'un calcaire lacustre (calcaire de Bouxwiller) et des roches pisolithiques en général. Bull Serv Carte Géol Als Lorr (Strasb) 2:99–105

Larqué P (1981) Mise au point sur les paléoaltérations rubéfiées à kaolinite des bassins tertaires de l'Est du Massif central. Existence de deux périodes d'altération rubéfiante dans le Paléogène du Velay. Sci Géol Bull (Strasb) 34:183–191

Logan BW (1987) The Mac Leod evaporite basin, western Australia. Holocene environments, sediments and geological evolution. Am Assoc Petrol Geol Mem 44:140 pp

Lombard A (1956) Géologie sédimentaire. Masson, Paris, 722 pp

Louail J (1984) La transgression crétacée au Sud du Massif armoricain. Cénomanien de l'Anjou et du Poitou, Crétacé supérieur de Vendée. Etude stratigraphique, sédimentologique et minéralogique. Mém Soc Géol Minér Bretagne 25:333 pp

Maglione G (1976) Géochimie des évaporites et des silicates néoformés en milieu continental confiné. Les dépressions interdunaires du Tchad, Afrique. Trav Doc ORSTOM, Paris, 50, 336 pp

Marshall WL, Warakomski JM (1980) Amorphous silica solubilities. II. Effect of aqueous salt solutions at 25 °C. Geochim Cosmochim Acta 44, 7:915–924

Martin-Serrano A (1991) El relieve del Macizo Hesperico y sus sedimentos asociados. In: Blanco JA, Molinar E, Martin-Serrano A (eds) Alteraciones y paleoalteraciones en la morphologia del Oeste Peninsular. Inst Tecn Geominero esp, Monogr 6:9–26

Millot G (1949) Relations entre la constitution et la genèse des roches sédimentaires argileuses. Géol Appl Prospec Min, Nancy, 2, 352 pp

Millot G (1957) Des cycles sédimentaires et de trois modes de sédimentation argileuse. CR Acad Sci Paris 244:2536–2539

Millot G (1964) Géologie des argiles. Masson édit, Paris, 499 pp

Millot G, Bonifas M (1955) Transformations isovolumétriques dans les phénomènes de latéritisation et de bauxitisation. Bull Serv Carte Géol Als Lorr (Strasb) 8:3–20

Millot G, Bocquier G, Paquet H (1976) Géochimie des paysages tropicaux. Recherche 7/65:236–244

Milon Y (1930) L'extension des formations sidérolithiques éocènes dans le centre de la Bretagne. CR Acad Sci Paris 194:1360–1361

Moretto R (1987) Etude sédimentologique et géochimique des dépôts de la série salifère paléogène du bassin de Bourg-en-Bresse (France). Mém Sci Terre, Nancy, 50, 252 pp

Nahon D, Millot G (1977) Géochimie de la surface et formes de relief. V. Enfoncement géochimique des cuirasses ferrugineuses par épigénie du manteau d'altération des roches mères gréseuses. Influence sur le paysage. Sci Géol Bull (Strasb) 30:275–282

Paquet H (1983) Stability, instability and significance of attapulgite in calcretes of mediterranean and tropical areas with marked dry season. Sci Géol Mém (Strasb) 72:131–140

Paquet H, Duplay J, Valleron-Blanc MM, Millot G (1987) Octahedral compositions of individual particles in smectite-palygorskite and smectite-sepiolite assemblages. Proc Int Clay Conf, Denver, 1985, Clay Minerals Society, pp 73–77

Platel J (1989) Le Crétacé supérieur de la plate-forme septentrionale du Bassin d'Aquitaine. Stratigraphie et évolution géodynamique. Doc BRGM, 164, 572 pp

Pouget M (1968) Contribution à l'étude des croûtes et encroûtements gypseux de nappe dans le Sud Tunisien. Cah ORSTOM, Sér Pédol, Paris, 6, pp 309–365

Roche E (1988) Sporopollinic biostratigraphy and Ypresian paleoenvironment. Bull Soc Belge Géol 97/3–4:373–383

Roulin F (1985) L'Eocène continental du synclinal d'Apt (Vaucluse, Sud-Est de la France); enchaînements silcrète-calcrète et argilogenèse. Thèse, Univ Lyon, 242 pp

Ruellan A, Nahon D, Paquet H, Millot G (1977) Géochimie de la surface et formes de relief. VI. Rôle des encroûtements et épigénies calcaires dans le façonnement du modelé en pays aride. Sci Géol Bull (Strasb) 30:283–288

Saugrin T (1982) Les calcaires lacustres de Touraine: sédimentation et diagenèse. Thèse, Univ Orléans, 244 pp

Sautereau JP, Decarreau A (1973) Genèse des minéraux argileux, géochimie des éléments majeurs, du chrome et du vanadium dans le Bartonien moyen du bassin de Paris. Thèse, Univ Paris-Sud, 79 pp

Schalch F (1922) Erläuterungen zu Blatt Griessen (Nr 157). Geol Spezialkarte Baden, 117 pp

Schuler M (1990) Environnements et paléoclimats paléogènes. Palynologie et biostratigraphie de l'Eocène et de l'Oligocène inférieur dans les fossés rhénan et rhodanien et en Hesse. Doc BRGM, 190, 503 pp

Simon-Coinçon R (1989) Le rôle des paléoaltérations et des paléoformes dans les socles: l'exemple du Rouergue (Massif Central français). ENSMP, Mém Sci de la Terre, Paris, 9, 290 pp

Simon-Coinçon R, Astruc JG (1991) Les pièges karstiques en Quercy: rôle et signification dans l'évolution des paysages. Bull Soc Géol Fr 162:595–605

Singer A (1984) Pedogenic palygorskite in the arid environment. In: Singer A, Galan E (eds) Palygorskite-sepiolite. occurrences, genesis, and uses. Developments in sedimentology 37. Elsevier, Amsterdam, pp 169–176

Sittler C (1965) Le Paléogène des fossés rhénan et rhodanien. Etudes sédimentologiques et paléoclimatiques. Mém Serv Carte Géol Als Lorr (Strasb) 24:392 Pp

Steckman W (1952) Der Braunkohlenbergbau in Nordhessen. Hessischen Lagerstättenarch 1:1–212

Steinberg M (1969) Sédimentologie des formations continentales de l'Eocène. Mém BRGM 69:353–357

Teichmuller R (1958) Die Niederrheinische Braunkohlenformation. Stand der Untersuchungen und offene Fragen. Fortschr Geol Rheinld Westf 2:721–750

Thiry M (1981) Sédimentation continentale et altérations associées: calcitisations, ferruginisations et silicifications. Les Argiles Plastiques du Sparnacien du Bassin de Paris. Sci Géol Mém (Strasb) 64:173 pp

Thiry M (1982) Les kaolinites des Argiles de Provins: géologie et cristallinité. Bull Minéral 105:521–526

Thiry M (1989) Geochemical evolution and paleoenvironments of the Eocene continental deposits in the Paris Basin. Palaeogeogr Palaeoclimatol Palaeoecol 70:153–163

Thiry M, Turland M (1985) Paléotoposéquences de sols ferrugineux et de cuirassements siliceux dans le sidérolithique du nord du Massif central (bassin de Montluçon-Domérat). Géol Fr 2:175–192

Thiry M, Cavelier C, Trauth N (1977) Les sédiments de l'Eocène inférieur du Bassin de Paris et leurs relations avec la paléoaltération de la craie. Sci Géol Bull (Strasb) 30:113-128

Thiry M, Delaunay A, Dewolf Y, Dupuis C, Ménillet F, Pellerin J, Rasplus L (1983) Les périodes de silicification au Cénozoïque dans le Bassin de Paris. Bull Soc Géol Fr (7)25:31–40

Thiry M, Simon-Coinçon R (1995) Tertiary paleoweatherings and silcretes in the southern Paris Basin. Catena 26:1–26

Trauth N (1977) Argiles évaporitiques dans la sédimentation carbonatée continentale et épicontinentale tertiaire. Bassins de Paris, de Mormoiron et de Salinelles (France), Jbel Ghassoul (Maroc). Sci Géol Mém (Strasb) 49:195 pp

Trauth N, Lucas J, Sommer F (1968) Etude des minéraux argileux du Paléogène des sondages de Chaignes, Montjavoult, le Tillet et Ludes (Bassin de Paris). Mém BRGM 59:53–76

Trauth N, Astruc G, Archanjo J, Dubreuilh J, Martin P, Cauliez N and Fauconnier D (1985) Géodynamique des altérations ferralitiques sur roches sédimentaires, en bordure sud-ouest crétacée du Massif central: paysages sidérolitiques en Quercy Blanc, Haut Agenais, Bouriane et Périgord Noir. Géol Fr 2:151–160

Triat JM, Trauth N (1972) Evolution des minéraux argileux dans les sédiments paléogènes du bassin de Mormoiron (Vaucluse). Bull Soc Fr Minér Cristallogr 95:482–494

Truc G (1975) Les encroûtements carbonatés liés à la pédogenèse; rôle important des microorganismes: biocorrosion et biosynthèse de la calcite en milieu pédologique confiné. IXe Congr Int Sédimentologie, Nice, Livret-guide, exc 2, pp 47–55

Valleron MM (1981) Les faciès calcaires du Lutétien à Planorbis pseudoammonius dans le Bas-Languedoc: argilogenèse et silicifications associées aux encroûtements calcaires. Thèse, Univ Strasbourg, 108 pp

Valleron MM, Dulau N, Pourzahed P, Saugrin T (1983) Calcitisations et opalitisations dans l'Eocène du Sud-Est de la France. Comparaison avec des faciès analogues d'Alsace et de Touraine. Bull Soc Géol Fr 25:11–18

Valleron-Blanc MM, Trauth N, Truc G (1985) Les calcrètes lutétiens de Laval-Saint-Roman (Gard). Géol Fr 2:161–173

Vernet JP (1969) Morphologie des cristaux de kaolinite des formations sidérolithiques. Schweiz Miner Petrogr Mitt 49:385–389

Wells SG, McFadden LD, Dohrenwend JC (1987) Influence of Late Quaternary climatic changes on geomorphic and pedogenic processes on a desert piedmont, eastern Mojaveo Desert, California. Quat Res 27:130–146

Wittmann O (1955) Bohnerz und preäozene Landoberfläche in Markgräflerland. Jahresh Geol Landesamt Baden-Württemb 1:267–299

12 On the Genesis of Sedimentary Apatite and Phosphate-Rich Sediments

Jacques Lucas and Liliane Prevot-Lucas

1 Introduction

Igneous rocks consist essentially of silicate minerals which are unstable at the earth's surface where they gradually undergo hydrolysis. When the constitutive chemical elements of these silicate minerals pass from the crystal state to that of ions in water, the relative affinities among the elements change, some of them moving away from each other, others becoming attracted towards each other. This concept of "convergence and divergence of elements" was precious to Millot (1964), who considered it to be one of the keys to understanding the organization and evolution of the geologic materials at the surface of the earth. Some elements in solution have a mainly inorganic geochemical behavior and are quickly incorporated into clay minerals. Other elements are involved preferentially in biological activity and are associated with organisms and organic matter before becoming constituents of authigenic minerals in sediments. Apatite is one of these authigenic minerals which, because of its relatively low solubility, is commonly preserved in its primary state. As a consequence, it has a far greater potential to reflect the original geochemical characteristics of its sedimentary environment than common biogenic minerals such as carbonates and sulfates.

Clay minerals constitute a major component of the sedimentary sequences in most sedimentary basins. Their capacity to exchange cations and their ability to be transformed into new minerals make them active competitors for scavenging and trapping trace elements, processes that commonly involve interactions with organic matter which is also modified. Consequently, clay minerals and non-silicate authigenic minerals in sediments provide complementary data for studying the mechanisms of sedimentary rock formation.

Clay minerals form a large family with a variety of members whose behavior, structure and genesis remain far from being fully understood. Phosphate minerals form another complex group of authigenic minerals which also raise many problems and controversies concerning their behavior, significance and genesis. The problems in both families have many similarities. Though a comparison of both groups seems a priori rather bold and artificial, it is very instructive since these two groups, although seemingly very dissimilar, are in fact complementary and often interactive.

Phosphate minerals are numerous. Nriagu (1984) described more than 300 species and emphasized the fact that they occur as accessory minerals in almost all types of known sedimentary, igneous and metamorphic rocks, in meteorites, in regoliths and as components of biological structures. In other words, phosphate minerals may form as easily in magmas controlled by the Gibbs's phase rule, as in liquids ruled by the law of mass action. This is a common characteristic of phyllosilicates which also occur in a wide range of settings, from micas originating in deep crustal rocks to clay minerals forming in surface environments. Among all phosphate minerals described, those of the apatite group, and more especially Ca apatites, are by far the most abundant, originating in igneous, sedimentary and purely biological systems.

Apatites crystallizing from a magma at high temperatures and those formed in aqueous media at low temperatures are as different as are micas and clay minerals. When apatites occur as accessory minerals in igneous rocks (or more rarely, as major constituents, e.g., deposits associated with carbonatites), they are of macroscopic size and are idiomorphic. These are generally fluorapatites which are relatively rich in Cl and are characterized by high contents of trace elements, especially of rare-earth elements (REE; Nash in Nriagu 1984). By contrast, apatites formed in surface environments are usually sub-microscopic and are barely visible, even with a scanning electron microscope (SEM). These apatites are usually carbonate fluorapatite (CARFAP) in sediments, or hydroxyapatite in vertebrate skeletons. They are also rich in a great variety of trace elements, including U, Cd and many other metals, when compared to their surrounding sediment.

Comparisons between the clay and apatite groups will not be pursued further since the two groups, despite the broad similarities discussed above, are fundamentally different. In surface environments, clay mineral formation is closely tied to inorganic geochemical processes, whereas the apatite group is typically formed by organically mediated processes, primarily because P plays a major role in biological activity. This difference is well shown by experiments made to synthesize the two mineral groups under pressure and temperature conditions typical of the earth's surface. The synthesis of clay minerals is difficult and results solely from inorganic reactions, whereas the synthesis of apatites is much easier and may be accomplished by either purely inorganic precipitation or by chemical precipitation resulting from bacterial mediation, microbes being directly or indirectly involved.

2 Apatite Synthesis

It is now well known that apatite can be obtained experimentally by several methods. Pure inorganic syntheses have been performed, in both fresh and seawater, by a double exchange reaction between a soluble salt of phosphate (Na_2HPO_4, $12H_2O$) and a Ca salt (e.g., $CaCO_3$; $CaSO_4$, $2H_2O$; $CaCl_2$) with the addition of Na fluoride (Nathan and Lucas 1976; Van Cappellen 1991). In seawater, apatite forms only when the medium is acid and when its Mg/Ca ratio has decreased from its normal value of 5 to around 1.

Apatite precipitation through bacterial mediation has been obtained in both water and seawater from organic P and a solid or dissolved Ca donor (Lucas and Prévôt 1984; Prévôt et al. 1989). The ribonucleic acid which was used in these experiments served both as a P donor and as a nutrient substrate for the microorganisms, without which the environment remained inactive, and it proved impossible to synthesize apatite. The details of the experimental procedures having been given in several publications, only the most significant results will be presented here.

Apatite obtained experimentally by bacterial mediation is very similar to that recovered from unaltered sedimentary phosphate deposits (i.e., where they have not been transformed by meteoric alteration or metamorphism). In both cases, the mineral is a CARFAP which displays the same structure (as determined by XRD) and has a broadly similar chemical composition with comparable CaO/P_2O_5 and F/P_2O_5 ratios, and CO_2 contents. CARFAP has been obtained both by direct precipitation from entirely dissolved products (Plate I, photo 1) and from solid carbonate-mineral precursors (Plate I, photo 2), including bioclasts (corals, sea urchins, foraminifera) and ooids (Lucas and Prévôt 1985). In the latter case, the primary structure of the object may be pseudomorphously preserved by a mechanism of dissolution-recrystallization on a sub-microscopic scale, the dissolved carbonate being replaced by the precipitation of apatite. In both cases, direct precipitation or replacement, the apatites are sufficiently well crystallized to give good XRD peaks, but the crystals are always very tiny and visible only with difficulty by SEM. The apatite generally occurs either as micrometer-scale spheroids with radial-fibrous internal structures, which are commonly aggregated to form "cauliflowers", or elongated grains interpreted as being mineralized microbial bodies.

The end products of the organically mediated experiments, both precipitated cryptogranular aggregates and replaced solid precursors, always incorporate organic matter which is difficult to remove, even by long contact with strong oxidizing agents such as concentrated H_2O_2. A very similar situation occurs in phosphorites which very commonly contain organic matter trapped in aggregates of apatite cryptocrystals. This organic matter cannot be hosted in the apatite lattice, which implies that it has to be adsorbed on the tiny apatite crystals which form an impermeable felt, causing pore waters and solvents to have almost no access to the enclosed nanopores. These nanopores probably also trap other constituents such as U which may become incorporated in amounts up to several weight percents. Such high concentrations are generally considered to be incompatible with structural incorporation into the apatite lattices. However, the latter is possible and has been demonstrated for Sr to Ca substitution and for partial Ba to Ca substitution (Lucas et al. 1990).

The well-known inhibiting role of Mg in apatite formation has been confirmed experimentally. In seawater, apatite only begins to form when the Mg/Ca ratio has decreased significantly due to the precipitation of a Mg mineral resulting from bacterial activity; in this case, the precipitated mineral is struvite, NH_4MgPO_4, $6H_2O$. Of course, this can only happen in a closed environment, insulated from a re-

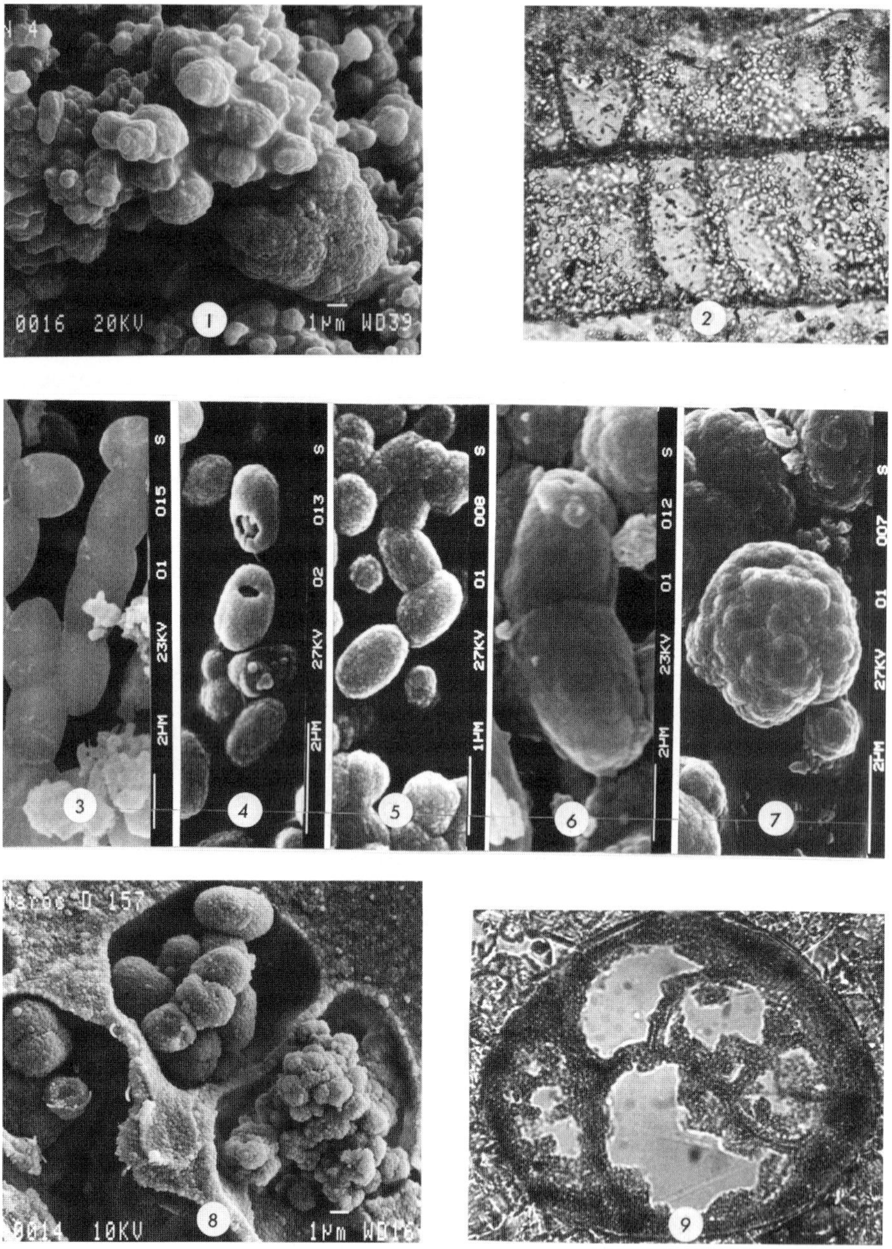

enrichment in Mg. It should be emphasized that Mg is only an inhibitor, it merely slows down considerably the formation of crystalline apatite. Eventually, with time, apatite will form (Gulbrandsen et al. 1984; Van Cappellen and Berner 1991).

All experiments were performed in an alkaline environment in which the pH was self-maintained at about 8. Such a high pH may seem inconsistent with the

Plate I. Examples of apatite micrograins obtained experimentally during the mineralization of cuttle bone. They are considered to represent successive stages of apatite growth with mineralized bacteria as nucleation sites. **1—7** Experimental products. **1** Scanning electron microscope (SEM) photomicrograph of coalescent cauliflower-like forms made by apatite precipitated from solution in the absence of a solid precursor. **2** Photomicrograph (PPL) showing the experimental replacement of aragonitic walls of cuttle bone by apatite; ´140. **3** Smooth bacteria chains. **4** Ragged "bacteria" connected by mycelian filaments. **5—7** Apatite growth, with the thickening masking the primary form and leading to cauliflower-like elements. **8** and **9** Natural products. **8** SEM photomicrograph of coalescent cocoon-like microspheres infilling foraminifera chambers (Dekakra, Ganntour, Morocco). **9** Photomicrograph (PPL) showing replacement of a calcitic foraminiferal test by microspherical apatite (borehole 285, Ganntour, Morocco); ´350. Note the similarity of these natural products with the experimental products (**1** and **2**)

dissolution of a solid carbonate precursor and with the necessity of an acidic environment for apatite precipitation, confirmed by purely inorganic apatite synthesis. This inconsistency is waived if we consider that bacteria cause an acidification of the closely surrounding microenvironment, by the lysis of RNA and the liberation of H_3PO_4, which is instantaneously buffered by the dissolution of nearby calcite and the in situ precipitation of apatite.

Apatite has been produced synthetically in aerobic, intermittently agitated environments, in disaerobic conditions (i.e., in uncovered, unagitated beakers) and in strictly anaerobic environments in closed bottles and under argon atmosphere. In all cases, the final products were very similar. The main difference between each of the experiments, which started with the same original composition, relates to the time needed for apatite formation: 2–3 weeks in the most oxygenated conditions and 5–6 months in anaerobic environments. Thus, apatite formation is not due to a specific type of microbe, though some species are more active, but depends mainly on the bacterial metabolism which controls the quantity and quality of enzymes that play a major role in the hydrolysis of the organic matter, and thus the availability of inorganic P. Experiments have demonstrated that the direct utilization of alkaline phosphatase leads rapidly (within 1 day) to the formation of apatite, even in the absence of bacteria. Again, the apatite occurred as microspheres and cauliflower-like aggregates.

The possibility that various Ca donors can be used for apatite synthesis has been well proved. In addition, the successful utilization of common mussel tissue in place of RNA (Hirschler 1990), has demonstrated that P can also be provided by more complex organic matter. The resulting apatite was similar to that of all previous experiments. Only the time necessary to obtain a precipitate under aerobic conditions was increased to about 1 year. Thus, it is not unreasonable to assume that any organic matter is a potential raw material for apatite formation.

The chemical inactivity of the sterilized preparations, the presence of recognizable bacterial forms in the mineralized material, and the formation of NH_4-containing struvite clearly demonstrate the presence of bacterial activity. The controls on apatite nucleation remain unresolved, although the morphology and location of the synthesized apatite suggest two possibilities (Hirschler 1990; Hirschler et al. 1990):

1. Crystallization of apatite in microenvironments close to bacteria. These microenvironments are modified by high concentrations of degradation products of metabolites, especially inorganic P, produced by bacteria. Nucleation may then occur on a solid carbonate substrate for apatite precipitation which acts afterwards as a pseudomorphic form for the dissolved carbonate. Nucleation may also start on the bacteria itself which acts as a passive substrate for apatite precipitation from surrounding solution. Thus, bacteria acquire rugged surfaces which later grow into coarser cauliflower-like aggregates.
2. Direct bacterial nucleation in the periplasm or cytoplasm. In this case, the bacteria are phosphatized from the interior of the cell and have a smooth external surface. However, they subsequently serve as nucleation sites for further apatite precipitation and lose their pristine aspect (Plate I, photos 3–7).

The first mechanism in which apatite precipitation occurs outside the bacteria seems to be the most frequent and the most efficient way by which most apatite is formed.

3 Genesis of Sedimentary Apatite

3.1 From Experiment to Nature

Can experimental results be transposed to the natural environment and can they be used to propose a model explaining the processes of apatite genesis in sediments? There are many similarities between the synthetic products which have been described above and the natural products, a representative selection of which will be discussed later.

The same radial-fibrous spheroids, cauliflower-like aggregates and elongated particles, which have been attributed to bacteria by many authors, have been observed in many sediments. The bacterial nature of these particles, though credible, has been challenged since there is no real proof that bacteria are able to self-phosphatize, and the morphological similarities cannot by themselves confirm the role of bacteria in phosphatization. Nevertheless, the importance of microbial or fungal activity in phosphogenic environments is amply demonstrated by the abundance of phosphatized organic forms which have been described (Baturin 1982; Soudry and Champetier 1983; Lamboy 1987; Soudry and Lewy 1988; Prévôt 1988, 1990), including filaments, tubes, and lamellae. Bacteria may have served as a substrate for apatite precipitation, as did many other biological or mineral particles. On the other hand, replacement of carbonate precursors is common. Many primarily calcitic or aragonitic bioclasts undergo pseudomorphism by apatite and sometimes the replacement structures are the same as those observed in experimental material. Vertebrate debris also serve as a substrate for apatite precipitation. In natural environments apatite can form, as in the experiments, either by direct precipitation from solution (Plate I, photos 1,8) or by replacement of a precursor (Plate I, photos 2,9).

It should be emphasized that both the natural and synthetic materials are very similar from both mineralogical and crystal habit viewpoints. CARFAP occurs in both cases as acicular cryptocrystals aggregated into particles lacking any organized internal macrostructure, so that for a long time sedimentary apatite has been considered to be amorphous and has been termed "collophane". These felt-like cryptocrystalline aggregates display a cryptoporosity. Single euhedral apatite crystals, although common in igneous and metamorphic rocks, are very rare in sediments where they result, for the most part, from secondary transformations caused by weathering, deep-burial diagenesis, or low-grade metamorphism.

The close association of preserved organic matter with phosphorites, even after intense weathering, is significant. This organic matter is generally located partly outside the apatite grains, and is responsible for the dark color of the unweathered rock matrix. However, light-colored phosphorites may also contain up to several percent of organic matter which, as in synthetic apatites, is trapped within cryptocrystalline apatite aggregates where it is difficult to be completely destroyed by weathering. This organic matter is composed of humic and fulvic material (Belayouni and Trichet 1979; Belayouni 1983; Benalioulhaj 1989) and of organic molecules which, as markers of bacterial activity (Meunier-Christmann et al. 1989), are useful in constraining phosphogenic models. As in synthetic apatites, other components such as U (Lucas and Abbas 1989), or crystals of NaCl may be trapped in the cluster of apatite crystals where their abundance and broad positive correlation with P may be stoichiometrically puzzling to the geochemist.

Mg, although abundant in seawater, is a minor constituent of both natural and synthetic apatites. The concentration of MgO in natural CARFAP is generally 0.05–0.5% (Lehr et al. 1968), but it may be much lower in synthetic apatites, many of which have been precipitated from solutions containing no Mg. Besides the macroscopic differences in morphology and scale which we will discuss later, these microscopic and chemical similarities are strong enough to allow the use, without temerity, of the experimental results to explain the genesis of natural apatite.

3.2 Natural Apatite

As most known phosphorites are marine sediments, it is essential to consider a model of apatite genesis in oceans. Dissolved phosphate is close to saturation with respect to CARFAP in seawater (Van Cappellen 1991). However, the phosphate content of the surface oceans depends on an equilibrium linked to biological activity. P is one of the limiting nutrients, which means in turn that it cannot increase markedly in normal marine environments where organisms able to use it are living. Consequently, P can only be concentrated in the organic matter resulting from biological activity and it is independent of the original P supply (such as guano, upwelling water, or continental weathering).

Among the constraints there is a strong necessity to extract most of the Mg out of the solution. It is not possible to significantly deplete unconfined seawater of its Mg. Thus, one has to consider temporarily closed volumes, or volumes with very restricted possibilities of exchange, in which precipitation of a high amount of the available magnesium is possible. These closed volumes could be the interstitial waters just below the water–sediment interface in which fluid circulation is already limited (Daumas 1976). Struvite, the transitory mineral which hosts Mg in the experiments, could then be a natural transitory mineral in the genesis of phosphorites. Subsequent dissolution of this highly unstable mineral could lead to the increase in Mg needed for the formation of more stable phases such as dolomite and Mg clays which are frequent components in phosphate-rich sediments. The question remains as to whether these Mg minerals are the rule and whether calcitic, non-Mg-mineral-rich phosphate deposits are invariably the result of dedolomitization, as has been documented for Morocco and Saudi Arabia (Lucas et al. 1979; Prévôt 1990; Sheldon, pers. comm., 1994), or whether, in some cases, apatite may precipitate in the absence of a Mg sink.

In the very upper part of the sediment, organic matter from pelagic organisms is deposited, and micro- and macrobenthic organisms are active, providing a further source of P. This sediment surface zone also contains abundant bacteria that transform organic matter. In this respect, our experimental results are well suited to the natural environment since we have been able to obtain apatite not only from various concentrations of RNA in solution, but also from desiccated mussel tissue with a P concentration (1.92% P_2O_5 dry weight) close to that of mean organic matter [2% P_2O_5, calculated from theoretical formula $(CH_2O)_{106}(NH_3)_{16}H_3PO_4$]. It is not unrealistic that, as in our experiments, bacteria destroy this organic matter, inducing mineralization of phosphate and precipitation of apatite outside the bacteria.

The experimental results, as well as similarities recognized between synthetic and natural apatite, suggest that the environments in which apatite forms are well isolated microenvironments which are accessible to microbes. In addition, the microenvironments are rich in organic matter, relatively depleted in carbonate, which would otherwise inhibit apatite precipitation, and probably oxygenated since aerobic bacteria seem the most active. These characteristics cannot be met all together in a single natural environment. Oxygenation of the uppermost sediments is achieved by agitation, which is not conducive to maintaining the necessary temporarily closed microenvironments for the removal of Mg by precipitation, which reduces the speed of apatite formation (Van Cappellen 1991). On the other hand, some anaerobic environments which may be favored by sedimentation of detrital clay particles could be totally asphyxiating, with the micropores being too small for bacterial access, and thus the organic matter would be partly protected. This would be the scenario for black shales which are often associated with phosphorites (Benalioulhaj 1991). The environments most conducive to apatite precipitation might be intermediate between these two extremes. They are commonly termed "oxygen-minimum", "oxygen-depleted" or "poorly oxygenated" zones, or "mildly reducing environments". Comparative studies of black shale and corresponding phosphorites (Meunier-Christmann et al. 1989) have shown that the destruction of organic matter follows similar pathways. In both cases, biomarker molecules prove the phytoplanktonic origin of the organic matter, the pristane/phytane ratio below unity points to a globally reducing environment, and the organic matter is immature. In black shales, the transformation of organic matter is slow, the humic and fulvic acids condensing into kerogen (Huc 1978) and mineralization being rare. By contrast, the transformation into phosphorites proceeds differently, phosphate being mineralized and apatite precipitating. Part of the remaining transformed organic matter is trapped in cryptopores within phosphate grains and then the evolution of this organic matter is blocked (Benalioulhaj 1989). The part of the organic matter which is not incorporated into the phosphate becomes increasingly oxidized, probably during periods of mild agitation, and is finally dispersed in solution.

The deposition of a sediment with high organic matter and low carbonate content depends on the seawater environment above the sediment–water interface. It appears that seawater needs to be enriched in nutrients leading to high productivity which, for reasons that are still to be understood (Westbroek 1991), produces mainly non-calcareous planktonic species. Interestingly enough, such favorable zones may occur either as huge surfaces of the continental shelf, such as in Morocco at the Cretaceous–Tertiary transition, or as isolated spots within a chalk basin like the Paris basin (Jarvis 1992) during the Late Cretaceous. The paleoenvironmental conditions which resulted in the formation of a dominantly chalk province on the northern side of the Tethys and a dominantly phosphate province to the southern side are in great need of more investigation.

Observations suggest that the phosphogenic sediments should be low in both detrital minerals and biogenic carbonate particles. This is strengthened if one thinks in terms of the budget of apatite production on oceanic shelves. This budget relies on the average chemical composition of organic matter [normally represented as $(CH_2O)_{106}(NH_3)_{16}H_3PO_4$] which contains 35.8 wt% organic C and 0.8% P. As 0.8% P corresponds to 5.6% of potential apatite, 100 g of sedimented organic matter corresponding to 35.8 g organic C may lead to 5.6 g apatite. According to Parsons et al. (1977), primary productivity of organic matter on the shelves in modern oceans is 60–80 g m^{-2} $year^{-1}$, of which 30–40% is trapped in the sediment, representing a deposition rate of 20–35 g m^{-2} $year^{-1}$. In areas of high productivity, the annual C deposition is much higher, as illustrated by the deposition of 1214 g m^{-2} $year^{-1}$, or even more off La Jolla, California. A high productivity zone of 1000 g m^{-2} $year^{-1}$ C (i.e., 2800 g m^{-2} $year^{-1}$ of organic matter) may thus provide 156 g m^{-2} $year^{-1}$ of apatite, which for a mineral density of 3, would correspond to 50 cm^3 m^{-2} $year^{-1}$ apatite. This is equivalent to an apatite layer being 0.05 mm thick. This estimate is of the same order of magnitude as that of Föllmi and Garrison (1991). To obtain this quantity of apatite, approximately 60 g of Ca are needed, which might be provided by microorganisms such as diatoms by their organic matter (Pérès and Devèze 1964 in Slansky 1980). It is understood that these are maximum values, since this calculation has the prerequisite that all of the deposited organic matter has been mineralized without any being recycled in the sediment and escaping in solution. In addition, any supply of detrital or biogenic minerals not involved in the mechanism of apatite formation will produce a diluting effect. Supplies of such material exist in most phosphate-rich series.

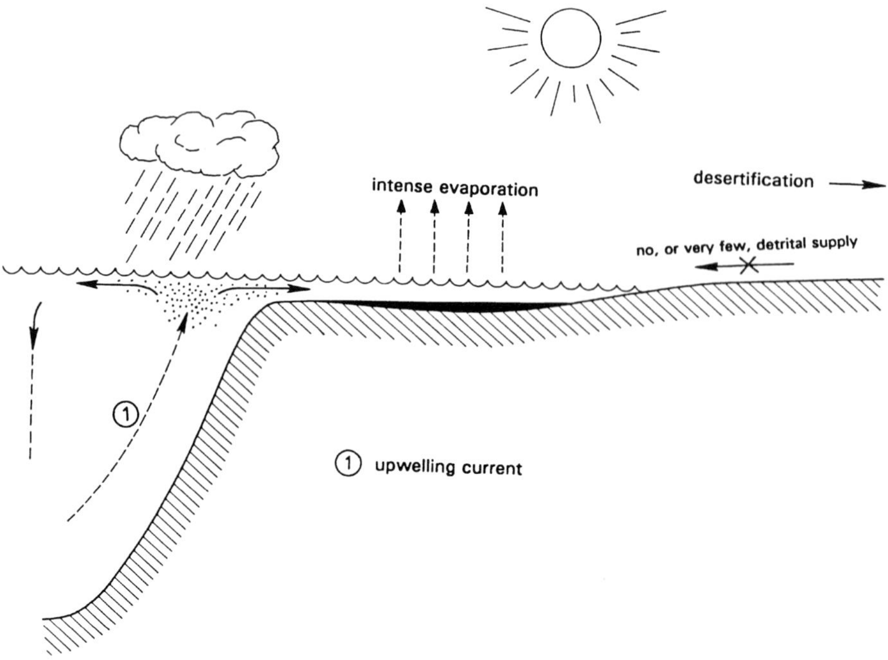

Fig. 1. Schematic diagram showing conditions favorable for the accumulation of organic matter which are needed for the mineralization of phosphorus and the crystallization of apatite

In summary, only microenvironments formed in sediment pores can support the bacterial processing of organic matter which is able to increase the dissolved phosphate content sufficiently for rapid apatite precipitation, and at the same time allow the dissolved Mg content to become depleted.

The deposition of a source bed with a high organic matter content requires a reduced detrital sediment supply. This means either a shelf fringing an arid continent, which is consistent with phosphogenesis being associated with an upwelling system (Fig. 1), or offshore shoals far from the coast. It also requires a dominance of organisms without carbonate shells in an environment which does not favor the precipitation of biogenic carbonate. High productivity in the water column has to be maintained by a nutrient flux, particularly a P supply. Given these requirements, the main supply of P is likely to be marine, and upwelling water provides the most satisfying source. Productivity is probably irregular because of its linkage with periodic upwelling (Eshet et al. 1994), biological blooms, mass mortalities and, on a longer time scale, global and local parameters including ocean opening, sea level changes, and climatic variation. Preservation of organic matter on the seafloor, albeit with partial oxidation to provide dissolved phosphate, requires a calm environment without high detrital clay input and moderately oxygenated, but not anoxic, conditions. Such intermediate environments cannot be expected to persist over long periods.

Though many questions remain to be addressed, the conditions of apatite genesis are slowly coming to light. We have a better insight into its paleoenvironmental conditions. It is likely that precipitation of tiny apatite crystals takes a very short time, comparable to that necessary to obtain it experimentally, which is instantaneous in geological terms. Experiments cannot help to explain the formation of phosphorites and phosphate-rich sequences. For that purpose, complementary observations are needed, not at the level of the microenvironment or of the pores of the source bed, but at the level of rocks, beds, and deposits.

4 Genesis of Phosphorites and Phosphate-Rich Sequences

The immense accumulations of phosphate in sedimentary deposits have been puzzling geologists for a long time. The main question to be addressed is the extraordinary disproportion between the crustal content of P (about 1000 mg g^{-1}), its content in sediments, most of which does not exceed 4000 mg g^{-1}, and the concentrations observed in phosphate deposits. For example, the P_2O_5 content of the raw phosphorite ore in Morocco is greater than 30%, and the estimated reserves in the Ganntour and Oulad Abdoun basins, which have been exploited for more than 60 years, are over 34 million tons. This estimate is certainly considerably less than the total phosphate content of the basins since associated phosphate-rich beds, containing insufficient amounts of phosphate to be considered worth exploitation, have not been taken into account. Furthermore, the total accumulation of phosphate in this specific Moroccan area of the North African Atlantic margin is even higher when the newly exploited deposits of Bou Craa and Meskala, which are still under preliminary study, will be included. Looking to the north and the south beyond the Moroccan phosphate area, it can be established that the whole Mediterranean basin, including large areas of Tunisia, Egypt and the Near East, and to a lesser extent Turkey, Greece and Albania, and the northwest margin of Africa including Mauritania, Senegal and Togo, broke into a "phosphate fever" during the late Cretaceous and early Tertiary (Prévôt 1990).

Since the Precambrian, several other "phosphatic fevers" have been detected as landmarks through geological time. They have produced sedimentary phosphorites displaying a great variety of characteristics: dark or light, friable or indurated, siliceous, calcitic or dolomitic, granular or nodular. Undoubtedly these characteristics do not all reflect primary states; some result from late diagenetic and even from metamorphic transformations, or from varying degrees of meteoric or hydrothermal alterations (Flicoteaux and Lucas 1984). It is remarkable that in all of these phosphorites originating from such a wide variety of ages and places, the apatite mineral differs little in composition. It is invariably a CARFAP with varying amounts of CO_2. This homogeneity of composition might be an indicator of similarities in the genesis of these apatites. But similarity is not unity and our purpose is not to claim a single invariant model. It is likely that each of the various phosphorite facies is a response to a globally similar model with specific characteristics due to local environmental factors.

In this contribution, we have been dealing primarily with granular phosphorites which represent the main constituent of most ancient sedimentary deposits. To address the problem of the genesis of the mineral apatite, it is important to study pristine phosphate sediments which have not been modified by burial diagenesis or weathering after formation. The choice is difficult and certainly partially arbitrary. Many studies have shown that ageing and weathering result in a decrease in the CO_2 content of apatite, and that weathering may lead to bleaching of the primary sediments by oxidation of the organic matter, to dedolomitization, and to other processes. We have chosen to investigate Cretaceous–Tertiary deposits of the Tethyan province in which, due to the important volume of phosphate-rich sediments, primary facies have been preserved locally and are distinguishable by their common characteristics.

Petrographically, phosphorites exhibit a great diversity reflected in a wide variety of phosphate grain types (Prévôt 1990). While many of their characteristics bear a close similarity to the ultrastructures observed in the experimental material, the natural grains are much bigger and are generally easily visible to the naked eye. Grain types include: (1) skeletal debris of vertebrate teeth and bones, (2) coated grains in which structured apatite coats (Plate II, photo 10) a nucleus of different mineralogy, (3) uncoated, structureless grains which are pure apatite, (4) entire or fragmented fecal pellets, and (5) microfossils, or debris of macrofossils, the originally calcitic or aragonitic tests of which have been phosphatized. Two characteristics of all phosphate grains which are particularly important to the genesis of phosphorites shall be highlighted: the uniform mineralogy of these various grains consisting in CARFAP suggests a common point in their history, while the widespread occurrence in phosphate sediments of silica and secondary silicification, of dolomite dominantly in small rhombs, and of clay minerals occurring as fragile bridges between all these constituents or as clusters around them, indicate a link between the parageneses of these different components.

With the exception of bone debris which retain their primary structure (after pseudomorphism), most other phosphate grains and matrices consist of cryptocrystalline aggregates. These polycrystalline apatite grains, even the smallest, clearly cannot have formed in a single pore of an ooze; several crystallization stages are required. We propose a mechanism with alternate opening and closing of the pore system corresponding to changes in the ooze structure. This could be compared to what happens in a jig. During the first stage, the ooze is structured and pores are created between the slightly compacted particles. In these pores, interstitial water is divided into separated closed microenvironments. During the second stage, the ooze is homogenized, particles are dispersed, insulated pore spaces disappear and the interstitial water forms a continuous and open system. When the pores are closed, bacteria transform the available organic matter and promote apatite precipitation. After a short time, equilibrium is reached in the pore and precipitation stops. Precipitation occurs again only after new nutrients are supplied for further bacterial activity, i.e., when fresh organic matter is introduced into the ooze. As diffusion seems to be far too slow a process to provide a sufficient supply of dissolved organic matter, the addition of organic matter probably occurs predomi-

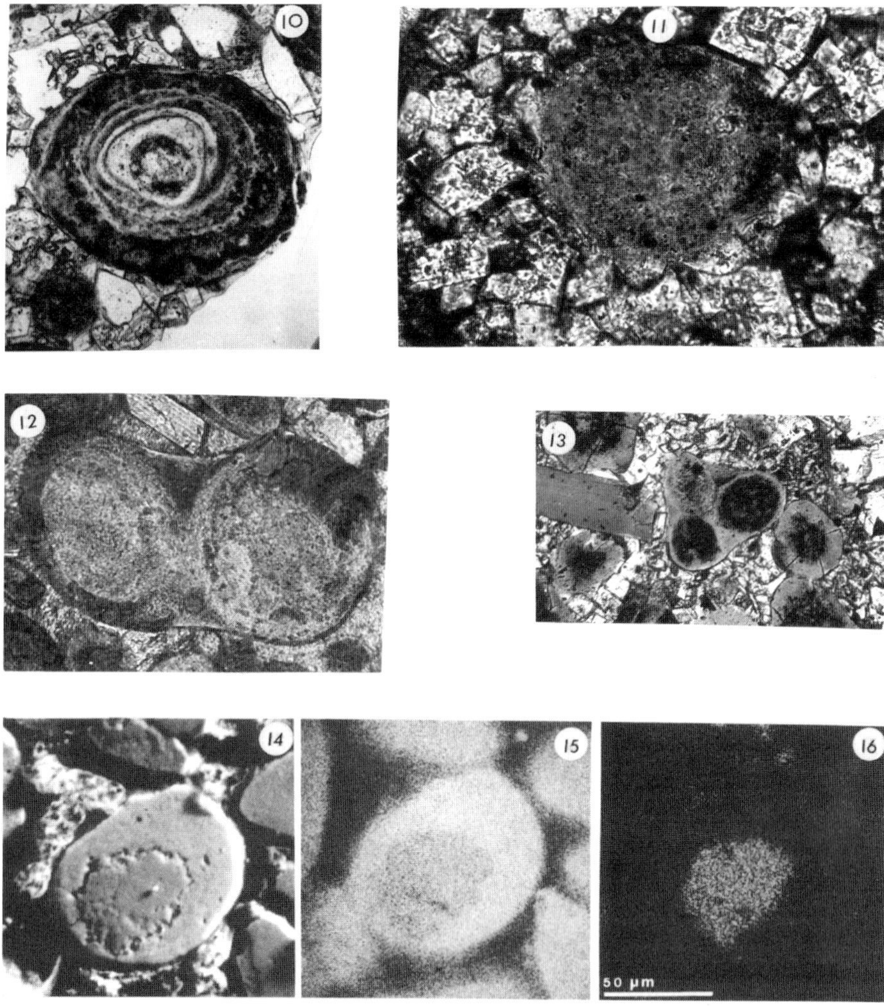

Plate II. Examples of natural apatite grains. **10** Grain showing several broadly concentric coatings (Bengrir, Ganntour, Morocco); ´140. **11** Apatite surrounded by dolomite rhombs. **12, 13** Association of two grains in a single apatitic envelope. **14—16** Dolomitic nucleus in an apatite grain (scanning electron microscope) and elemental distribution maps for Ca (**14**) and Mg (**15**)

nantly when the pores become reconnected during homogenization of the ooze. Repetition of many structuring-homogenization events is likely to produce apatite in varying amounts, depending on how long the system is operative.

During pore-insulation episodes, apatite precipitates as dispersed tiny crystallites. During homogenization, this insoluble apatite remains, while struvite and any other transitory Mg minerals dissolve, providing solutions with high Mg contents which are available to form dolomite. Each of these pore opening–closing successions result in the formation of new apatite crystals which progressively aggregate to form grains. Dolomite, on the other hand, precipitates by the overgrowth of early nuclei, forming abundant small dispersed rhombs (Plate II, photo 11).

This mechanism solves the apparent contradiction between the necessity for a closed environment to precipitate apatite and a continuous supply of material. When this sedimentary material is the result of mainly non-calcareous productivity, it has a low Ca carbonate content but is high in organic matter and Si. The two former constituents are involved in apatite formation and consequently are depleted as the proportion of apatite in the sediment increases, whereas Si continues to accumulate, giving the phosphate-rich sediment its typically siliceous character.

The above mechanism also helps to explain the change of scale between apatite precipitation in a microenvironment within a short time and the enormous apatite accumulations in phosphorite deposits, which took several million of years. Thus, it helps to explain how the stability of the environment of apatite precipitation can be maintained over such long periods, while studies of modern occurrences appear to indicate a lack of environmental stability. Moreover, such instabilities are actually required for the structuring-homogenization process. Some potential causes of instability include: the release of gas (e.g., CO_2, NH_4) resulting from microbially mediated decay of organic matter, burrowing by benthic organisms, variable bottom currents inducing local sediment mixing, seasonal variations of upwelling currents leading to an irregular supply of sediment, and catastrophic events such as storms and earthquakes. All of these "perturbing events" may participate in the proposed mechanism. It is easy to imagine that such regular, widespread and frequent structuring-homogenization episodes would not fundamentally modify the regional or global environment which would appear to remain stable over long periods.

Periods of repeated apatite crystallization leading to phosphate grain formation are interrupted by larger scale winnowing and reworking, which are periods of non-sedimentation. The previously formed apatite grains are redeposited and are again placed in an environment suitable for apatite formation. This second order alternation is responsible for the variety of petrographical aspects of grains, including the multiple coating of grains (Plate II, photo 10), multiple grains wrapped in the same envelope (Plate II, photos 12, 13) and the occurrence of dolomite rhombs forming the nuclei of apatite grains (Plate II, photos 14–16).

The story of phosphate-rich sediments is still not complete. Many deposits display rhythmic sequences in which beds of almost pure apatite alternate with beds of dolomite, silica, clay, and sometimes black shales. These complex successions are linked either to environmental changes or to changes in the dynamics of the deposit. It seems tenable that unconsolidated phosphate-rich sediments may be transported over varying distances during periods of higher energy, and that during winnowing different minerals which formed together in a common environment become fractionated and are redeposited separately after sorting. Such winnowing, which is proved by an extraordinary mixing of fossil species (Soncini and Rauscher 1988; Soncini 1990), probably leads to friable phosphate beds with very high apatite content, alternating with beds which are sometimes totally lacking in phosphate.

5 Conclusion: A Model for Phosphogenesis

The above observations explain why we are inclined to consider phosphorites as one type of bioproductite. Other types with which phosphorite is sometimes associated include black shales, dolomite, chert, chalk and glauconite. The physicochemical and biological factors that lead to the preferential formation of one or another of these bioproductites is far from being well understood, but from our current results we propose a model for the formation of phosphorites which can be described as a four-stage system (Fig. 2):

1. The first stage concerns the P trap. As indicated earlier, we link the trapping of P to its concentration in organic matter, caused by the proliferation of organisms in response to an increased nutrient supply. One of the best known and most efficient ways of supplying nutrients is by upwelling water. We know that there are various types of upwelling generated by various causes and occurring in various paleogeographic situations. The important point is that they all favor high organic productivity.
2. The second stage concerns the trapping of the organic matter in areas of relatively shallow water, such as on continental shelves. In deeper water areas much of the organic debris is recycled or dissolved before reaching the sea bed. In this box, the type of biogenic material depends on the species dominating the biomass (Fig. 3), which in turn depends, through still largely unexplained proc-

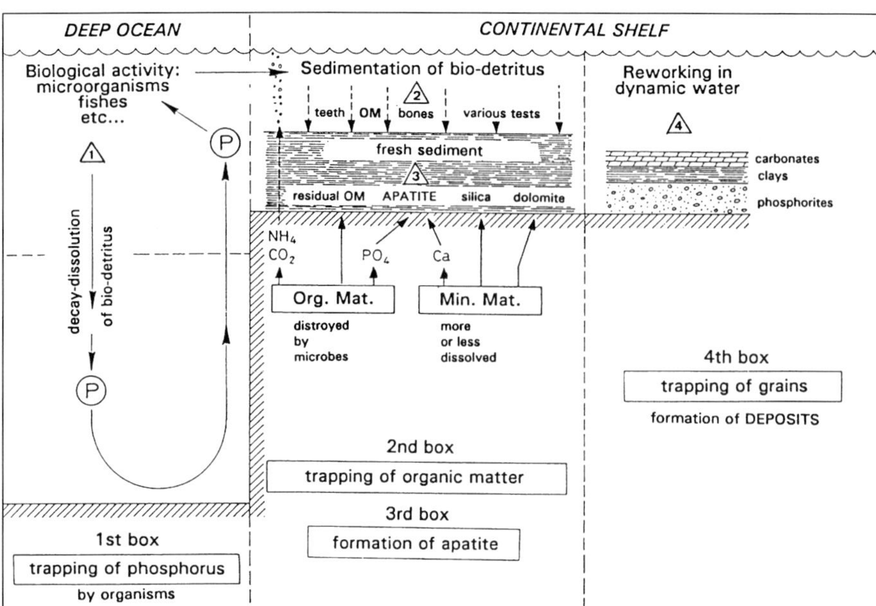

Fig. 2. Model for the formation of marine sedimentary phosphate deposits

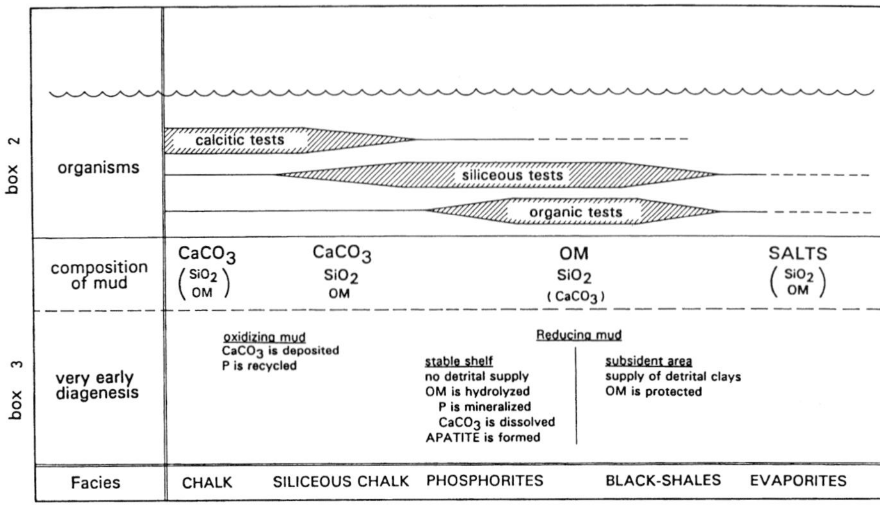

Fig. 3. Variations of the organisms living on the continental shelf as a function of environmental conditions, and consequent variations of early diagenetic facies

esses, on the characteristics of the aquatic environment (temperature, salinity, oxygenation, turbidity, exchange with the deep ocean, freshwater supply).

3. The third stage represented by the ooze located immediately below the sediment–water interface (Fig. 3), is the domain of early diagenesis in which authigenic minerals are formed during a succession of microenvironmental closures alternating with homogenizing episodes.
4. The fourth stage concerns the place where the minerals formed in the third stage are collected, sorted and concentrated, and where apatite is definitively locked up within the sedimentary sequence. This stage is the domain of hydrodynamic influences.

For the processes invoked in the model to be more easily understood, the four stages have been described as being separated in time and space, although they are probably alternating and recurrent. At the same time, stage 1 may govern conditions in one spatial area, while stage 4 may be active somewhere else. During the repeated alternation of the four stages at one place, the duration and intensity of the processes occurring in each stage are undoubtedly variable. Thus, the very fundamental concepts of changes in scale are dealt with. The spatial scale ranges from microenvironments in which the crystallites of apatite precipitate to immense accumulations in the ore deposits. The time scale records the occurrence, over a long period, of many short episodes of precipitation.

Thus, phosphorites are not the result of a continuous process, but are rather the products of several mechanisms, some continuous, some discontinuous, some occasional. During stage 1, the supply of P to the environment is a continuous process, even if modulated by seasonal variations. During stage 3, apatite precipitation and growth are necessarily discontinuous: in space within the pore-water system of the sediment, and in time due to the opening and closing of the system. Finally, even if these conditions for apatite formation are attained, only in occasionally favorable circumstances are the precipitated minerals concentrated and immobilized on the shelf, since winnowing currents may also disperse them toward the open ocean.

Each stage of the formation of a deposit needs a favorable conjunction of specific conditions. The rapid growth of the marine biomass (stage 1) and more especially the preferential growth of the most P-concentrating biomass, depends on many factors including climate, the CO_2 content of the atmosphere and of seawater, and the location and duration of upwelling currents. This is the stage where possible relations between phosphorites and glaciations or volcanism are intriguing. These relations are not direct connections, but are different results of the same state of the earth.

The trapping of organic matter (stage 2) depends on the morphology of the basins and on their relation with the ocean, in other words, on global paleogeography and local tectonics. The diagenetic formation of apatite (stage 3) depends on the stability of the basin floor, in which calm conditions result in stratification of the water column and the formation of temporarily oxygen-depleted bottom waters with a reduced supply of detrital material. In such basins, which appear to be tectonically stable for long time periods, catastrophic episodes provoke mechanical reworking, leading to the sorting of minerals by winnowing and to the deposition of the different beds in a reduced parautochthonous sequence.

The group of conditions which are linked to both global and local events and circumstances is encountered locally at different stages of the general evolution of the earth, displaying an apparent rhythmicity of phosphogenic processes, as described by Cook and McElhinny (1979). Again, there is a spatial and temporal discontinuity in the formation of phosphate deposits which does not necessarily correspond to a discontinuity in the P supply to the oceans. In other words, major variations in the P content of the oceans are not necessary to explain the formation of phosphate deposits. We believe that all geological periods supply P almost equally, but that only some were able to concentrate part of it, while others kept it regularly disseminated in sediments in which its low content is not striking. As a matter of fact, even economic deposits contain only a small part of the total P locked in the sedimentary record, most P occurring in disseminated form in phosphate-poor sediments.

Acknowledgments. This very condensed presentation owes much to the numerous contacts and discussions with members of IGCP 156 and 325. The networks created by these projects have been very efficient for the former and still are for the latter. The names are too numerous to all be mentioned here but we express our gratitude to all involved. The brevity and incompleteness of the list of references was a concession made in order to avoid an unreadable text. The reader will be able to find the main references in the bibliographies of the cited references. Finally, we are much indebted to Yaacov Nathan and the other reviewer of the manuscript for their pertinent comments, and to Ian Jarvis for the valuable revision of our English.

References

Baturin GN (1982) Phosphorites on the sea floor. Origin, composition and distribution. Developments in Sedimentology, vol 33. Elsevier, Amsterdam, 343 pp

Belayouni H (1983) Etude de la matière organique dans la série phosphatée du bassin de Gafsa-Metlaoui (Tunisie): application à la compréhension des mécanismes de la phosphatogenèse. Thèse, Univ Orléans, 202 pp

Belayouni H, Trichet J (1979) Glucosamine as a biochemical marker for dinoflagellates in phosphatised sediments. Adv Org Geochem Phys Chem Earth 12:205–210

Benalioulhaj N (1991) Les formations à phosphates et à schistes bitumineux du bassin des Oulad-Abdoun et du Bassin de Timahdit (Maroc): pétrographie, minéralogie, géochimie et environnement de dépôt. Thèse, Univ Strasbourg, 238 pp

Benalioulhaj S (1989) Géochimie organique comparée des séries du bassin phosphaté des Oulad-Abdoun et du bassin de schistes bitumineux de Timahdit (Maroc) Implications dans la phosphatogenèse. Thèse, Univ Orléans, 257 pp

Cook PJ (1970) Repeated diagenetic calcitisation, phosphatisation and silicification in the Phosphoria Formation. Geol Soc Am Bull 81:2107–2116

Cook PJ, MacElhinny (1979) A reevaluation of the spatial and temporal distribution of sedimentary phosphate deposits in the light of plate tectonics. Econ Geol 74:315–330

Daumas RA (1976) Modifications des constituants organiques dans la couche superficielle des sédiments marins: minéralisation et diagenèse. Bull Centre Rech Pau 10:149–158

Eshet Y, Almogi-Labin A, Bein A (1994) Dinoflagellate cysts, paleoproductivity and upwelling systems: a late Cretaceous example from Israel. Marine Micropal 23:231–240

Flicoteaux R, Lucas J (1984) Weathering of phosphate minerals. In: Nriagu JO, Moore PM (eds) Phosphate minerals. Springer, Berlin Heidelberg New York, pp 292–317

Föllmi KB, Garrison RE (1991) Phosphatic sediments, ordinary or extraordinary deposits? The example of the Miocene Monterey Formation (California). In: Müller D, Mackenzie JA, Weissert H (eds) Controversies in modern geology. Academic Press, London, pp 55–89

Gulbrandsen RA, Roberson CE, Neil ST (1984) Time and the crystallization of apatite in seawater. Geochim Cosmochim Acta 48:213–218

Hirschler A (1990) Etude de l'intervention des microorganismes dans la formation de l'apatite. Thèse, Univ Strasbourg, 142 pp

Hirschler A, Lucas J, Hubert JC (1990) Bacterial involvement in apatite genesis. FEMS Microbiol Ecol 73:211–220

Huc AY (1978) Géochimie organique des schistes bitumineux du Toarcien du Bassin de Paris. Thèse, Univ Strasbourg, 138 pp

Jarvis I (1992) Sedimentology, geochemistry and origin of phosphatic chalks: the upper Cretaceous deposits of NW Europe. Sedimentology 39:55–97

Lamboy M (1987) Genèse de grains de phosphate à partir de débris de squelette d'échinodermes: les processus et leur signification. Bull Soc Géol Fr III:759–768

Lehr JR, McClellan GH, Smith JP, Frazier AW (1968) Characterization of apatites in commercial phosphate rocks. Coll Int Phosphates Minéraux Solides, Toulouse, 1967, vol 2. Masson, Paris, pp 29–44

Lucas J, Abbas M (1989) Uranium in natural phosphorites: the Syrian example. Sci Géol Bull (Strasb) 42:223–236

Lucas J, Prévôt L (1984) Synthèse de l'apatite par voie bactérienne à partir de matière organique phosphatée et de divers carbonates de calcium dans les eaux douce et marine naturelles. Chem Geol 42:101–118

Lucas J, Prévôt L (1985) The synthesis of apatite by bacterial activity: mechanism. Sci Géol Mém (Strasb) 77:83–92

Lucas J, El Faleh EM, Prévôt L (1990) Experimental study of the substitution of Ca by Sr and Ba in synthetic apatites In: Notholt AJG, Jarvis I (eds) Phosphorite research and development. Geol Soc (Spec Publ), Lond 52:33–47

Lucas J, Prévôt L, El Mountassir M (1979) Les phosphorites rubéfiées de Sidi Daoui. Transformation météorique locale du gisement de phosphate des Ouled Abdoun (Maroc). Sci Géol Bull (Strasb) 32:21–37

Meunier-Christmann C, Lucas J, Albrecht P (1989) Organic geochemistry of Moroccan phosphorites and bituminous shales. A contribution to the problem of phosphogenesis. Sci Géol Bull (Strasb) 42:205–222

Millot G (1964) Géologie des argiles. Masson, Paris, 499 pp (English translation in 1970, Geology of clays. Springer, Berlin Heidelberg New York, 425 pp)

Nathan Y, Lucas J (1976) Expériences sur la précipitation directe de l'apatite dans l'eau de mer: implication dans la genèse des phosphorites. Chem Geol 18:181–186

Nriagu JO (1984) Phosphate minerals: their properties and general modes of occurrence. In: Nriagu JO, Moore PM (eds) Phosphate minerals. Springer, Berlin Heidelberg New York, pp 292–317

Parsons TR, Takahashi M, Hargrave B (1977) Biological oceanographic processes. Pergamon, New York, 332 pp

Prévôt L (1988) Géochimie et pétrographie de la formation à phosphates des Ganntour (Maroc). Utilisation pour une explication de la genèse des phosphorites Crétacé-Eocène. Thèse, Univ Strasbourg, 325 pp

Prévôt L (1990) Geochemistry, petrography, genesis of Cretaceous-Eocene phosphorites; the Gantour deposit (Morocco): a type example. Mém Soc Géol Fr 158:230 pp

Prévôt L, El Faleh M, Lucas J (1989) Details on synthetic apatites formed through bacterial mediation. Mineralogy and chemistry of the products. Sci Géol Bull (Strasb) 42:237–254

Slansky M (1980) Géologie des phosphates sédimentaires. Mém BRGM 114:90

Soncini MJ (1990) Palynologie des phosphates des Oulad Abdoun (Maroc). Biostratigraphie et environnement de la phosphatogenèse dans le cadre de la crise Crétacé-Tertiaire. Thèse, Univ Strasbourg, 243 pp

Soncini MJ, Rauscher R (1988) Associations de dinokystes du Maastrichtien-Paléocène phosphaté au Maroc. Bull Centres Rech Explor Prod Elf-Aquitaine 12:427–450

Soudry D, Champetier Y (1983) Microbial processes in the Negev phosphorites (southern Israel). Sedimentology 30:411–423

Soudry D, Lewy Z (1988) Microbially influenced formation of phosphate nodules and megafossil moulds (Negev, southern Israel). Palaeogeogr Palaeoclimat Palaeoecol 64:15–34

Van Cappellen P (1991) The formation of marine apatite, a kinetic study. PhD, Yale Univ, New Haven, 240 pp

Van Cappellen P, Berner RA (1991) Fluorapatite crystal growth from modified seawater solutions. Geochim Cosmochim Acta 55:1219–1234

Westbroeck P (1991) Global Emiliana modelling initiative (GEM). Int Worksh on a coccolithophore and global climate. Terra Nova 3:572–574

13 Clay Mineral Sedimentation in the Ocean

Hervé Chamley

1 Introduction

Georges Millot's thesis (1949) was on the «Relationships between the nature and genesis of argillaceous sedimentary rocks». His last scientific contribution was published in early 1991 a few months before he died and was entitled «About the abundance of smectite minerals in common marine sediments deposited during high sea level stages in late Jurassic-Paleogene times» (Chamley et al. 1990). Georges Millot's first and last reviews were, therefore, devoted to clay sedimentation in marine environments where he defined the widely used terms of heritage, transformation and neoformation. This underscores the interest and energy that he deployed throughout his scientific life in order to better understand clay mineral sedimentation in the ocean in the course of the earth's geological history. His scientific and didactic approaches are expressed in all the publications he was part of or wrote by himself, especially in the outstanding book that he published in 1964 (English version in 1970).

The aim of the present contribution is to pay tribute to Georges Millot by summarizing data published during the last few decades on the clay mineral marine sedimentation. Largely based on a textbook (Chamley 1989, 1992), this review takes into account a few additional publications issued more recently. We consider successively clay mineral origin in relation to detrital supply, clay destruction and its authigenesis close to the sediment–seawater interface, and the evolution of clay materials in the marine paleoenvironments during post-sedimentary diagenesis. The present contribution follows the content of the textbook by focusing on the major lessons provided by hundreds of references dealing with specific chemical environments from coastal to deep sea with detrital versus diagenetic constraints in past geological series, and with paleoenvironmental signatures by the clay mineral during the course of geologic times.

2 Sedimentary Formation and Destruction of Clay Minerals

2.1 In Alkaline Evaporitic Environments

Evaporative conditions develop in lagoon and shelf environments characterized by hot arid climate and chemical confinement. Such conditions do not determine conspicuous or noticeable clay mineral formation in recent coastal sediments. By contrast, past peri-marine environments show massive Mg clay formation of smectite and/or palygorskite groups, at different periods between the late Devonian and the Tertiary. The chemical formation of clay occurred during some geologic stages in a similar way within peri-marine and continental sediments, especially during early Paleogene times (e.g. Trauth 1977; Thiry 1981; see also Singer and Galan 1984). In addition, chemical clays are proved to form through pedologic processes, especially by authigenesis in ancient calcareous soil crusts (see Paquet and Ruellan, this Vol.).

The convergence observed in the nature of pedologic, river–lacustrine and coastal clays sheds light on some genetic conditions that must have been similar in the three types of environments (see Blanc-Valleron and Thiry, this Vol.). Salinity appears to have played a minor role, while strong alkalinity, water stagnation, intense evaporation and warm climate were probably critical. This implies poor drainage and flat relief to prevent the evacuation of ionic solutions and to permit chemical concentration. Such climatic, morphological and geochemical controls in the formation of Mg clays under evaporative conditions cast doubt on the possibility of crystallization of the same minerals in non-hydrothermal deep-sea environ-

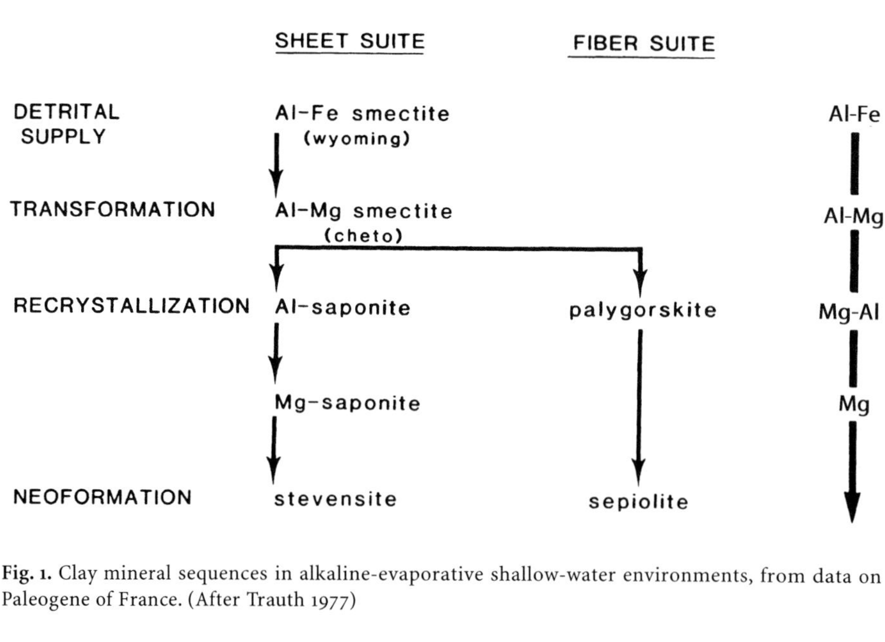

Fig. 1. Clay mineral sequences in alkaline-evaporative shallow-water environments, from data on Paleogene of France. (After Trauth 1977)

ments characterized by cold, slightly alkaline or chemically concentrated and mobile water. Deep-sea Mg clays appear to be able to develop only in a few, very restricted environments marked by local ion concentrations, such as those determined by halmyrolysis of crustal basalts or peridotites (e.g. Karpoff et al. 1989).

Minerals preferentially originating in alkaline evaporitic environments include Al-Mg beidellite, Al and Mg saponite, stevensite, palygorskite and sepiolite. A noticeable part of peri-marine and lacustrine smectites consists of Al–Fe beidellite and is detrital (Trauth 1977); they chiefly accumulated in sediments deposited on the borders of the evaporative basins. These minerals were transformed into Al-Mg smectites (i.e. saponite) and palygorskite in intermediate sedimentary environments. Stevensite (Mg smectite) and sepiolite appear to have proceeded from true neoformation under evaporitic conditions that prevailed in the central part of basins or in the upper levels of chemical sequences. The genesis of Mg sheet and chain silicates seems to have followed two independent geochemical paths (Fig. 1).

Other evaporative clay minerals reported in the literature include mainly Permo–Triassic chlorite and corrensite which appear to depend on hypersaline conditions and perhaps also on diagenetic control by geothermal gradient. Furthermore, recent investigations and experiments stress the possibility of illitization through evaporation by repeated wetting and drying cycles (e.g. Deconinck 1992).

2.2 In Organic Environments

Living organisms appear to be rarely able to modify the mineral composition of clay associations by bioturbation and nutrition activities. There are a few examples of a slight degradation of chlorite and mixed-layer clays through sediment ingestion by some shallow-water benthic crustaceans and annelids and of the transformation of mineral standards through digestion by some copepods living in nearshore seawater (see references in Chamley 1989). Alteration of Al–Si detrital minerals by living organisms appears to be restricted to shallow-water fecal pellets and to be of little quantitative importance, as proved by the fact that marine clay mineral assemblages fundamentally resemble those identified in river sediments from adjacent landmasses.

Sapropels and sapropelic marls developing temporarily during late Cenozoic time in the Mediterranean Sea, especially in eastern basins, display a wide range of relations between organic and mineral sedimentary constituents. In some sapropels, clay minerals tend to be diversely degraded according to the initial mineral composition, the abundance of organic matter, and the bathymetric and topographic situation (Fig. 2). Clay degradation tends to increase toward the central and deepest parts of sapropelitic areas, in depressed zones and at the bottom of organic layers, and to affect palygorskite, smectite, chlorite, mixed layers and illite in decreasing order. Kaolinite appears to be rather stable in Mediterranean sapropels. Correlative changes occur in inorganic geochemistry and in other sedimentary constituents such as organic matter, sulfides, and biocalcareous and biosiliceous debris. Late Cenozoic Mediterranean sapropels represent unusual marine environments in which clay degradation occurred close to the sediment–water interface

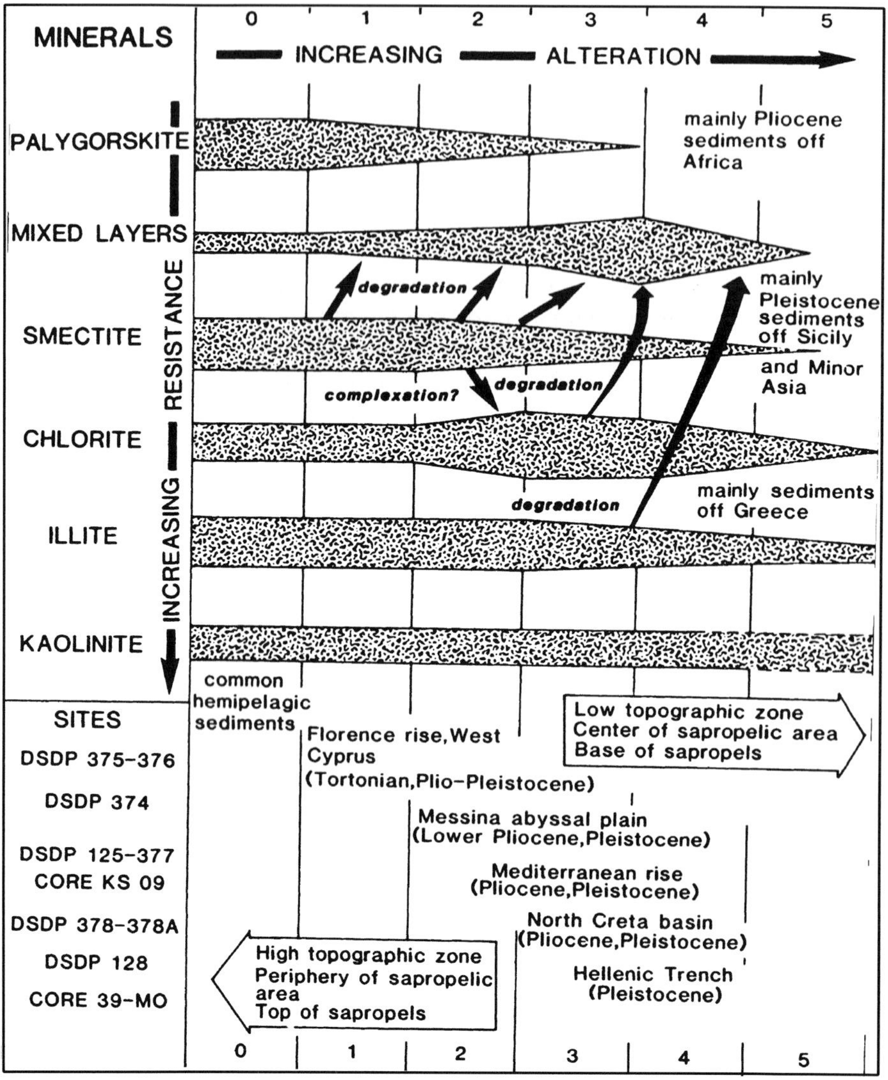

Fig. 2. Interpretation of submarine clay-mineral alteration in Pliocene—Pleistocene sapropels of the eastern Mediterranean Sea. (After Sigl et al. 1978)

mainly through the action of organic acids, in a manner similar to that characterizing continental podzolic soils and some underclays.

By contrast, Cretaceous black shales in the Atlantic domain contain diversified clay associations that mostly vary independently of the organic matter content. Smectite, which is the most common mineral, is well preserved in all black-shale sedimentary facies, as are other clay species including palygorskite and chlorite. In addition to the near absence of in situ clay alteration, Atlantic black shales differ from Mediterranean sapropels by their great extension and thickness, their small dependence on submarine bathymetry and topography, their diverse lithologies, and their often terrigenous, moderately abundant, and immature organic matter.

2.3 In Clay Granular Environments

Green Fe-clay granules represent typical marine particles that form close to the sediment–water interface in the upper part of continental margins and submarine topographic heights, in semiconfined microenvironments (Odin 1988). Green granules dominantly form in marine regions marked by reduced terrigenous supply, essentially through neoformation processes. The genesis of green granules is favored by high sea level stands, but it does not necessarily correlate with transgressive stages, sedimentation gaps, or turbulent water. Iron-rich clay granules develop from various substrates, the most suitable being calcareous microfossils and fecal pellets. The rapidity and intensity of greening noticeably depend on the nature and size of the host substrates and on the microchemical environment, as shown by various cases summarized by Odin (1988).

Green clay granules in modern sediments are of two mineralogical types. Verdine consists mainly of a 7 Å Fe-clay mineral called odinite (Bailey 1988) and is identified only in late Pleistocene–Holocene sediments. It develops in shallow-water zones of intertropical regions (maximum 50–80 m depth), close to river mouths that provide abundant dissolved Fe. The second type, referred to as glaucony, is a group characterized by minerals of various types: Fe smectite, glauconitic smectite and mixed layers, and glauconite (Fe–K illite). Glaucony statistically develops in deeper and cooler marine areas than verdine. Glaucony granules could have formed during at least the Holocene, at latitudes as high as 50° and at water depths of 1000 m. However, its maximum production occurs between 150 and 300 m in temperate warm to equatorial regions (Fig. 3). Glaucony formation proceeds in several stages, marked by a strong enrichment successively in Fe and K, by an increase in initial volume causing external cracks and by the disappearance of the original shape of the substrate. Glaucony also has film and diffuse habits, in contrast to verdine.

The chemical and mineralogical evolution of green clay granules stops either after long exposure at the sediment surface (about 10^5–10^6 years for glaucony) or after significant burial (usually a few decimeters). Most past Fe-rich clay granules appear to have formed, like recent ones, in semi-confined environments located at the sediment–water interface. But ancient glaucony minerals often display better crystallinity than recent ones, without any noticeable chemical difference. Ancient Fe-rich clay granules are reported from fresh–saline water-mixing zones to deep offshore. The formation processes of old glauconitic peloids and of oolitic ironstones are sometimes still poorly understood (see van Houten and Purucker 1984).

The formation of green clay granules is the first important step in the incorporation of terrigenous Fe in mineral structures within the marine environment. Continental shelves and upper slopes marked by glauconitization and incidently by verdine formation serve as geochemical filters during the transfer of dissolved

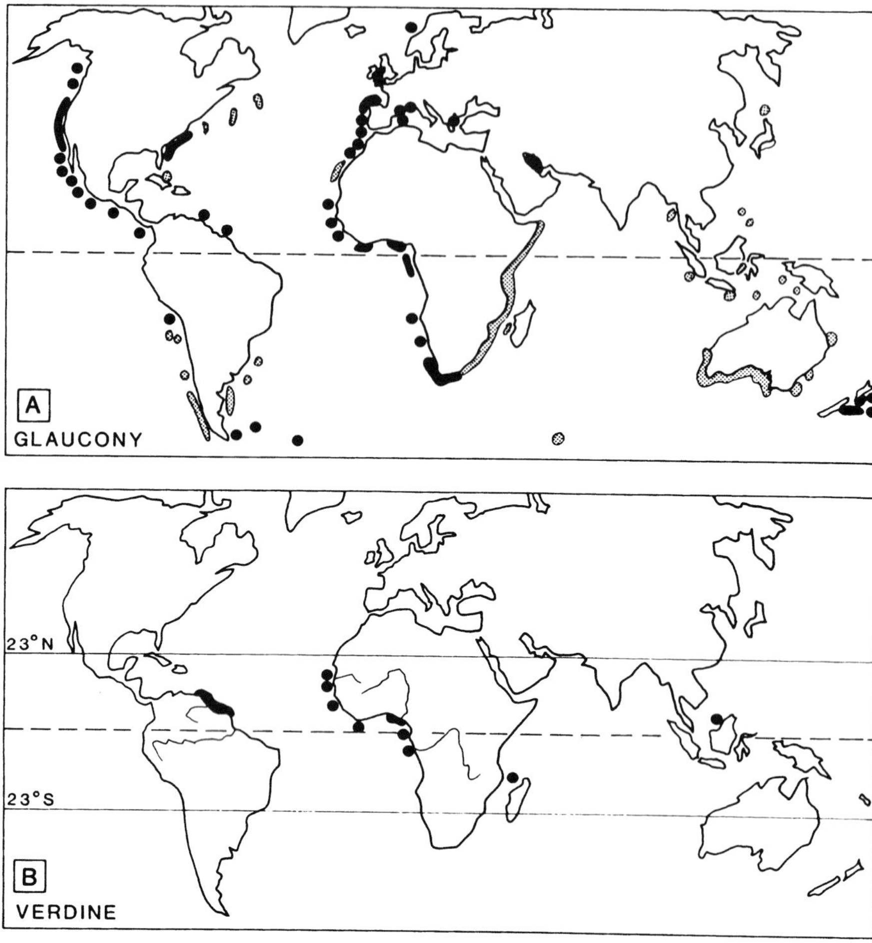

Fig. 3. A Distribution of glaucony and **B** verdine in surface sediments of the World Ocean (after Odin 1988). *Hatched areas* without detailed mineralogical data

elements from land to sea, a similar role being played in the deeper ocean by metalliferous sediments. Note that the celadonite-bearing facies, despite its mineral, chemical, morphological, and environmental particularities related to volcanic dependence, presents Fe and K enrichments, restricted conditions and greening mechanisms comparable to those typical of Fe-rich sedimentary clay.

2.4 In Deep-Sea Metalliferous Clay Environments

Deep-sea metalliferous clay which represents a major step in the fixation of Fe in marine sediments, appears to form essentially close to the sediment–seawater interface and, therefore, results largely from hydrogenous processes. Authigenic aluminosilicates in deep-sea brown clay mainly comprise nontronite-like Fe-smectites, which contain minor but significant amounts of Al and Mg (Hoffert 1980). Metal oxi-hydroxides form the dominant constituents in these environments, and are frequently associated with zeolites of phillipsite type. Very little evidence exists for the hydrogenous formation in detectable amounts of palygorskite or sepiolite in non-hydrothermal deep-sea sediments.

Despite their frequent occurrence, authigenic smectites are often only an accessory component of metal-bearing deep-sea clays, dominant Fe and Mn oxi-hydroxides forming abundant independent particles, aggregates, micronodules or nodules. Some «brown clays» are almost devoid of clay minerals, particularly in the South Pacific and south Indian Oceans where metal oxides, zeolites and biogenous debris are the main sedimentary constituents (Fig. 4). Some subsurface metalliferous clays in the South Pacific virtually lack clay minerals, in contrast to surface deposits in the same domain, raising the question of the stability of Fe-smectites under burial conditions. The central and South Pacific deep-sea brown oozes affected by intense authigenic processes, contain less clay minerals, less fine-grained fractions and less volcanic materials than North Pacific deep-sea brown oozes receiving dominantly terrigenous input (in Chamley 1989). In general, these Cenozoic clays are characterized by a very low sedimentation rate (often <1 mm/10^3 year) and small thickness (less than a few tens of meters) and they have few true equivalents in the geologic records.

The mechanism of smectite hydrogenous formation in deep-sea brown clays is only partly understood. The most probable process in the Pacific Ocean consists of a low-temperature interaction between biogenic silica providing Si and perhaps a proportion of Al, Fe-oxi-hydroxides providing Fe, and seawater providing Mg and accessory elements. The negative anomaly of Ce, which is registered in many authigenic smectite-bearing sediments, underlines the influence of seawater (e.g. Hoffert 1980; Bonnot-Courtois 1981). An alternative mechanism is suggested for some brown, siliceous clays of the Indian Ocean where amorphous Fe–Si complexes are possible precursors for smectite genesis (Fröhlich 1982). Whatever the mechanism, it appears to depend little on direct volcanic activity or direct hydrothermal input. The relative importance of synsedimentary and early diagenetic contributions in clay formation is often difficult to assess, because the extremely low rate of deposition of pelagic brown clay favors long-lasting exchanges with the open sea environment, as well as bioturbation. Early-burial effects generally appear to be of little importance, since most old-brown clays from typical authigenic environments apparently show no significant increase in smectite abundance (e.g. south equatorial Pacific and south Indian Oceans). In addition, deep-sea metalliferous clays set up difficult questions about the transition between autochthonous and allochthonous influences. Because smectites, for instance, may issue from various continental and submarine sources and may be reworked from one marine environment into another one, the identification of their initial origin constitutes a difficult task. The combination of sophisticated chemical, microchemical, micromorphologic and isotopic investigations in both sedimentary and interstitial environments, together with modelling, may greatly help to characterize genetic

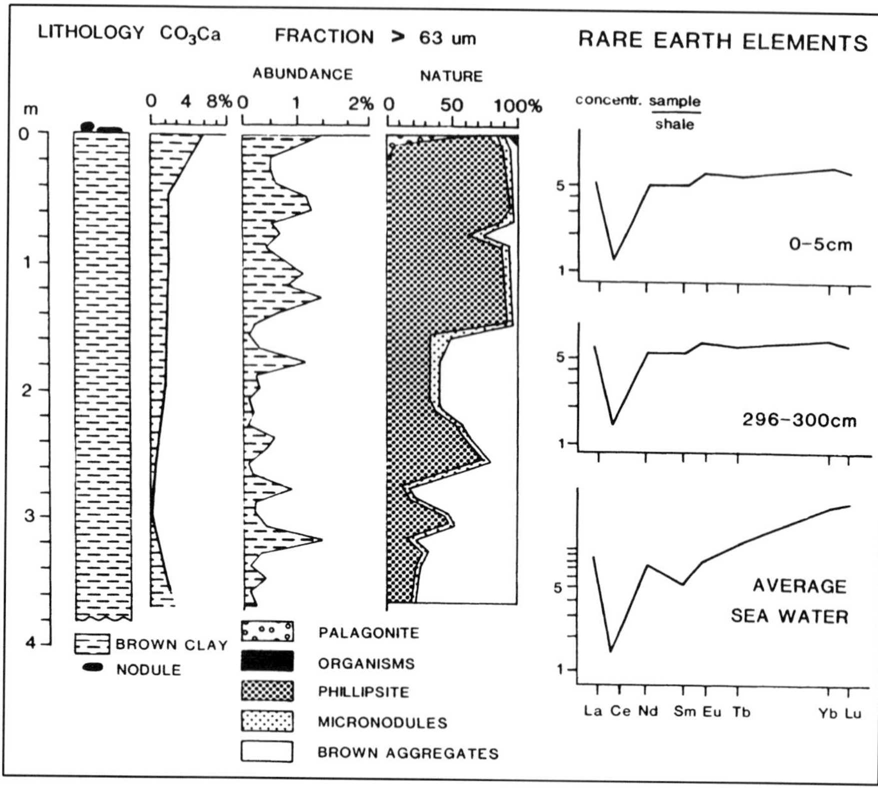

Fig. 4. Sedimentological and rare-earth-element data on metalliferous clay in a core of the central south equatorial Pacific Ocean. (After Hoffert 1980)

conditions. Fruitful approaches are provided by rare-earth element measurements on selected mineral species and grain-size fractions and by detailed Rb–Sr isotope analyses.

2.5 In Volcanic and Hydrothermal Environments

The submarine alteration of volcanic glass classically gives way to the formation of various types of smectites, as does the subaerial alteration of these unstable, poorly crystalline materials (see Fisher and Schmincke 1984; Chamley 1989). The influence of hydrothermalism on the clay mineral formation is less often investigated in the literature (e.g. Buatier et al. 1989). Internal hydrothermal processes include alteration of the oceanic crust by high-temperature fluids and metamorphism of sediments intruded by basalt sills. Surficial processes include low-temperature alteration and precipitation of hydrothermal material in the deep sea. Low-temperature alteration of glass is called palagonitization and is characterized by successive formations of hydrated glass, Mg smectite (saponite), metal oxides and zeolites (phillipsite); the final stage of palagonitization corresponds to the development of exclusive authigenic minerals intermixed in complex aggregates. Low-temperature alteration of crystalline basalt determines the formation of various minerals: oxidizing conditions favor the formation of Mg–Fe smectites relatively depleted in Mg (<6% MgO) and of K-rich celadonite, carbonates, and Fe–Mn oxides; reducing conditions lead to Mg-rich smectites (>20% MgO) and Fe sulfides. A large part of hydrothermal alteration phenomena depends on the action of seawater migrating within the upper oceanic crust, which corresponds to the first phases of the submarine hydrothermal cycle.

Hydrothermal deposits are fundamentally characterized by the precipitation of abundant Fe–Mn oxi-hydroxides relatively depleted in minor transition elements (Co, Cu, Ni). Deposition of massive sulfides and other minerals may also take place near hot-water vents, as the fluids emerge. Clay minerals are often not abundant in hydrothermal deposits, perhaps because of an excess of Fe or depletion of Si or Al. When noticeably present, clay minerals basically consist of nontronite, a typical dioctahedral Fe smectite that usually precipitates fairly close to the vent areas. Nontronite is considered as an end-member characterized by very abundant Fe and little Al and Mg (Fig. 5). Several other chemical types of smectite occur in the deep ocean, in vent areas, and may have various origins extending from hydrothermal sources to typically soil-derived sources (McMurty et al. 1983).

Nontronite may further evolve into glauconitic minerals during the very early diagenetic history. For instance, the hydrothermal mounds of the Galapagos Spreading Center contain green clay layers interbedded with common pelagic oozes; these green clays display an increase in K contents in the oldest layers, and a development of lathed clay particles made up of rigid, straight, and dislocation-free glauconite (Buatier et al. 1989).

Unstable trioctahedral Fe smectite (Fe^{2+} saponite) locally occurs adjacent to vents in Red Sea depressions filled with warm brines (Badaut et al. 1983). Fibrous clays of both palygorskite and sepiolite types may also form in a hydrothermal

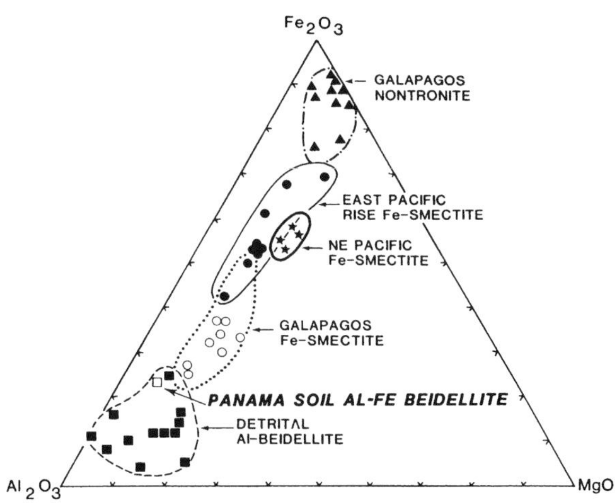

Fig. 5. Al_2O_3—Fe_2O_3—MgO variation diagram for East Pacific deep-sea smectites. (After McMurtry et al. 1983)

environment; they appear to result preferentially from transformation of smectite-rich sediments located in the immediate vicinity of basalts and subjected to the combined influence of hydrothermal fluids, basalt and seawater reacting under semi-confined conditions. The pelagic sediments intruded by or rapidly blanketing hot basalt sills are baked for a few centimeters thickness and experience a very local contact metamorphism; the resulting products consist of regular mixed-layer clays such as corrensite (chlorite–smectite) or of chlorite and high-temperature zeolites (Desprairies and Jehanno 1983).

The progressive dilution of hydrothermal plumes in seawater determines a transition from hydrothermal to hydrogenous environments. It is not easy to recognize the dominant influences and genetic processes occurring in sediment transitions between true hydrothermal and true hydrogenous or terrigenous environments, or to identify hydrothermal signatures in ancient oceanic sediments. The great abundance of Fe and Mn relative to Al and the depletion of Co, Cu, and Ni relative to Zn are fairly reliable indicators of a dominantly hydrothermal impact relative to a hydrogenous influence characterized by more abundant minor metals and to a terrigenous influence marked by less amorphous metal oxides (e.g. McMurtry et al. 1983). Schematically, pure hydrothermal smectite is very rich in Fe (nontronite), whereas pure hydrogenous smectite contains noticeable amounts of Al and Mg (Al-Mg-bearing Fe smectite; Fig. 5). By contrast, terrigenous smectite is frequently of an Al–Fe beidellite type, especially when formed in warm climatic regions and it is often accompanied by various other clay minerals. But such a distribution characterizes end-member clays. Additional information is expected from data on chemical elements from hydrothermal sources (As, Se, some rare earth elements, etc).

3 Detrital Supply and Diagenesis of Clay Minerals

3.1 Terrigenous Input in Modern Oceans

Terrigenous supply represents the dominant agent responsible for the constitution of clay suites in many recent sediments of the world ocean. This is deduced from the latitudinal distribution of clay minerals in many oceanic regions, which parallels their distribution in terrestrial soils, from the scarcity of isotopic and chemical equilibrium of marine clays with the oceanic environment, from the correspondence between the age of most marine clays and the average age of rocks cropping out on the adjacent continents and from evidence for very high particulate terrigenous flux to the ocean relative to the dissolved flux (e.g. Chamley 1989; Weaver 1989). Additional evidence is provided by the absence or scarcity of clay mineral changes at the land–sea transition, the essentially physical character of sorting and settling processes during marine transportation and the frequent similarity of fine-grained mineral compositions in soils, river sediments, sestons, aerosols, marine suspended matter and surface sediments of a given region.

Illite, chlorite, associated quartz, feldspars and various dense minerals, commonly called primary minerals, typically constitute terrigenous species. Kaolinite, random mixed layers and vermiculite minerals are mostly characteristic products of chemical weathering and pedogenesis developed on exposed landmasses. Less common sheet silicates such as talc, pyrophyllite and serpentine also result essentially from the erosion of continental rocks.

Smectites are largely supplied to the sea by rivers and coastal runoff from lowest to highest latitudes and can issue from various soils, weathering complexes, sedimentary rocks, and plutonic and volcanic rocks. They do not suffer from river and marine transportation and are often carried to the marine environment farther than other minerals because of their high buoyancy. Both non-volcanogenic and volcanogenic smectites derived from terrigenous sources add to marine-derived smectitic minerals, from which they can be distinguished only by detailed geochemical and micro-morphological investigations. Fibrous clays, especially palygorskite but also sepiolite, tolerate transport constraints much better than was classically believed and can constitute detrital components of various marine sediments (in Chamley 1989). This is the case in coastal muds (e.g. southwestern Mediterranean, Persian Gulf), turbidites (Gulf of Oman, eastern Atlantic), hemipelagites (eastern Atlantic, Mediterranean), and pelagites (Arabian Sea). Palygorskite bundles survive reasonably both the eolian and the subaqueous transport. Like smectites, fibrous clays set a problem of boundary between detrital and authigenic sources in marine environments.

The present-day zonation of terrigenous clays in the ocean is basically controlled by the climate. Kaolinite forms mainly through chemical weathering in soils of humid low-latitude regions, whereas chlorite and illite mostly derive from the physical weathering of magmatic and metamorphic rocks that crop out widely in high-latitude and desert regions. Random mixed layers, vermiculite minerals and poorly-crystallized smectite mainly characterize temperate regions; well-crystallized Al–Fe smectites largely form in soils of warm arid and poorly drained regions (Millot 1970). This soil- and climate-driven zonation is particularly well reflected by Atlantic Ocean sediments, and indicates the accessory influence of north-south currents on the clay distribution relative to latitudinal influences (in Chamley 1989).

The climate, however, is only one of the factors responsible for the distribution of terrigenous minerals in modern oceans. The average petrography of source regions becomes essential when there is little chemical weathering, which theoretically determines the erosion and deposition in the sea of any mineral occurring in exposed rocks. Thus kaolinite, smectite, and other so-called low-latitude minerals may be supplied in noticeable amounts in Arctic or peri-Antarctic seas through the erosion of ancient sediments and soils where they have been stored during past hydrolyzing periods. The same reasoning explains the active supply of paleosol-derived kaolinite in marine basins located off present desert or arid regions (e.g. Saharian range). Transportation by nearshore surface or density currents also modifies the original clay zonation induced by climate, as does long-distance advection by marine or eolian currents. Such differentiation processes allow the identification of the hydrological regimes and their seasonal controls in a given area. For instance, the clay mineral flux issuing from the Columbia river is characterized by abundant smectite in a transect along the western coast of the United States, while the south Oregon and north California rivers provide abundant chlorite and associated illite (Karlin 1980). In winter time, discharge of the Oregon and California coastal streams is maximum and associated with northwards marine currents, which determine the supply and transportation of abundant chlorite. In summer time, the Columbia river discharge is maximum and determines a smectite-rich turbid plume associated with a southwestwards current off the river mouth (Fig. 6).

3.2 The Origin of Smectite in Old Common Marine Sediments

3.2.1 Classical Potential Origins

Many clay mineral assemblages deposited under open, not chemically-driven conditions in marine basins since the Early Mesozoic, are comparable to modern assemblages and largely result from terrigenous input without noticeable diagenetic changes. This is potentially the case for all clay mineral groups (Chamley 1989; Weaver 1989).

Nevertheless, the smectite group raises a particularly difficult question since it includes mineralogical varieties which suggest diversified possible origins. The chemistry and shape of the individual particles, and some other geochemical characters (e.g. isotope composition) may help to recognize the precise origin of smectites, but such an aim is often complicated by the coexistence of mineral varieties that proceed from different sources. In addition, some smectites yield variable chemical compositions in a given genetic environment. It is, therefore, sometimes difficult to distinguish easily and accurately between the following possible origins (Fig. 7):

1. Reworking from soils developing under hydrolyzing conditions in poorly drained areas, or from past sediments that resulted from reworking of such soils (mainly Al–Fe beidellites).
2. Reworking of subaerial alteration products from exposed volcanoes (Al to Mg or Fe smectites).
3. Submarine evolution of volcanic material, through hydrothermal influence (often Mg-smectites) or not (often Fe-smectites).
4. *In situ* precipitation, or reworking of smectites from evaporative, more or less alkaline basins (diverse Mg-smectites originating in ancient lakes or perimarine basins).
5. Preservation of smectites in some epimetamorphic rocks, especially in carbonate-rich types.

One should add the possibility of reworking degradation smectites of various chemical compositions that have widely formed since the Miocene in soils and weathering profiles from temperate regions, as well as the occurrence in Pacific-like regions of hydrogenous Fe-smectites.

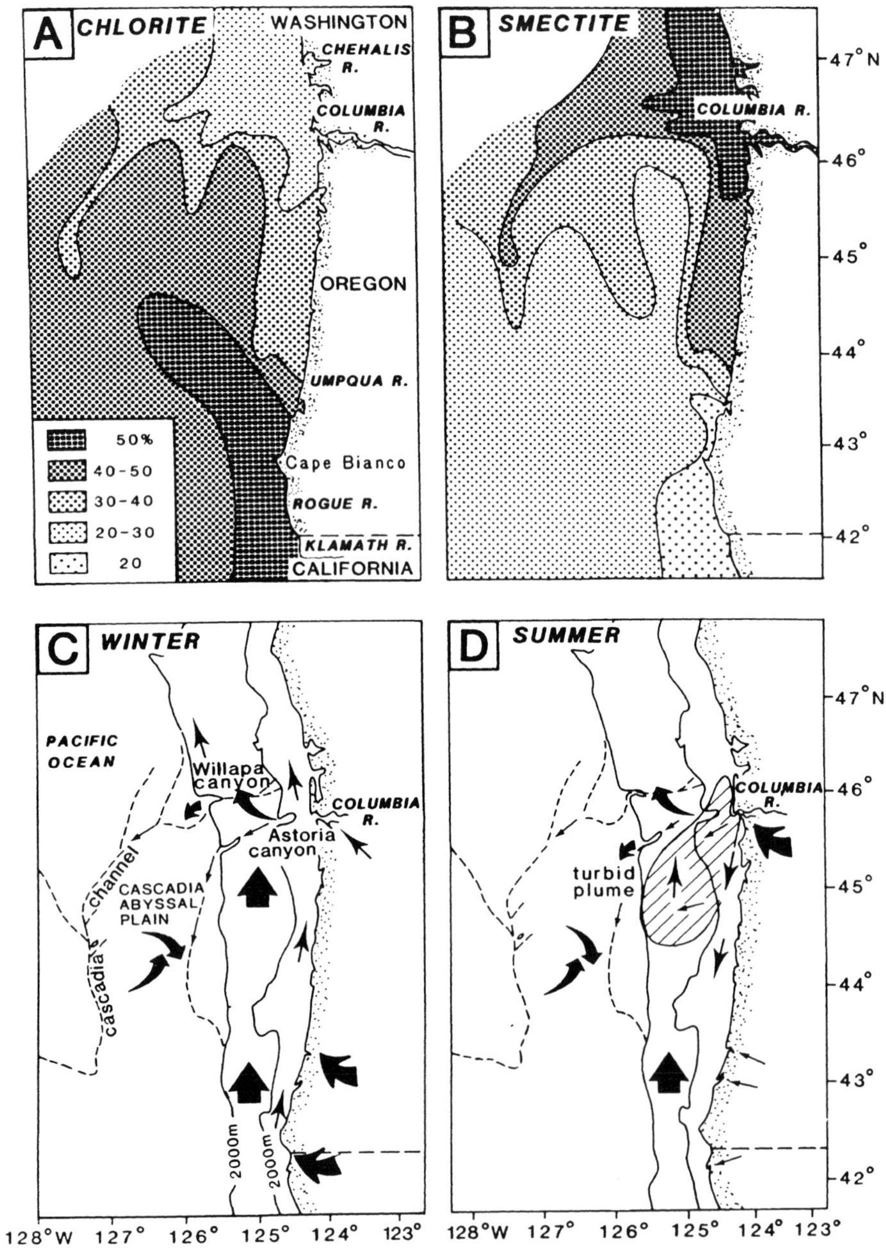

Fig. 6. A Distribution of chlorite and **B** smectite in surficial sediments off the Oregon coast. Transport pathways inferred from the clay mineral distribution patterns, **C** in winter and **D** summer. (After Karlin 1980)

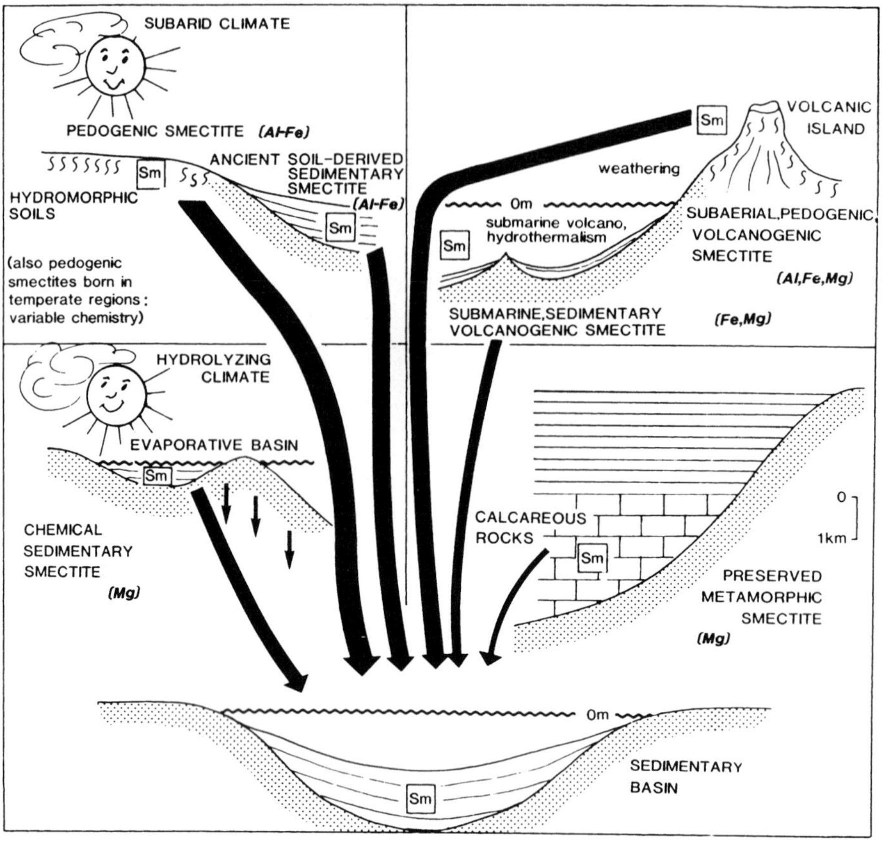

Fig. 7. Six classical origins for smectites in sedimentary basins. (After Chamley 1989)

3.2.2 A Case Study: The Smectites of the Atlantic Ocean

Common Atlantic deposits of late Mesozoic–early Cenozoic age constitute favorable materials to test the hypotheses of either volcanogenic, early diagenetic or detrital origin of smectite, for the following reasons (in Chamley 1989).

1. The Atlantic Ocean was a fairly young ocean marked by an important volcanic activity, the expression of which should be easily recognized in contemporaneous marine sediments.
2. These sediments have usually undergone a moderate depth of burial (less than 2 km) and were subjected to moderate geothermal gradients (around 30 °C/km).
3. They mostly comprise biogenic to clayey components allowing the ready recognition of secondary, *in situ*-formed constituents.

4. They were deposited at noticeable sedimentation rates (usually >1 cm 10^3 year), which *a priori* precluded appreciable hydrogenous genesis at the sediment–seawater interface.
5. Their richness in smectite allows a precise identification of the mineralogy, geochemistry and morphology of this mineral species (see Chamley 1989).

In most Atlantic series, the relative abundance of smectite does not depend on the main lithologic successions. The mineral content varies independently of the dominance of either reddish clay, organic-rich clay, limestone, chalk, chert, silt, pelagic or hemipelagic ooze. This general independence, combined with the absence of control of smectite abundance by burial depth, suggests that the mineral formation and evolution are not noticeably controlled by early diagenesis. Of course, this general behavior does not preclude the possibility of minor lithologic control, such as local degradation in organic environment, or dependence on short-scale changes like limestone–marl alternations.

Thousands of X-ray diffraction analyses made on Cretaceous–Paleogene sediments from the Atlantic domain indicate that most smectites contain few illite-like layers and constitute dioctahedral minerals. X-ray diffraction characters vary little from one area or geologic period to another. On transmission electron micrographs, smectites mostly display flaky outlines very similar to those encountered in continental soils. Some laths occur sometimes at the periphery of smectite flakes (Holtzapffel and Chamley 1986), but they are not associated with any change of the quantitative mineralogical composition and they display only slight chemical adjustments relative to non-lathed smectite particles (Steinberg et al. 1984).

Microprobe investigations on insulated particles show that smectites from common Cretaceous-Paleogene shale-to-limestone rocks consist of beidellites marked by the presence of Al in both the tetrahedral and the octahedral positions, and by a partial substitution of Fe^{3+} and Mg for octahedral Al (Debrabant et al. 1985). The relative amounts of Al, Fe, and Mg may vary in a moderate manner, which characterizes a rather well-defined group of Al–Fe beidellites. The geochemical composition is very similar for flaky smectite particles and lathed overgrowths. This composition is comparable to that of smectites originating in continental soils of warm regions or reworked from such soils in peri-continental deposits (e.g. Table 1). By constrast, the chemical composition of most Atlantic smectites differs greatly from that of smectites derived from alteration of volcanic glass or basalt, or from hydrothermal or hydrogenous precipitation, which are much richer in Mg or Fe. Only local examples of smectite–bentonite layers clearly derived from *in situ* transformation of volcanic glass exist (e.g. Deconinck et al. 1991).

Sediments from the eastern Atlantic Ocean enriched in clay fraction (60–90%) and containing abundant smectite, have been studied for their REE compositions from both the stratigraphic and the geographic points of view (Bonnot-Courtois 1981). The vertical distribution of the REEs displays a very homogeneous pattern in most sediments deposited from early Cretaceous to Quaternary. When normalized to continental shales, the curves appear rather flat. The REE are devoid of both enrichment in heavy elements (HREE) and depletion in Ce. Only a slight enrichment in light elements (LREE) is usually recorded. All these characters suggest a

	Tetrahedra		Octahedra				Interlayers			
Samples	Si	Al	Al	Fe^{3+}	Mg	Ti	Na	K	Mg	Ca
Senegal Basin, Kafoutine										
N_0 620, Maastrichtian	3.12	0.88	1.00	0.50	0.30	0.02	0.62	0.42	-	0.33
N_0 1190, Santonian	3.22	0.78	1.31	0.44	0.23	0.02	0.32	0.30	-	0.20
Cape Verde Basin, DSDP 367										
19-1-100, late Cretaccous	3.74	0.26	1.23	0.43	0.44	0.02	0.03	0.17	-	0.06
22-6-91, late Albian	3.31	0.69	1.07	0.69	0.32	0.02	0.05	0.40	0.05	0.07
23-2-46, Albian	3.57	0.43	1.47	0.38	0.18	0.02	0.04	0.16	0.08	0.04
Mormoiron Basin, SE France										
Al-Fe smectite (Wyoming)	3.92	0.08	1.21	0.40	0.30	0.03	?	0.21	-	0.16
West African Soils (parent rock)										
M-5D (gneiss w/amphibole)	3.67	0.33	1.07	0.58	0.47	0.03		?		
GB 92 (granite)	3.38	0.62	1.43	0.49	0.21	0.06		?		
LIV-2-3 (Eocene clay)	3.61	0.39	1.37	0.46	0.24	0.05		?		
Godola (granite)	3.30	0.70	1.29	0.68	0.19	0.05	?	0.09	?	0.01

Table 1. Structural formulae of smectites from clay to calcareous sediments of the late Cretaceous of tropical northeastern Atlantic. Comparison with Paleogene detrital Al—Fe smectite (Wyoming) of the Mormoiron Basin, SE France, and with smectites from recent West African soils. (In Chamley 1989)

strong continental influence and the absence of noticeable impact of volcanic activity or seawater influence on smectite formation.

Smectites of diverse grain sizes from Cenomanian and Paleocene sediments of the North Alantic Ocean were studied by Clauer et al. (1990) for their Rb–Sr compositions. For instance, Paleocene oozes at DSDP (Deep Sea Drilling Project) site 386, southeast of Bermuda, display $^{87}Sr/^{86}Sr$ ratios of 0.70860±0.00040 for the 0.2–0.4 μm fraction, and of 0.70899±0.00043 for the <0.2 μm fraction. These values are much higher than those characteristic of Paleocene marine depositional environments (0.70783±0.00028), but slightly lower than those typical of pure detrital origin. The apparent ages indicated by Rb–Sr isochrones range between 167±5 and 200±6.10^6 years, instead of 56±3.10^6 years which corresponds to the Paleocene stage. These results indicate that Atlantic smectites were dominantly inherited from old, presumably terrigenous parent-rocks and were only slightly modified by early-diagenetic changes.

In the Atlantic sediments, smectite is often accompanied by clinoptilolite and opal CT (cristobalite-tridymite series), which suggests a mineral paragenesis developed through diagenetic processes. In fact, the distribution of the three mineral groups is largely independent. If clinoptilolite- and/or opal-rich sediments frequently contain abundant smectite, smectite-rich sediments are often devoid of clinoptilolite and opal CT. Clinoptilolite and opal develop mainly in late Cretaceous to Paleogene sediments (Stonecipher 1976), while smectite-rich deposits occur from late Jurassic to late Paleogene. In addition, the smectite composition and relative abundance do not depend on the presence or absence of zeolite or opal.

Volcanic glass is seldom recognized in significant amounts in late Jurassic to Cenozoic sediments of the Atlantic domain, where smectite is abundant. It is somewhat intriguing that marine sediments whose abundant smectite is attributed by some authors to diagenetic transformation of volcanic glass are virtually devoid of obvious volcanic remains. Most authors supporting the hypothesis of a volcanic origin for smectite in common sediments imply that initial volcanic glass has been altered and destroyed through smectite formation. A complete lack of glass is nevertheless difficult to understand, knowing that the potential alteration of volcanic material greatly differs with size, shape, porosity and composition, and that non-altered volcanic debris usually still exist in deposits rich in unquestionably volcanogenic smectite. In addition, the sediments blanketing the oceanic basalt crust generally reveal a decrease in smectite abundance relatively to illite, chlorite, and kaolinite (e.g. Robert 1987). Albian sediments deposited close to the North American continent contain abundant smectite, whereas contemporary deposits located close to the Atlantic ridge and in the volcanically active Cape Verde basin are depleted in this mineral (Chamley and Debrabant 1984).

In summary, the characters and distribution of most late Mesozoic-to-Cenozoic Atlantic smectites appear to depend little or not at all on burial depth, lithology, associated minerals and volcanic activity. On the contrary, these characters and the distribution are largely compatible with land-derived minerals, suggesting that smectite is only little controlled by post-sedimentary processes and notably generated on the landmasses adjacent to the Atlantic Ocean; this differs from the somewhat theoretical approaches proposed in the literature (e.g. Thiry and Jacquin 1993). Smectite-bearing formations eroded on landmasses may have been very diverse and may comprise various weathering profiles, including those developing at the expense of volcanic rocks (e.g. Kimblin 1992).

3.2.3 Smectite and Sea-Level Changes

Recent data on Jurassic-to-Paleogene deposits from Atlantic and Western Tethyan Oceans show that the highest amounts of the smectite group frequently characterize geological periods marked by either long-term or short-term high sea-level stages (e.g. Chamley et al. 1990; Debrabant et al. 1992; Deconinck 1992). Such a relationship is demonstrated during periods and at locations as diversified as the Jurassic sediments drilled in the northern Paris Basin (Fig. 8), the Cretaceous–Paleogene deposits cropping out in central Italy near Gubbio, and the Cretaceous series deposited in the Moroccan Atlas Gulf.

The main periods of high sea-level stages result notably from an increase in the oceanic spreading and subduction rates, as well as in thermal fluxes from asthenosphere to lithosphere, and correlate with warm, often humid, climates on landmasses. Such conditions prevailed during late Jurassic to Paleogene times and may have determined the increase in smectite abundance in marine sediments during four distinct processes:

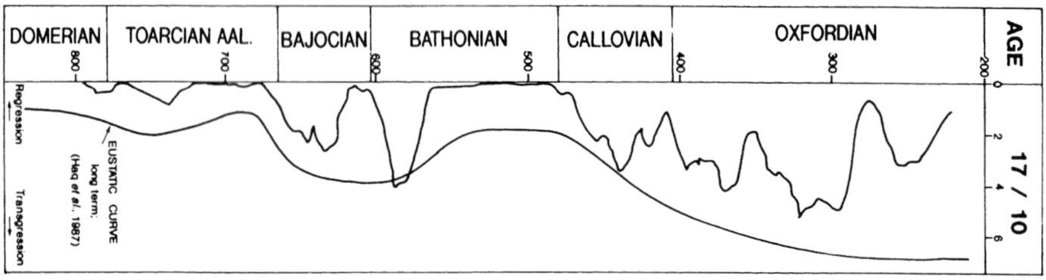

Fig. 8. Smectite/illite ratio (17/10 Å, X-ray diffraction of glycolated samples) in the Liassic to Oxfordian series of borehole ANDRA A 901, north of Paris Basin (after Debrabant et al. 1992). Comparison with the long-term eustatic curve. (After Haq et al. 1988)

1. The pedogenic formation and reworking towards the ocean of Al–Fe beidellites was enhanced when the climate was warm and the sea level high, since the chemical weathering expanded laterally and vertically and the erosion mainly affected the soil blankets. The preservation of detrital smectites in marine sediments is supported by the lack of noticeable change after the deposition (see above).
2. The smectitization of volcanic products, a mechanism widely quoted in the literature, increased when the volcanic activity due to active spreading increased. Nevertheless, the scarcity of volcanic debris and of volcanic clay mineral shapes and chemistry in common marine sediments, as well as the lack of mineralogical gradient around the oceanic ridges, weaken the reliability of such an explanation (see above).
3. The increase in hydrothermal fluxes contemporary with the expansion of ridges determined an increased supply of the ions necessary for the smectite growth (Si, Fe, Ca, etc.). Such a hypothesis is supported by a combined increase in both Mn and smectites in some sediments deposited during high Mesozoic sea-level stages (Accarie et al. 1989). However, the smectites formed in recent hydrothermal environments present morphological and geochemical patterns different from those of common Mesozoic-to-Paleogene beidellites (e.g. Buatier et al. 1989).
4. Early diagenetic processes in open-sea sediments were favored when the terrigenous ion output increased due to active chemical weathering, and the exchanges between particles and interstitial fluids or seawater were enhanced by increased temperature of bottom water masses. Unfortunately, there is a lack of present-day equivalents to Mesozoic conditions, and most clay minerals in recent Atlantic-type deposits are hardly sensitive to early diagenetic influence.

In summary, there is a convergence in the four geodynamical mechanisms able to favor the development of beidellite minerals in common late Mesozoic–early Cenozoic sediments, despite the fact that it is difficult to quantify each of these mechanisms. The detrital supply constitutes the best-documented process and the *in situ* smectitization of volcanic products constitutes the less documented influence at our present state of knowledge. Whatever the precise causes, early diagenetic processes were enhanced by ionic output from terrigenous or/and oceanic sources, as well as by the warm temperature of water masses, which are dominantly controlled by climatic factors. Al–Fe smectites appear to represent specific minerals possibly formed at the surface of the Earth in both the continental and the marine environments, during fast spreading high sea-level and warm-climate stages. Under surficial conditions, they represent reliable markers of processes driven by internal geodynamic constraints (Chamley et al. 1990).

3.3 Diagenesis with Increasing Burial Depth

Knowledge of clay behavior after burial has increased greatly in the last few years, mainly because of the development of high resolution electron microscope and microprobe techniques and of *in situ* borehole measurements:

1. The successive steps in the decrease in expandability of smectite minerals and in the formation of illite species represent a reliable and easy marker of depth effect that can be observed on X-ray diffraction diagrams (Fig. 9). Illitization appears to result mainly from micro-environmental, chemical exchanges involving the K and Al released by the degradation of smectite and also of coarser-sized feldspars.

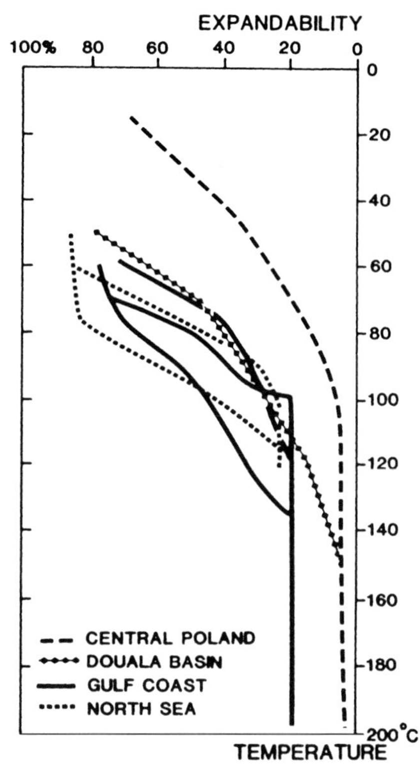

Fig. 9. Smectite-to-illite conversion in shales from different basins, as a function of temperature. (After Srodon and Eberl 1984)

2. Modifications affecting the clay minerals usually occur at burial depths exceeding 2 km and appear to be less progressive than was formerly thought. In series marked by normal geothermal gradients (about 30 °C/km), the major ordering processes develop at between 2.5 and 3.5 km and do not progress much beyond 5–6 km.
3. The residence time at a given temperature in a buried series appears to be more important than the absolute, instantaneous temperature values. Borehole experiments show that some deeply buried sediments are less affected by clay diagenetic changes than shallower sediments of the same lithology and age subjected to comparable temperatures. This points to the dominant control of kinetics relative to chemical conditions.
4. The physical reorganization of clay particles and aggregates subjected to progressive compaction during early diagenesis, is followed by a physico-chemical reorganization during late diagenesis. Both types of modifications converge to determine a subparallel to parallel orientation of clay particles and clay layers.

Some classical concepts on burial diagenesis of clay should be modified in view of recent results (see Chamley 1989):

1. Chlorite does not usually result from the transformation of smectite in a Mg environment. Diagenetic chlorite is often not abundant (less than 10% in Gulf Coast sediments) and is of a Fe type; Mg chlorite develops commonly only during metamorphism. Common diagenetic chlorites could be by-products of illitization resulting from significant dissolution and recrystallization.
2. Kaolinite is not progressively altered to chlorite or illite with increased depth; it is either stable, growing, or simply destroyed, according to the chemical environment.
3. True regular mixed layers such as allevardite (illite/smectite) or corrensite (chlorite/smectite) do not frequently form as transitional minerals during the burial history of smectite. Regular mixed layers are only rarely identified in sediments buried at less than 4–6 km and often appear to be related to specific lithologies.
4. The concept of mixed layering during the illitization process appears questionable in view of high-resolution electron microscope data. Mixed-layer spacings occur only locally on ultra-thin clay sections cut normal to the layer planes. Many observations indicate the existence of packets of single clay minerals (smectite, illite, chlorite, kaolinite) rather than random or regular layer alternations (e.g. Ahn and Peacor 1986, 1987). This raises the question of the actual significance of X-ray diffraction patterns attributed to random or ordered mixed-layer minerals. Some authors have thought that they result from inter-particle instead of intra-particle effects, suggesting that illitization at depth results from dissolution–recrystallization processes (Nadeau et al. 1984). Other authors describe the coexistence of smectite and illite packets, the latter increasing at depth and intergrowing through transformation of the former (Ahn and Peacor 1986).

5. Diagenetic processes do not depend on the absolute age of the buried series. Some very old sediments contain very diverse clay assemblages similar to those of recent detrital deposits, and late Tertiary series may have experienced pervasive diagenetic changes. More important than geologic age and lithostatic pressure are the geothermal gradient and residence time at a diagenetically active temperature. Notice that some thick sedimentary piles deposited rapidly may lack appreciable diagenetic clay change despite an overburden of about 5 km, as recorded in the Pliocene–Pleistocene mudstones accumulated on both sides of the permanently uplifted Central Range of Taiwan island (Chamley et al. 1993).

The mechanisms of clay evolution during burial diagenesis largely differ according to lithology, fluid nature and pressure, and geothermal gradient. The major control appears to be the sediment permeability, which determines the water/rock ratio and the importance of ion exchanges. In most sediments, chemical migrations occur only over short distances (i.e. less than a few meters). In almost closed systems such as compacted argillaceous series, illitization results chiefly from transformation of adjacent detrital smectite, while chlorite and kaolinite may form locally in small amounts along processes involving dissolution and precipitation. In more open systems, allowed by coarser lithology or high interstitial water flows, diagenetic illite forms newly after dissolution of smectite and K feldspars. All intermediate situations may exist.

The diagenetic history of buried clay series closely parallels that of organic matter, which has led to extensive investigations of the mineral characteristics and evolution in relation to oil generation, migration, accumulation, preservation and exploitation. Abundance of smectite in initial sediments favors the production of hydrocarbons because this mineral is able to adsorb abundant organic compounds, to release large amounts of water which may act as a carrier, to increase the permeability through its degradation, and to act as a catalyst. The major stage of smectite--illite ordering during clay diagenesis often occurs fairly shortly before oil generation and migration, indicating close organic–inorganic interactions.

3.4 Other Diagenetic Constraints

Lateral clay diagenetic changes in mountain belt areas reflect the attenuate effects of metamorphism linked to orogeny. Particularly well documented in the western Alpine range, the clay mineral modifications recorded in Cretaceous deposits outcropping from outer to inner parts of mountain chains roughly parallel the modifications observed vertically at increasing burial depths. Clay mineral suites tend to be enriched in illite and chlorite, the latter being often dominant (e.g. Levert and Ferry 1988; Fig. 10). Overburden of sedimentary formations by tectonic nappes may also induce mineral modifications similar to those resulting from progressive burial. Less permeable and chemically confined sediments may resist the diagenetic constraints induced by depth of burial or tectonics. This is the case of the limestone–marl alternations subjected to vertical or lateral diagenesis, in which the marly interbeds are preserved much better and longer than calcareous beds

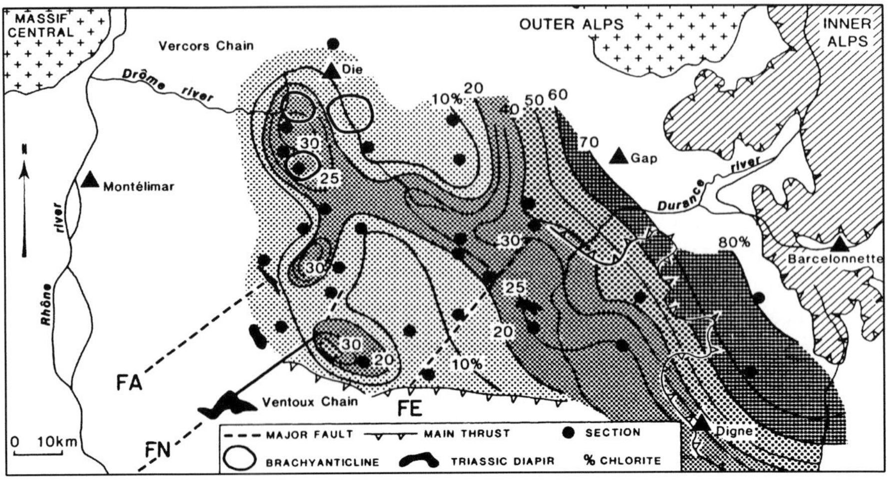

Fig. 10. Percentage distribution of chlorite in Kimmeridgian isochronous limestone—marl alternations from the Vocontian basin, southeastern France. (After Levert and Ferry 1988) *FA* Aygues fault; *FE* Eygalayes fault; *FM* Ménée fault; *BA* brachyanticlines

(Deconinck 1987). Another case of resistance to thermodynamic changes concerns smectites, which may be preserved locally and display enhanced crystallinity in alkaline, often magnesian environments of late diagenetic–early metamorphic zones.

Highly permeable sediments favor horizontal circulation of groundwaters and induce specific diagenetic changes that interfere strongly with burial effects. This is especially the case for coarse to medium-sized sandstones, the major constituents of which are detrital and may have been modified or complemented by variable amounts of various secondary minerals. Clay minerals are the species most often formed diagenetically in permeable sandstones. They mainly comprise kaolinite, illite and chlorite, along with mixed layers and other minerals. The habits, mineral suites and chemical compositions of diagenetic clays in sandstones display great diversity (massive or hexagonal crystals, blades, rims and cements; e.g. Huggett 1984). Diagenetic clay minerals in permeable sedimentary rocks are actively studied from environmental, textural and kinetic points of view because their formation or inhibition dramatically affects hydrocarbon reservoir properties. Occupancy of pore spaces by clay decreases the rock permeability, which is of great importance if diagenetic processes occur in an early phase. Infilling of sandstones by oil or gas prevents further argillization, whereas migration of saline waters favors late diagenetic illitization.

Volcaniclastic sediments are particularly sensitive to diagenetic modifications because of their usually high permeability, the high reactivity of their amorphous constituents to chemical exchanges, and their frequent association with hydrothermal exhalations. A mineral sequence commonly encountered at increasing depth comprises smectite and zeolites of the clinoptilolite, heulandite and analcite type, successively. Analcite also constitutes local accumulations (analcimolites) derived from early diagenetic alteration of fine vitric ash in alkaline lacustrine deposits and, subsequently, possibly altered into various clay minerals. Bentonites are thin, widespread beds of clay-rich sediments that derive chiefly from early diagenetic subaqueous alteration of fine volcanic ash. Bentonites are characterized by either smectite (bentonites s.s. or S-bentonites), kaolinite (K-bentonites or tonsteins), or illite (I-bentonites). Smectite-rich bentonites are particularly frequent and result mostly from alkaline, submarine alteration (e.g. Fisher and Schmincke 1984; Deconinck et al. 1991).

4 Preservation and Destruction of the Paleoenvironmental Record of Marine Sedimentary Clay

4.1 Paleoclimate

Clay minerals preferentially form through weathering and pedogenesis at the surface of the earth. The widespread, easy and nearly continuous erosion of soft pedogenic blankets results in dominant participation of soil-derived clay minerals in the formation of recent and past sediments. As clay minerals apparently experience none or only slight diagenetic modifications in most oceanic sedimentary series buried to less than 2–3 km and devoid of significant volcano-hydrothermal impact, they may be useful markers of past continental climates. The correlations observed in the variation of mineral, biogenic and isotopic markers demonstrate the usefulness of detrital clays in reconstructing past climate successions in Meso–Cenozoic sediments. Clay minerals basically express the intensity of weathering, especially that of hydrolysis, in landmasses adjacent to sedimentary basins. Pedogenic minerals fundamentally integrate the combined effects of temperature and precipitation and sometimes provide additional data on seasonal rainfalls or drainage conditions. Furthermore, clay mineral records sometimes provide useful information on indirect climatic changes such as those affecting marine currents, sea level stands and ice blankets on land or sea.

Paleoclimatic reconstruction from clay mineral data can be done for different geological periods (in Chamley 1989). The middle to late Quaternary alternation of glacial and interglacial periods generally correlates with alternately dominant physical and chemical alterations of clay assemblages. At temperate latitudes, cold marine periods indicated by microfauna or oxygen isotopic data, usually correspond to sedimentary levels enriched in well-crystallized illite, chlorite and smectite, and in feldspars, which indicates an increasing supply of rock-derived minerals and, therefore, an average cold and dry climate on land. By contrast, periods characterized by warm marine water often correspond to enhanced amounts of kaolinite, poorly crystallized expandable minerals (smectite, random mixed layers) mainly reworked from weathering profiles. These variations thus express the dominant supply to the sea of materials alternately eroded from rocky substrates and from currently forming soils.

Cenozoic sedimentary series frequently display a step-by-step increase in slightly weathered and rock-derived minerals, namely illite, chlorite, random mixed layers and feldspars, at the expense of pedogenic Al–Fe smectite and often kaolinite (in Chamley 1989). This general trend, whose accelerating and slackening stages roughly parallel the cooling and warming periods indicated by climatic markers such as oxygen isotopes, is particularly well documented in the Atlantic range characterized by stable margins and little tectonic instability (Fig. 11). This general clay mineral change is attributed to the irregular transition from non-glacial to glacial conditions at the earth's surface. The development of ice caps since the late Eocene and the correlative latitudinal climate differentiation resulted in a

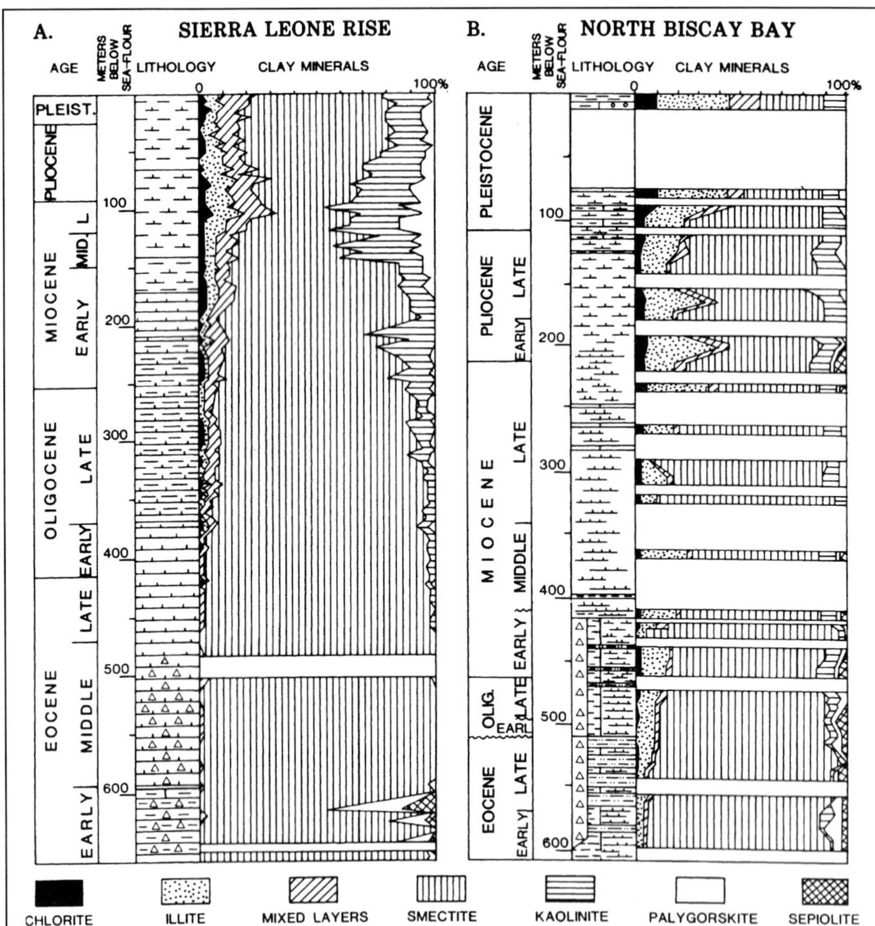

Fig. 11. A—B Clay stratigraphy in Cenozoic sediments. **A** At DSDP Site 366 on the Sierra Leone Rise, north-equatorial eastern Atlantic (after Robert 1982). **B** At DSDP Site 400 in the north Biscay Bay, northeastern Atlantic. (After Debrabant et al. 1979)

decrease in chemical weathering in the course of time and an increase in physical alteration. Other modifications affected the marine circulation, which may also have been expressed by suspended clay mineral assemblages (e.g. Robert and Chamley 1991).

Pre-Cenozoic series offer a wide field for paleoclimate research based on clay data, and this area is still little investigated. Cretaceous times often correlate with smectite-rich clays which mainly result from erosion of thick pedogenic blankets developed under high temperature and seasonal variations in humidity. On the other hand, Jurassic sediments usually display large amounts of kaolinite in clay fractions, suggesting warm conditions combined with constant annual humidity. Both Jurassic and Cretaceous stages appear to be characterized by little variation in climatic conditions over time, at various latitudes. The differentiation in soil development is noticeable only from late Cretaceous and especially from Paleogene time. Paleoclimate reconstructions from clay stratigraphic data may be used in series as old as the late Precambrian, especially in regions where there were no significant postsedimentary overburden and tectonic structuration and where clay-rich sediments were actively and regularly deposited at the periphery of variously weathered landmasses (in Chamley 1989).

4.2 Past Continental Sources, Paleocirculations and Tectonics

Many examples are available in the literature on the use of clay assemblages as indicators of detrital sources in recent and ancient sediments from various depositional environments. Progress has still to be made on the precise identification of terrigenous origins, since a given clay-mineral group may result from different sources in a single region. For instance, smectites in recent sediments of the northern Indian Ocean may derive from common soils developed under a hydrolyzing climate on the Asian continent, from soils and rocks cropping out in Australia or East Africa, from the subaerial alteration of volcanic rocks (Indonesia), from altered submarine volcaniclastites or basalts, or from *in situ* growth (Bouquillon et al. 1989). The combination of mineralogical, micro-morphological and micro-chemical data helps greatly in reconstructing recent and past sources.

Extensive investigations also report the use of clay suites as paleocurrent markers. Reliable studies exist on the Quaternary and late Tertiary periods, for which advected minerals may serve as indicators of the different superimposed water masses, of their variations in volume and reaction relative to climate changes, of the respective influence of wind and water circulation, or of the existence of marine barriers such as divergence zones (e.g. Debrabant et al. 1993). Further investigations now concern older Cenozoic periods. For instance, Early Paleogene deposits at DSDP and ODP (Ocean Drilling Program) sites of the Atlantic and southern Oceans are characterized by increased amounts of palygorskite (and/or sepiolite) at low latitudes and of kaolinite at high latitudes principally (Robert and Chamley 1991). Good correlation of clay mineral and oxygen isotope data indicates a dominant control of climate. Results suggest that the global warming which culminated near the Paleocene–Eocene boundary favored evaporation and formation of fibrous clays in low latitude coastal areas, while ascendance of warm air resulted in precipitation and kaolinite genesis in continental hinterlands. Enhanced poleward transfers of heat involving surface marine currents and atmospheric circulation warmed high latitude areas which also experienced high precipitation leading to kaolinite genesis (Fig. 12).

Tectonic phases generally determine important and long-lasting modifications of detrital clay assemblages, which differ from most effects induced by changes in climate or currents. As tectonic rejuvenation usually impedes the development of continental soils and favors the direct removal of material eroded from various rocky substrates, the resulting deposited clays often displaying a mineral diversification. Because of their large dispersal properties, clays may reflect rejuvenation phases far away from tectonized areas in sedimentary basins never filled by coarse terrigenous fractions. Because of their dependence on meteoric conditions, pedogenic clays are highly sensitive to slight tectonic changes; they are able to reflect some epeirogenic stages that determine small morphologic modifications only, but noticeable changes in the development of surficial soils. A decrease in epeirogenic activity tends to be followed by the development of new soils, which may form in a few hundred years and whose erosion products accumulating in sedimentary basins bear modified paleoenvironmental messages.

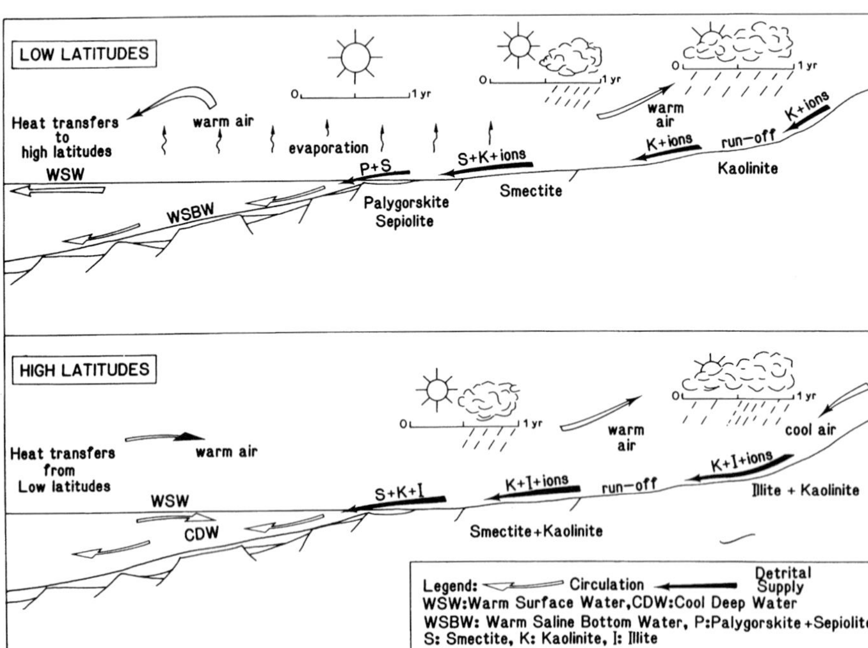

Fig. 12. Interpretation of Early Eocene weathering conditions, thermal exchanges and paleocirculation in the Atlantic range, as inferred from clay mineral data. (After Robert and Chamley 1991)

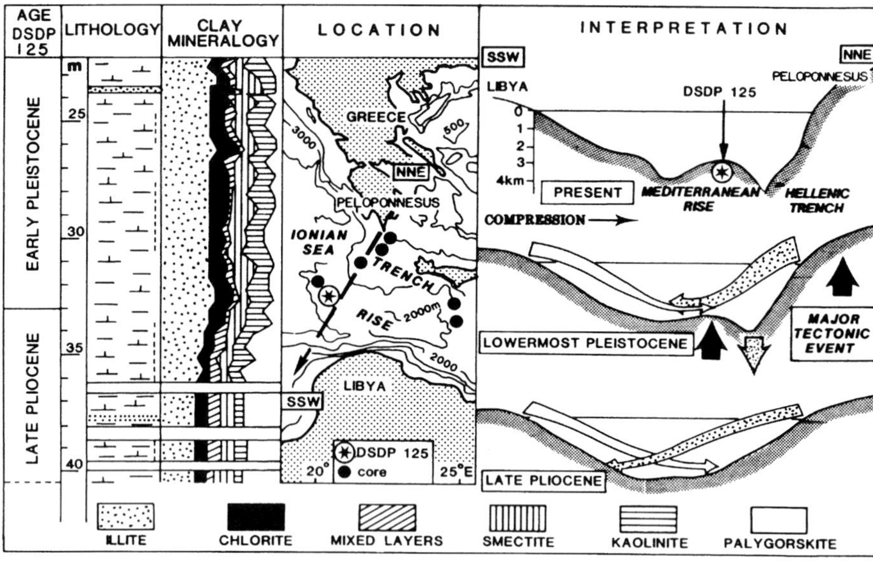

Fig. 13. Clay mineral stratigraphy and interpretation at the Pliocene—Pleistocene transition on the eastern Mediterranean rise, Ionian Sea. (In Chamley 1989)

Sedimentary clay suites reflect the tectonic events occurring on exposed landmasses, especially the uplift movements determined by compression phases or by the positive reaction of continental crust to subsidence in adjacent marine basins. Clay successions may also reflect tectonic events occurring in the basins, such as those resulting in the formation of a submarine barrier (uplift) or of a trough or trench (downlift); in both cases, the dispersion of clays by marine currents tends to be impeded, which often modifies the composition of detrital assemblages. For instance, late Cenozoic sedimentary sections drilled on the Mediterranean rise and Hellenic trench, eastern Mediterranean, display a decrease in Africa-derived palygorskite and an opposite increase in Greece-derived illite and associated minerals, close to the Pliocene–Pleistocene boundary (Fig. 13). Sedimentological, palynological and geophysical arguments suggest that this clay mineral change results both from an acceleration of the Peloponnesus uplift responsible for increasing erosion in southern Greece, and from a combination of uplift of the submarine Mediterranean rise and of tectonic deepening of the Hellenic trench (in Chamley 1989).

4.3 Progressive Obliteration of Paleoenvironmental Messages by Diagenesis

The comparison of contemporaneous series deposited under different geodynamic and diagenetic conditions allows the recognition of the respective parts of syn- and post-sedimentary records visualized by clay mineral assemblages. For instance a multidisciplinary approach including clay mineralogy, chemistry and morphology has been carried out on Cretaceous–Paleogene sediments at site Kafountine 1, which was drilled on land in the Senegal coastal basin and at DSDP site 367 located in the adjacent Cape Verde deep-sea basin (NE Atlantic). It allows identification of the respective controls on the sedimentary record of diagenesis relative to margin subsidence related to ocean opening, tectonic instability compared to climatic changes, sources, transport and depositional conditions (Chamley et al. 1988).

The diagenetic effects are mainly determined by the temperature driven by depth of burial, and they, therefore, occur only in the Senegal coastal basin which experienced very high sedimentation rates (5387 m drilled at the Kafountine site, Valanginian to Miocene). Diagenetic changes related to depth are marked by an increase in clay grain size, increase in crystalline chlorite, vermiculite mixed layers, Al–Mg illite, quartz and Na feldspar abundance, decrease in kaolinite, and disappearance of smectite after a downward decrease in its deficit in interlayer charges. Such changes do not exist at deep-sea site 367 (penetration 1153 m), which points to the importance of lithostatic pressure effects compared to hydrostatic ones (Fig. 14). The comparison of the downhole variation in the amount of illite, chlorite and mixed layers at both sites, clearly reflects the distinct behavior of clay assemblages (Fig. 15). Comparisons between coastal and deep-sea mineralogical and geochemical records indicate that the deepest diagenetic level studied in the coastal basin (5200 m) lies significantly above the top of the anchizone. Diagenetic effects appear to be very low down to 2000 m, low down to 3500/3900 m, and stronger below 4500 m (Fig. 16). Clay diagenetic modifications related to sediment–interstitial water exchanges seem quantitatively negligible. They mainly affect the deep-sea basin, where some clay-rich sediments show iso-mineralogical and nearly iso-geochemical micro-morphological changes (recrystallization of flaky smectite into lathed smectites). Pore water circulation did not significantly modify the sediments in either basin, which is demonstrated by the lack of correspondence between mineralogy or geochemistry and lithology, even in the permeable Late Cretaceous sandstones of western Senegal.

A strong mineralogical break identified in the Cretaceous series (diminution of the illite group, Fig. 15) suggests that tectonic activity significantly decreased after the early–middle Cenomanian, close to the time at which the subsidence rate and crustal thinning decreased on the West African margin (Latil-Brun and Flicoteaux 1986). After that time, moderate sedimentation rates developed in both coastal and deep-sea basins, which resulted in the disappearance of important modifications of detrital sedimentary components. As a consequence, the comparative study of synchronous long-range sedimentary records in both coastal and deep-sea domains allows identification of the combined effects of internal and surficial geodynamic modifications occurring at the continent–ocean boundary (Fig. 16).

Fig. 14. Clay minerals schematic zonation since the early Cretaceous at Sites DSDP 367 (Cape Verde basin) and Kafountine 1 (Senegal basin). Results are given as average values for successive time intervals. (After Chamley et al. 1988)

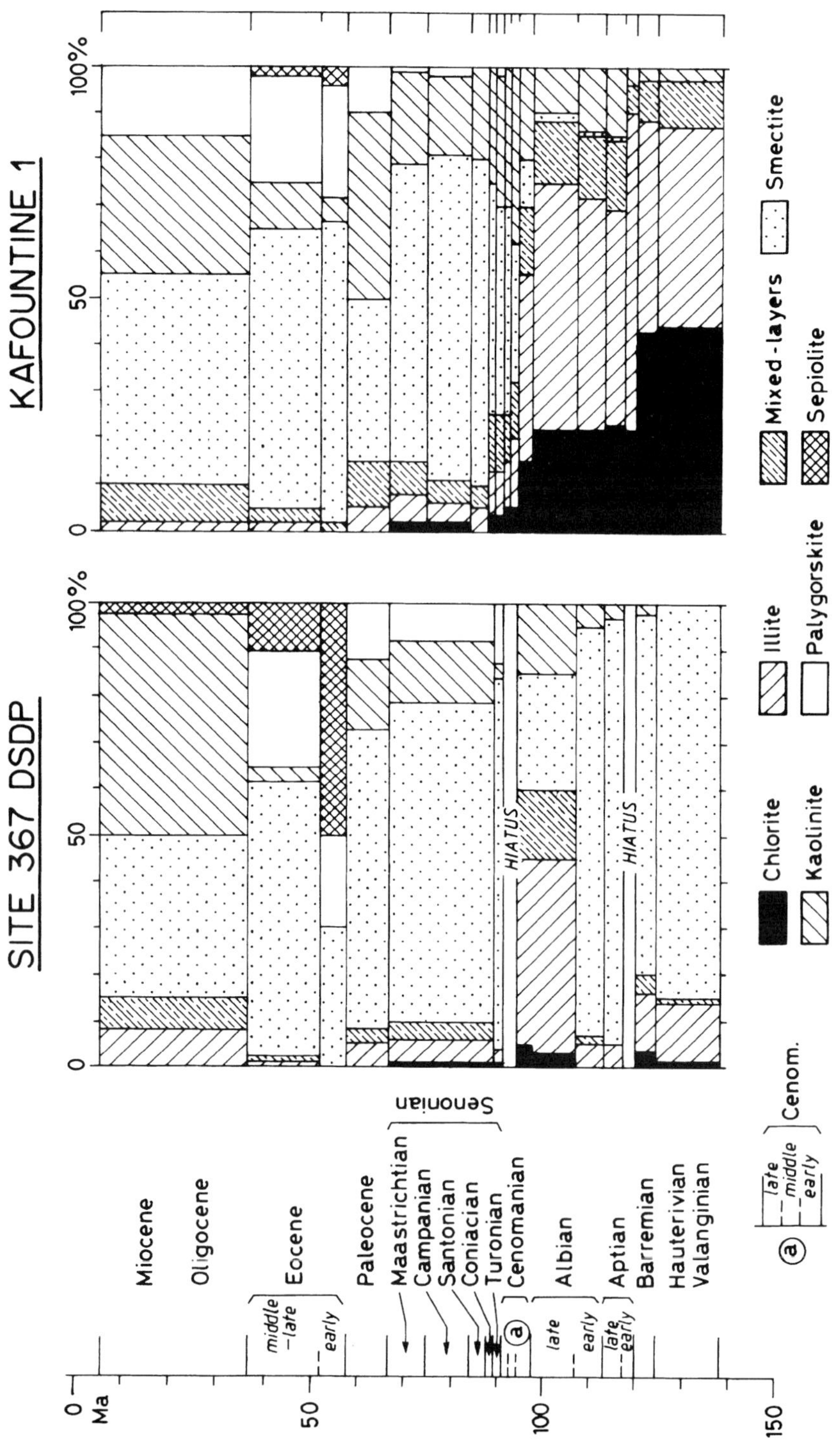
KAFOUNTINE 1
SITE 367 DSDP
0
50
100%
HIATUS
HIATUS
Chlorite
Kaolinite
Illite
Palygorskite
Mixed-layers
Sepiolite
Smectite
0 Ma
50
100
150
Miocene
Oligocene
Eocene
middle
late
early
Paleocene
Maastrichtian
Campanian
Santonian
Coniacian
Turonian
Cenomanian
Senonian
Albian
late
early
Aptian
late
early
Barremian
Hauterivian
Valanginian
a
late
middle
early
Cenom.

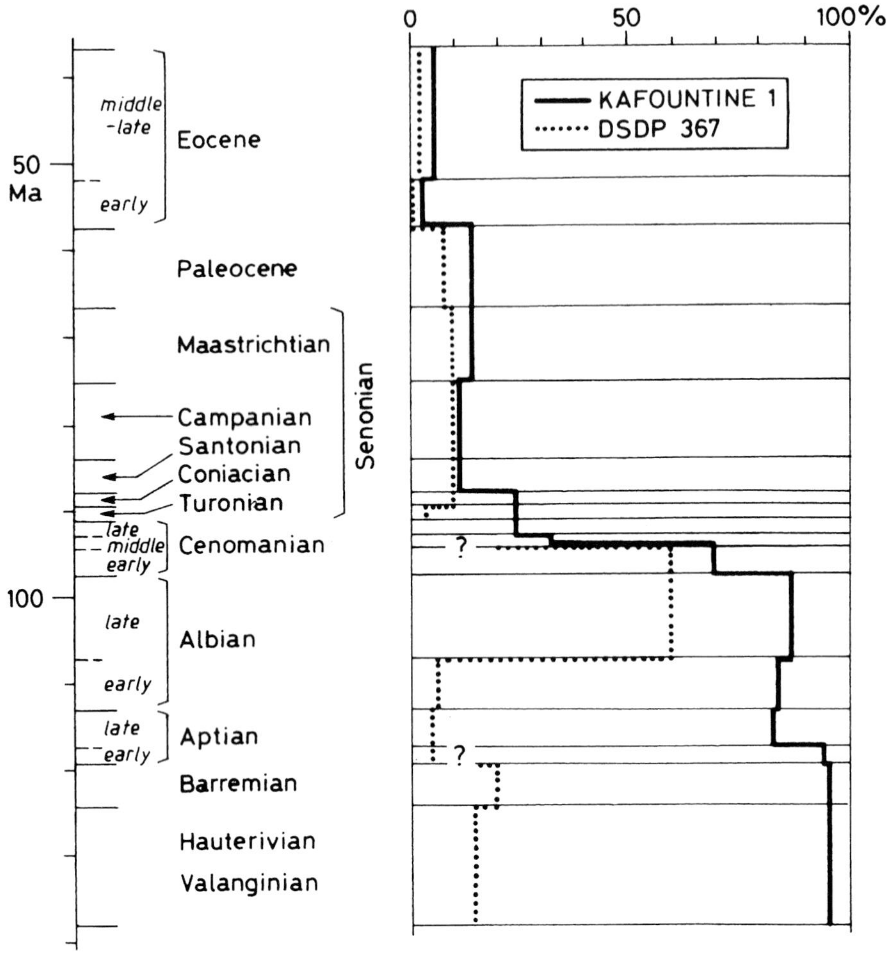

Fig. 15. Relative abundance of illite + chlorite + random mixed layers at sites 367 and Kafountine during the Valanginian to Eocene time interval. (After Chamley et al. 1988)

The mineralogical modification produced by the tectonic instability prevented the easy identification by clay assemblages of other paleoenvironmental changes, which are more easily recognized during the further relaxation stages. Comparison of coastal and deep-sea basin records confirms the warm and rather humid climatic conditions indicated by Cretaceous–Paleogene soil-derived detrital assemblages (smectite). Minor changes in climate are better recognized in the proximal area (Senegal basin), suggesting drier conditions in the uppermost Cretaceous and parts of the Paleocene–Eocene. The African provenance of the main terrigenous sources is confirmed by the presence of similar mineralogical and geochemical assemblages at both sites. The increased distance from sources is marked by a diminution of kaolinite supply, reduced compositional variations and stronger

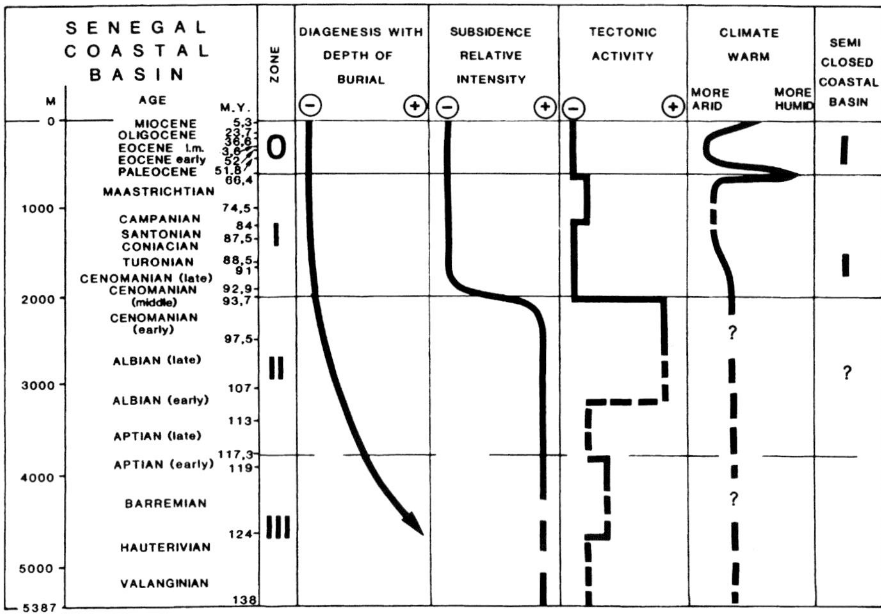

Fig. 16. Cretaceous to Paleogene clay mineral expression of diagenesis and paleoenvironment in the Senegal coastal basin based on comparisons with data from the Cape Verde deep-sea basin. (After Chamley et al. 1988)

reworking from evaporitic areas of semi-closed coastal basins (palygorskite). Thus the comparison between coastal and deep-sea sedimentary records points to close genetic relationships in terrigenous supply and significant differences under hydrodynamic and topographic conditions.

References

Accarie H, Renard M, Deconinck J-F, Beaudoin B, Fleury J-J (1989) Géochimie des carbonates (Mn, Sr) et minéralogie des argiles de calcaires pélagiques sénoniens. Relation avec l'eustatisme (Massif de la Maïella, Abruzzes, Italie). CR Acad Sci Paris 309/II:1679–1685

Ahn JH, Peacor DR (1986) Transmission and analytical electron microscopy of the smectite- to-illite transition. Clays Clay Min 34:165–179

Ahn JH, Peacor DR (1987) Transmission electron microscopic study of the diagenesis of kaolinite in Gulf coast argillaceous sediments. In: Schultz LG, Van Olphen H, Mumpton FA (eds) VIII Proc Int Clay Conf, Denver 1985, The Clay Minerals Society, Bloomington, pp 151–157

Badaut D, Besson G, Decarreau A, Rautureau R (1983) Occurrence of a ferrous trioctahedral smectite in recent sediments of Atlantis II deep, Red Sea. Clay Min 20:389–404

Bailey SW (1988) Odinite, a new dioctahedral-trioctahedral Fe^{3+}-rich 1:1 clay mineral. Clay Min 23:237–247

Bonnot-Courtois C (1981) Géochimie des terres rares dans les principaux milieux de formation et de sédimentation des argiles. Thèse Doc es-Sci, Univ Paris-Sud, 230 pp

Bouquillon A, Chamley H, Fröhlich F (1989) Sédimentation argileuse au Cénozoïque supérieur dans l'Océan Indien nord-oriental. Oceanol Acta 12:133–147

Buatier M, Honnorez J, Ehret G (1989) Fe smectite-glauconite transition in hydrothermal green clays from the Galapagos spreading center. Clays Clay Min 37:532–541

Chamley H (1989) Clay Sedimentology. Springer, Berlin Heidelberg New York, 623 pp

Chamley H (1992) Clay Sedimentology. Encycl Earth Syst Sci 1:485–502

Chamley H, Debrabant P (1984) Paleoenvironmental history of the North Atlantic region from mineralogical and geochemical data. Sediment Geol 40:151–167

Chamley H, Debrabant P, Flicoteaux R (1988) Comparative evolution of Senegal and eastern central Atlantic Basins, from mineralogical and geochemical investigations. Sedimentology 35:85–103

Chamley H, Deconinck J-F, Millot G (1990) Sur l'abondance des minéraux smectitiques dans les sédiments marins communs déposés lors des périodes de haut niveau marin du Jurassique supérieur au Paléogène. CR Acad Sci Paris 311/(II):1529–1536

Chamley H, Angelier J, Teng L (1993) Tectonic and environmental control of the clay sedimentation in the late Cenozoic orogen of Taïwan. Geodyn Acta 6:135–147

Clauer N, O'Neil JR, Bonnot-Courtois C, Holtzapffel T (1990) Morphological, chemical, and isotopic evidence for an early diagenetic evolution of detrital smectite in marine sediments. Clays Clay Min 38:33–46

Debrabant P, Chamley H, Foulon J, Maillot H (1979) Mineralogy and geochemistry of upper Cretaceous and Cenozoic sediments from North Biscay Bay and Rockall Plateau (eastern North Atlantic), DSDP Leg 48. In: Montadert L, Roberts DG et al. (eds) Init Rep Deep Sea Drill Proj, 48. US Gov Print Office, Washington, pp 703–725

Debrabant P, Delbart S, Lemaguer D (1985) Microanalyses géochimiques de minéraux argileux de sédiments prélevés en Atlantique Nord (forages du DSDP). Clay Min 20:125–145

Debrabant P, Chamley H, Deconinck J-F, Recourt P and Trouiller A (1992) Clay sedimentology, mineralogy and chemistry of Mesozoic sediments drilled in the northern Paris Basin. Sci Drilling 3:138–152

Debrabant P, Fagel N, Chamley H, Bout V, Caulet J-P. (1993) Neogene to Quaternary clay mineral fluxes in the Central Indian Ocean. Palaeogeogr Palaeoclimatol Palaeoecol 103:117–131

Deconinck J-F (1987) Identification de l'origine détritique ou diagénétique des assemblages argileux: le cas des alternances marne-calcaire du Crétacé inférieur subalpin. Bull Soc Géol Fr 3:139–145

Deconinck J-F (1992) Sédimentologie des argiles dans le Jurassique-Crétacé d'Europe occidentale et du Maroc. Mém Habilit Univ Lille I, 226 pp

Deconinck J-F, Amedro F, Desprairies A, Juignet P, Robaszinski F (1991) Niveaux repères de bentonite d'origine volcanique dans le Turonien du Boulonnais et de Haute-Normandie. CR Acad Sci Paris 312/(II):897–903

Desprairies A, Jehanno C (1983) Paragenèses minérales liées à des interactions basalte- sédiment-eau de mer (sites 465 et 456 des legs 65 et 60 du D.S.D.P.). Sci Géol Bull (Strasb) 36:93–110

Fisher RV, Schmincke HU (1984) Pyroclastic rocks. Springer, Berlin Heidelberg New York, 472 pp

Fröhlich F (1982) Evolution minéralogique dans les dépôts azoïques rouges de l'océan Indien. Relations avec la stratigraphie. Bull Soc Géol Fr 14:563–571

Haq BU, Hardenbol J, Vail PR (1988) Mesozoic and Cenozoic chronostratigraphy and cycles of sea-level change. SEPM Spec Publ 42:71–107

Hoffert M (1980) Les «argiles rouges des grands fonds» dans le Pacifique centre-est. Sci Géol Mém (Strasb) 61:257 p

Holtzapffel T, Chamley H (1986) Les smectites lattées du domaine Atlantique depuis le Jurassique supérieur: gisement et signification. Clay Min 21:133–148

Huggett JM (1984) Controls on mineral authigenesis in Coal Measures sandstones and mudstones in the Westphalian Coal Measures using back-scattered electron microscopy. Clay Min 21:603–616

Karlin R (1980) Sediment sources and clay mineral distributions off the Oregon coast. J Sediment Petrol 50:543–560

Karpoff A-M, Lagabrielle Y, Boillot G, Girardeau J (1989) L'authigenèse océanique de palygorskite par halmyrolyse de péridotites serpentinisées (marge de Galice): ses implications géodynamiques. CR Acad Sci Paris, 308/(II):647–654

Kimblin RT (1992) The origin of clay minerals in the Coniacian chalk of London. Clay Min 27:389–392

Latil-Brun MV, Flicoteaux R (1986) Subsidence de la marge sénégalaise, ses relations avec la structure de la croûte. Comparaison avec la marge conjuguée américaine au niveau du Blake-Plateau. Bull Centre Rech Pau 110:64–82

Levert J, Ferry S (1988) Diagenèse argileuse complexe dans le Mésozoïque subalpin révélée par cartographie des proportions relatives d'argiles selon des niveaux isochrones. Bull Soc Géol Fr 4:1029–1038

McMurtry GM, Wang C-H, Yeh H-W (1983) Chemical and isotopic investigations into the origin of clay minerals from the Galapagos hydrothermal mounds field. Geochim Cosmochim Acta 47:475–489

Millot G (1949) Relations entre la constitution et la genèse des roches sédimentaires argileuses. Thèse, Univ Nancy, Géol Appl Prosp Min 2:1–352

Millot G (1964) Géologie des argiles. Masson, Paris, 499 pp

Millot G (1970) Geology of clays. Springer, Berlin Heidelberg New York, 425 pp

Nadeau PH, Tait JM, McHardy WJ, Wilson MJ (1984) Interstratified XRD characteristics of physical mixtures of elementary clay particles. Clay Min 19:67–76

Odin GS (ed) (1988) Green marine clays. Developments in sedimentology 45. Elsevier, Amsterdam, 445 pp

Paquet H (1970) Evolution géochimique des minéraux argileux dans les altérations et les sols des climats méditerranéens et tropicaux à saisons contrastées. Mém Serv Carte Géol Als Lorr (Strasb) 30:212 pp

Robert C (1982) Modalité de la sédimentation argileuse en relation avec l'histoire géologique de l'Atlantique Sud. Thèse, Univ Aix-Marseille II, 141 pp

Robert C (1987) Clay mineral associations and structural evolution of the South Atlantic Jurassic to Eocene. Palaeogeogr Palaeoclimatol Palaeoecol 58:87–108

Robert C, Chamley H (1990) Paleoenvironmental significance of clay mineral associations at the Cretaceous-Tertiary passage. Palaeogeogr Palaeoclimatol Palaeoecol 79:205–219

Robert C, Chamley H (1991) Development of early Eocene warm climates, as inferred from clay mineral variations in oceanic sediments. Palaeogeogr Palaeoclimatol Palaeoecol 89:315-331

Sigl W, Chamley H, Fabricius F, Giroud D'Argoud G, Muller J (1978) Sedimentology and environmental conditions of sapropels. In: Hsü KJ, Montadert L et al. (eds) Init Rep Deep Sea Drill Proj, 42 A. US Gov Print Office, Washington, pp 445–464
Singer A, Galan E (1984) Palygorskite-sepiolite. Occurrences, genesis and uses. Developments in sedimentology 37. Elsevier, Amsterdam, 352 pp
Srodon J, Eberl DD (1984) Illite. In: Bailey SW (ed) Micas. Rev Miner 13, Miner Soc Am, Washington, pp 49–544
Steinberg M, Holtzapffel T, Rautureau M, Clauer N, Bonnot-Courtois C, Manoubi T, Badaut D (1984) Croissance cristalline et homogénéisation chimique de monoparticules argileuses au cours de la diagenèse. CR Acad Sci Paris 299/II:441–446
Stonecipher SA (1976) Origin, distribution and diagenesis of phillipsite and clinoptilolite in deep-sea sediments. Chem Geol 17:307–318
Thiry M (1981) Sédimentation continentale et altérations associées: calcitisations, ferruginisations et silicifications. Les argiles plastiques du Sparnacien du Bassin de Paris. Sci Géol Mém (Strasb) 84:173 pp
Thiry M, Jacquin T (1993) Clay mineral distribution related to rift activity, sea-level changes and paleoceanography in the Cretaceous of the Atlantic Ocean. Clay Min 28:61–84
Trauth N (1977) Argiles évaporitiques dans la sédimentation carbonatée continentale et épicontinentale tertiaire. Bassins de Paris, de Mormoiron et de Salinelles (France), Jbel Ghassoul (Maroc). Sci Géol Mém (Strasb) 49:203 pp
Van Houten FB, Purucker ME (1984) Glauconite peloids and chamositic ooids. Favorable factors, constraints, and problems. Earth Sci Rev 20:211–243
Weaver CE (1989) Clays, muds, and shales. Developments in sedimentology 44. Elsevier, Amsterdam, 819 pp

14 Revisited Isotopic Dating Methods of Sedimentary Minerals for Stratigraphic Purpose

Norbert Clauer and Sam Chaudhuri

1 Introduction

Cormier (1956) and Wasserburg et al. (1956) initiated many isotopic investigations of sedimentary minerals and whole rocks for stratigraphic purposes. Since this pioneering period, different approaches have been evaluated to identify the isotopic signatures of mineral components which may set reasonably narrow limits to the time of deposition of sediments. Many contradictory opinions have been expressed about the merits of these approaches and the significance of the various isotopic signatures in relation to the stratigraphic ages of the studied minerals. It is obvious from analysis of the available literature, that questionable dates were often generated without the necessary efforts to delineate clearly the origin of the analysed materials and to evaluate the potential impact of mineral impurities on the final dates.

Recent years have witnessed improved analytical capabilities for gentle separation and accurate characterization of the different types of mineral phases present in the separated size fractions, as well as emergence of new potential through Sm–Nd isotopic analyses for definition of major periods of clay mineral authigenesis. As a result of these recent developments, isotopic dating of clay minerals for stratigraphic purposes has been more promising than ever. Other non-silicate mineral phases, such as carbonates (Faure 1982), salts (Brookins 1980; Baadsgaard 1987), sulphates (Shanin et al. 1968; Blanco et al. 1982) and phosphates (Kolodny and Luz 1992) have also been evaluated. For all these materials, the major limiting factor in any successful stratigraphic dating by isotopic means remains the information gained about the origin of the mineral components of the analysed materials and about their post-depositional history.

Clay minerals can be of varied origins in sedimentary rocks as: (1) inherited detrital particles; (2) neoformed crystals by ionic precipitation from a liquid; (3) transformed precursor particles through ion exchange with the surrounding aqueous medium but without major changes in the structural framework; and (4) recrystallized crystals resulting from a simultaneous dissolution–precipitation with structural modifications of the parent material. Knowledge derived from independent means about the growth of clay minerals is critically needed to make reasonably valid assumptions about the isotopic compositions of the analysed authigenic mineral phases. Such information can be provided by X-ray measurements including determination of illite crystallinity indices and polymorphic forms, electron microscope observations, stable isotope or trace element (REE) geochemistry. Information about the isotopic compositions of the aqueous medium in which the minerals were formed or modified by the processes outlined above, is also of importance to calculate accurate isotopic ages. Authigenic clay minerals in isotopic equilibrium with their immediate environment are supposed to incorporate elements whose isotopic compositions reflect this environment. It is clear that during authigenesis in a marine environment, any newly formed clay mineral phase will, at least, incorporate Sr, Nd and Pb with isotopic compositions that should match the values of co-precipitating chemical mineral phases such as carbonates, oxides, sulphates or phosphates. Knowledge of the chemical and isotopic composition of the formational environment becomes very critical in the cases of formation reactions, such as transformation and sometimes dissolution–precipitation, which may occur in nearly closed systems. In these cases, marine fluids may have had no significant influence on the isotopic composition of the minerals at the time of their formation and the use of isotopic compositions of contemporaneous chemical mineral phases of marine origin is not appropriate at all.

Since the middle of the 1970s, many leaching experiments of clay particles have been made and refined in order to trace and reconstruct the Sr and, later, the Nd isotope compositions of the formational environments of these minerals (e.g. Clauer 1976; Clauer 1982a; Ohr et al. 1991; Bros et al. 1992; Clauer et al. 1993). From single leaching with dilute HCl to sequential leaching with distilled water, dilute HAc and dilute HCl, and from determination of only the Sr isotope composition to a complete chemical analysis together with Nd and Sr isotope measurements of each leachate (Stille and Clauer 1994), the technique has proved its potential of chemically and isotopically characterizing the formational environments in which clay particles formed during synsedimentary conditions, or later. The reliability of this approach will be discussed below.

Another determining parameter for isotopic dating of sediments is that, at their time of growth, authigenic mineral components have to have nearly identical isotopic compositions across a finite stratigraphic interval. Such spatial isotopic homogeneity, which has to be recognized for the use of stratigraphic isotope dating, will also be evaluated here using the available studies, provided evidence can be furnished that this isotopic uniformity occurred in a time period closely coincident with that of the deposition.

2 Some Fundamentals

The most critical step in isotopic dating of sediments is the selection of appropriate materials that are either known for their synsedimentary origin or likely to have the potential to have been able to form shortly after sedimentation. Clay sediments are known to consist of various fractions of detrital and early-to-late diagenetic mineral components, and this requires extreme caution in the separation of the clay materials from these rocks, to make sure that only the authigenic component is dated. In this respect, classical crushing and grinding methods should be avoided because of the possibility of creating artificially some clay-sized materials of detrital origin . Liewig et al. (1987) demonstrated that a cryostatic disaggregation procedure greatly minimizes the influence of artificial production of clay particles so highly prevalent in classical disaggregation methods. Other methods of gentle disaggregation have also been tested. Among other potential mineral phases suitable for isotopic stratigraphic dating, at least carbonates and salts are known to be sensitive to alteration and recrystallization processes which easily erase the original isotopic signatures.

2.1 Analytical Aspects

In theory, low-temperature minerals or whole rocks may be dated by one of the commonly used isotopic dating methods, provided the material suffered no loss or gain of either the presumed radioactive parent or the corresponding radiogenic daughter isotope since formation. Uncertainties about the initial isotopic conditions and the post-depositional effects on the isotopic budget of a sedimentary material are a major hindrance to stratigraphic dating by any isotopic method, as already mentioned. Evidence has, therefore, to be produced that the analysed materials meets the criteria for isotopic dating presented hereunder.

A common method of Rb–Sr, Sm–Nd or U–Pb isotopic dating of geological materials is the isochron method as applied to a suite of cogenetic materials that had the same initial isotopic composition for the daughter element but various ratios of parent to daughter elements. Historically, it was first used in Rb–Sr isotopic studies of cogenetic plutonic materials, describing a linear array for the analytical data in a rectangular co-ordinate diagram with $^{87}Sr/^{86}Sr$ as the ordinate and $^{87}Rb/^{86}Sr$ as the abscissa (Nicolaysen 1961). The slope of the linear array was shown to be a function of time and, hence, to be related to the formation period of cogenetic materials. The intercept of the line, with the ordinate at a zero value, is the initial isotopic composition of the considered element. One may consult Faure (1986) or any of several other text books for further details about this method of dating, but it should be kept in mind that linear trends for a suite of analysed materials may also arise from mixing two mineral components of different origins, such as detrital and authigenic K-bearing clay minerals. The slope of such a line cannot, consequently, be an expression of the time of formation of the analysed mineral fractions. Various critical tests are, therefore, needed to ensure that an apparent isochron is not a mixed line. In the case of an isochron with the true indication of the time of formation for the analysed materials, the initial isotopic composition may shed light on the history of formation of the materials.

Comparisons between the Sr isotopic composition of carbonate phases associated with a clay unit and of the leachable Sr from clay particles, are necessary to evaluate accurately the size of the chemical system in which clay authigenesis took place. Similar isotopic compositions for the marine carbonates and the clay leachates will incline towards an evolution in an open chemical system, while significant different values are rather indicative of an authigenesis in a restricted system unrelated to open sea. It will be seen later that such differences may introduce more or less pronounced biases in the calculated isotopic ages. As a general rule, it can be suggested that clay-type rocks behave preferentially as closed chemical systems, while sandy and silty units are more frequently open. In terms of clay genesis, crystal growth generally needs open chemical systems, whereas transformation often occurs in restricted rock volumes. Dissolution–precipitation is more difficult to link to a chemical volume, as it has been described in both open and closed chemical systems.

Alternatively, recent years have seen considerable analytical refinement of secular variations in the Sr isotopic composition of seawater during the Phanerozoic from analyses of fossils in marine carbonate rocks. This raised the prospect of dating young marine sedimentary rocks by analysing Sr isotopic compositions of carbonate or phosphate mineral phases in sedimentary rocks. Such a method of dating rests on the facts that chemical mineral phases precipitate in strict isotopic equilibrium with the open sea and that the secular variation curve of the marine Sr isotopic composition has been constructed with stratigraphically well-ordered mineral phases. Based on a curve relating the time-dependent variation in the Sr isotopic composition of open-seawater, it requires careful evaluation of whether the analysed mineral or rock has suffered alteration of the primary isotopic signature. In view of the high probability that diagenetic alteration is more pervasive in old carbonate rocks than in young ones and that the trend of the isotopic variation for very ancient seawaters is poorly defined, pre-Phanerozoic carbonate minerals or rocks cannot be accurately dated despite an increased knowledge due to recent studies by Derry et al. (1989), Burns et al. (1994), Kaufman et al. (1994), and Derry et al. (in press), whereas very detailed age information may be provided for recent sequences (Stille et al. 1994).

2.2 Mineral Phases Suitable for Isotopic Dating

Among materials that have been commonly targeted for isotopic analyses for the purpose of stratigraphic dating, glauconites have received much attention because of their occurrence in many sedimentary rocks and of their textural appearance suggesting that these minerals evolved pene-contemporaneously with the deposition of the host sediment. Many different accounts of geochronological information about these minerals have been given in Odin (1982). More recently, Clauer et al. (1992a,b) and Stille and Clauer (1994) have elaborated on chemical and isotopic evolution of these glauconites, as will be outlined below. Other K-bearing clay minerals such as illite and illite/smectite mixed layers, especially from Paleozoic and older rocks, have also been analysed (Clauer 1974, 1976). As many of these clay minerals are known to evolve from episodes of burial history also, their isotopic dates can be considered to be the minimum dates for the deposition time of sediments. Since burial-related diagenetic evolution for these minerals could occur within a few million years after deposition of the sediments, the isotopic dates may closely approximate the time of deposition of the rocks containing these minerals in early Paleozoic and older rocks. A limited number of studies has also shown that other clay minerals such as smectite and palygorskite can be useful for isotopic dating (Clauer 1976; Clauer et al. 1990).

Smectite clays with various amounts of illite layerings have also proved useful in some instances for stratigraphic dating by isotopic means. As such minerals often underwent post-depositional recrystallizations, the conclusions from isotopic data sets may suffer from some uncertainty unless the results are constrained by independent lines of evidence. Such lines of evidence are based on X-ray diffraction, electron microscopy, and trace-element geochemistry. Zeolites (Bernat et al. 1970; Clauer 1982a) and siliceous cherts (Brueckner and Snyder 1979) have been analysed, beside clay minerals, for K–Ar or Rb–Sr isotopic compositions to determine their usefulness as materials for stratigraphic dating.

Whole-rock samples were frequently examined in early Rb–Sr studies on dating sediments (e.g. Compston and Pidgeon 1962; Bofinger et al. 1968). Cordani et al. (1978, 1985) maintained that Rb–Sr isotopic dates of whole-rock samples can provide information about the time of sedimentation; they argued that silty-to-sandy sediments may become homogeneous in their isotopic compositions through a combined process of continental alteration and riverine transport, while differing in their Rb/Sr ratios. This claim has been questioned by Clauer (1982b) who argued that whole-rock sediments mainly consisting of detrital components of various origin are unlikely to define linear horizontal trend in a Rb–Sr isochron diagram at the time of deposition. This argument is evident from almost all studies of modern sediments (Clauer and Chaudhuri 1995). On the other hand, Stille and Clauer (1986) published a Sm–Nd isochron based on dates of several shale whole-rock samples and one clay fraction intercalated in authigenic banded iron units of the Proterozoic Gunflint Formation in Canada. The Sm–Nd date was 2.08±0.59 Ga with an initial $^{144}Nd/^{143}Nd$ ratio close to that of the known contemporary marine environment. On the basis of geological considerations and independent isotopic dates on volcanic rocks intruding into the basement of these banded iron formations, the authors interpreted the isochron age as a reasonable approximation to the presumed deposition time. The K–Ar and Rb–Sr dates obtained on the same samples were significantly lower at 1.47±0.10 and 1.46±0.53 Ga, respectively, probably due to later diagenetic activity which altered the Rb–Sr and K–Ar systems, while leaving the Sm–Nd system unchanged. Such a conclusion has not been reached by Nägler et al. (1992) in a similar study on pelite-type whole rocks, as they found a Sm–Nd isochron age of 1.52±0.07 Ga for Ordovician rocks. The reason why these authors could not date the early diagenesis in whole-rock samples is again due to the different lithofacies: the shales intercalated in almost pure chemical Fe units were mainly of chemical origin, while the pelites in the second case were mainly detrital. Interpretation of the line obtained by Nägler et al. might be twofold: either it results from an essentially bimodal mixing of detrital components as the data points of the samples yield also a linear trend in a $^{143}Nd/^{144}Nd$ vs 1/Nd diagram, or it represents reworked material with a homogeneous REE reservoir and in this case the calculated age could reflect the age of provenance, as claimed by the authors. Actually, the

Sm–Nd method seems to have some potential in whole-rock isotopic dating when applied to rocks whose components are mainly authigenic and contain enough REEs.

2.3 Influence of Particle Size and Temperature on Isotopic Dates

The fundamental aspect of retention of radiogenic isotopes, especially of radiogenic ^{40}Ar, in relation to the crystallinity and particle size of clay minerals has been investigated in several studies. A common notion that small particle size, frequently attendant with poor crystallinity, favours increased loss of radiogenic isotopes, has been advocated in many circumstances when isotopic dates of clay-type material are lower than expected. However, this notion remains largely unsupported by evidence. For instance, Clauer et al. (1984) found that K–Ar and Rb–Sr dates were identical for detrital <0.2 µm smectites, which means that no preferential loss of radiogenic ^{40}Ar could be detected in these poorly crystallized clay particles relative to radiogenic ^{87}Sr. Results of many other studies reported in the literature have similar conclusions. Ultimately, Clauer et al. (1995) have made K–Ar isotopic determinations on <0.03 µm "fundamental" illite/smectite particles of bentonites which represent the smallest possible clay crystals. The K–Ar dates were found to be either similar to those obtained on larger particles, or even higher, demonstrating that no preferential loss of radiogenic ^{40}Ar can be expected due to reduction of the grain size.

Isotopic dates of clay minerals may also be significantly influenced by temperature as enhanced temperature effect may induce preferential loss of radiogenic isotopes, especially ^{40}Ar relative to the non-radiogenic Ar isotopes. Odin and Bonhomme (1982) made experiments on glauconite grains with different K contents and they found that critical temperature for starting loss of radiogenic ^{40}Ar under vacuum was about 250 °C for poorly crystallized grains, while it was about 300 °C for well-crystallized ones. In nature, the temperature increase occurring with progressive burial of sedimentary sequences may or may not be high enough to alter the budget of radiogenic isotopes in the clay fractions of the buried sediments. Hunziker et al. (1986) reported that temperatures of about 260±30 °C are needed to cause total Ar diffusion of illitic particles of smaller than 2 µm size (Fig. 1). The effective temperature for such Ar losses is equivalent to that of 5–6 km burial at an average geothermal gradient. Bath (1977) analysed the Rb–Sr isotopic systems of clay minerals that were subjected to simulated hydrothermal interactions with brines at temperatures between 315 and 360 °C and observed no preferential loss of radiogenic ^{87}Sr relative to the radioactive ^{87}Rb.

A critical review of these results allows the conclusion that temperature has less diffusional effect on Rb–Sr ages than on K–Ar ages of clay minerals, probably because of structural accommodations for Sr and not for Ar. Of course, any crystallization of new clay phases after sedimentation will provide lower isotopic dates for the mixtures of both generations. Therefore, it is needed, especially when the K–Ar method is chosen for stratigraphic dating, to take carefully into account the burial history of the sedimentary units under consideration. Simulation programs of Ar

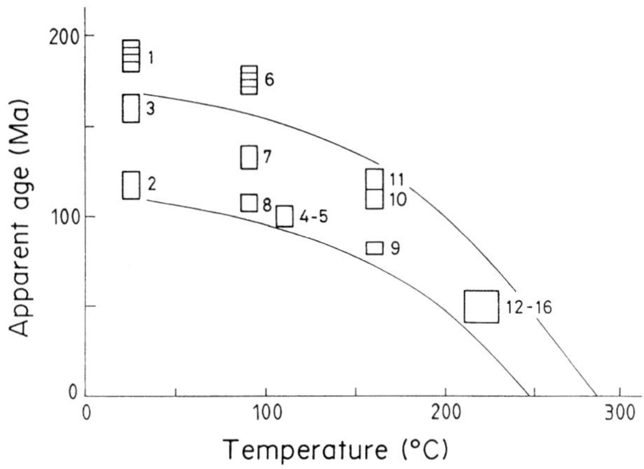

Fig. 1. K—Ar dates of detrital illite as a function of temperature measured in drill holes. (After Hunziker et al. 1986)

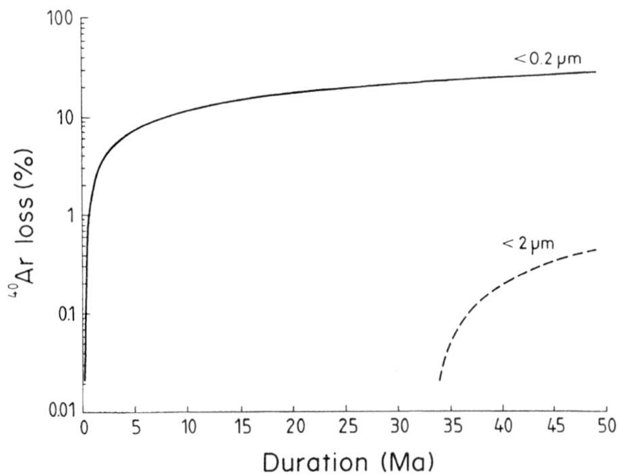

Fig. 2. Calculated Ar loss in percent of clay particles of different sizes depending on the duration of a 150 °C thermal pulse in millions of years

diffusion out of clay structures may efficiently complete this control in helping to examine critical situations. Knowledge of the burial histories of the sedimentary sequences allows the evaluation of maximum temperatures and time spans during which they were maintained at these temperatures. These parameters can be integrated in diffusion codes to determine the amounts of radiogenic ^{40}Ar loss depending on particle size, temperature and duration (Fig. 2; Zwingmann 1995). It can be seen on this diagram that a temperature of 150 °C, which corresponds to a burial of about 4000 m under a geothermal gradient of 35 °C/km, maintained during 50 million years may induce a loss of about 30% of radiogenic ^{40}Ar out of <0.2 μm clay particles. This loss becomes almost negligible, about 1%, for <2 μm particles under the same conditions.

2.4 Concept of Isotopic Homogenization

As outlined in the above discussion, clay-mineral phases of a sedimentary rock are datable only if all crystals dispersed in a rock volume were isotopically homogeneous for a considered isotopic system at a given moment of their evolution. This isotopic homogeneity is a prerequisite for stratigraphic dating. Odin and Matter (1981), for instance, summarized the criteria for glauconites that are suitable for stratigraphic dating and presentation of these criteria will help to understand how this isotopic homogenization may occur during sedimentation, and why it may not. According to the authors, the glauconitization process occurs in a micro-environment in which Fe-smectites evolve into glauconies and then into glauconites by progressively incorporating K. Based on electron microprobe analyses, Birch et al. (1976) suggested that the Fe enters the mineral structures very early in the glauconitization process, probably independently of the K incorporation. Odin and Dodson (1982) also proposed, on the basis of XRD results, that authigenic glauconitic minerals probably form by crystal growth from interstitial fluids. Odin et al. (1979) found from studies of Holocene glauconites, that the higher the K content of the grains, the closer are their K–Ar apparent ages to the true stratigraphic age (Fig. 3). They thought that this decrease was due to the loss of radiogenic ^{40}Ar during the process of glauconitization accompanied by the uptake of K_2O. Amouric and Parron (1985) observed structural discontinuities between smectitic

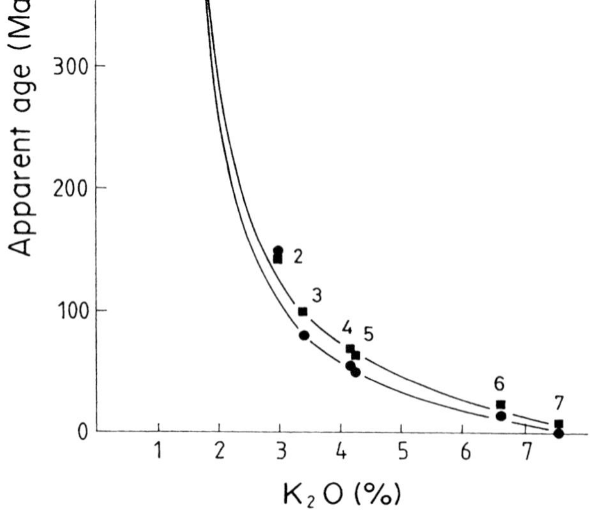

Fig. 3. K—Ar and Rb—Sr dates of recent glauconite-type faecal pellets and of the associated clay material from host mud in the Gulf of Guinea, as a function of their K_2O contents. (After Clauer et al. 1992b)

Fig. 4. Evolution of the $^{87}Sr/^{86}Sr$ ratios and the $\delta^{18}O$ values of recent glauconies from the Gulf of Guinea as a function of their K_2O contents. (After Clauer et al. 1992b)

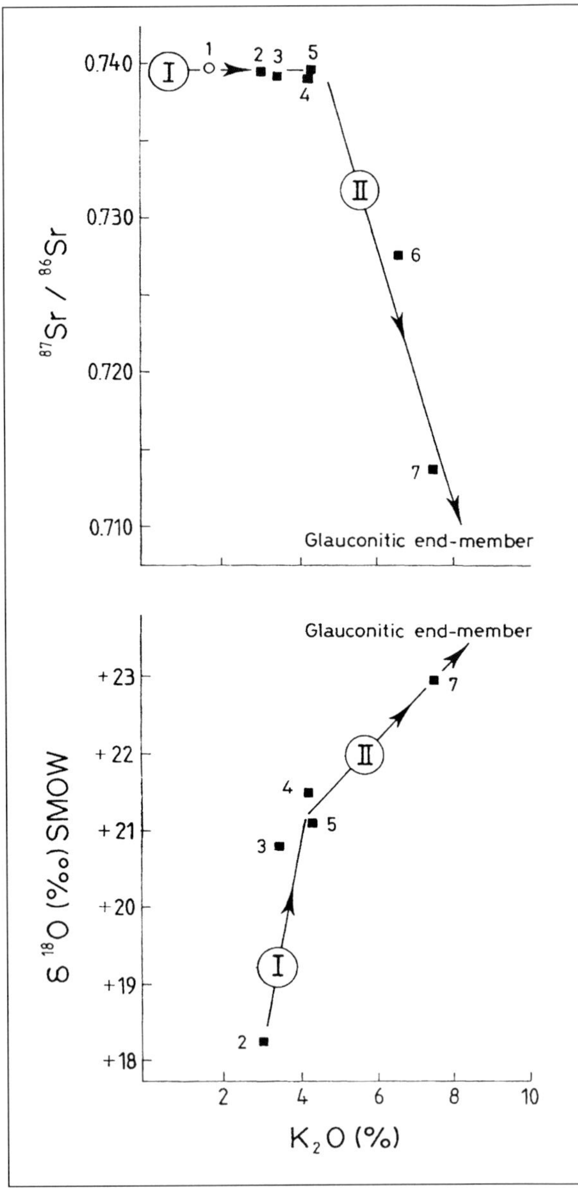

layering and glauconitic layering in the same particle and concluded that glauconitization had to proceed by dissolution–precipitation.

Recent Sr, Nd and oxygen isotopic studies on present-day to Holocene glauconitic faecal pellets of the Gulf of Guinea provided more insights into the glauconitization process (Keppens and O'Neil 1984; Clauer et al. 1992a,b). Rb–Sr and Sm–Nd analyses were made on pellets leached with 1 N HCl to remove the non-silicate mineral phases (carbonates, phosphates and oxides) and the adsorbed elements which could have modified the isotopic compositions and the amounts of elements in question. The Rb–Sr dates of the glauconite residues decreased when the amount of K_2O (and that of Rb) increased, so did the K–Ar dates (Odin 1975; Odin et al. 1979; Fig. 3). The comparison between the $^{87}Sr/^{86}Sr$ ratios and the K_2O amounts of the clay fraction from mud and from glauconitic pellets showed that there was no significant change in the $^{87}Sr/^{86}Sr$ ratio until the glauconitic pellets contained about 4.5% K_2O (Fig. 4), while the K–Ar and Rb–Sr apparent ages decreased from 473 to 50 Ma and from 415 to 65 Ma, respectively. Stille and Clauer (1994) refined the leaching technique by performing a sequential removal first of the adsorbed elements with distilled water, then of the carbonates and part of the oxides with dilute HAc, then of the oxides and part of the phosphates with dilute HCl, and finally of the remainder of the soluble phases with another leaching using dilute HCl. They could demonstrate that three components of different origin were involved in the glauconitization process: a detrital one, a marine one and a fluvial one.

These results suggest that there is persistence of the isotopic signature of the detrital clay fraction of the mud in the glauconitic pellets, at least at the beginning of the process. A genetic link between the two mineral phases, as proposed by Keppens and O'Neil (1984) on the basis of oxygen isotope data, therefore, exists and only when the amount of K_2O is above 4.5%, do the $^{87}Sr/^{86}Sr$ values of the glauconitic pellets shift toward that of the environmental seawater (Fig. 4). The glauconitization process can be decomposed into two steps: the first could be a progressive dissolution–crystallization of the detrital clay material of the grains, as suggested by Odin and Matter (1981) and described by Amouric and Parron (1985), but occurring in a closed environment without interference of seawater, and the second is a crystal growth from a solution with significant influence of seawater Sr. These two steps defined by the Sr isotopic evidence were confirmed by Nd isotopic data (Clauer et al. 1992b; Stille and Clauer 1994). This modelled evolution explains why only glauconies with K_2O contents higher than at least 6.5% should be used for stratigraphic dating, as they are the only ones having completed the glauconitization process, which means that the detrital signature is progressively erased and that isotopic equilibrium with the marine environment can be established. Another model leading to Nd isotopic homogenization has been discussed by Stille et al. (1993) in considering an interaction between clay mineral phases, detrital feldspar-like components and organic matter.

Additional understanding of the mechanism of isotope homogenization of clay minerals soon after deposition comes from a study of Clauer et al. (1990) which focused on mineralogy, morphology, geochemistry, and Rb–Sr and oxygen isotope geochemistry of smectite-rich size fractions of Albian–Aptian and Paleocene black shales. TEM observations revealed three distinct morphologies for the clay particles: flake-type particles with diffuse, irregular and often rolled-up edges, elongated lath-type particles with straight edges, and intermediate particles consisting of the two former types. The chemical and XRD analyses showed that both the flakes and the laths are Al–Fe smectites. Point countings have emphasized that the lath-type particles are concentrated in the fine size fractions (<0.2 μm) and that the flake-type particles are concentrated in coarser fractions (1–2 μm). The authors suggested that the laths were of authigenic origin and that the flakes were of detrital origin.

Several size fractions of different black-shale samples were systematically leached with dilute HCl (1 N) to analyse separately the Sr trapped in the silicate structures (clays) during crystallization and the Sr adsorbed onto the particles or trapped in soluble mineral phases (carbonates, oxides, phosphates). Depending on the sizes of the smectite particles and, therefore, on the relative amounts of lath-type and flake-type particles, the Rb–Sr apparent ages of the mixing lines fitted through the data points of the untreated fractions, the leachates and the residues were different. The finest fractions, those enriched in lath-type particles, yielded the lowest apparent ages but not low enough to correspond to the time of deposition. To determine the recrystallization period of the clay minerals, the authors compared the Rb–Sr apparent ages of the different size fractions to the amounts of lath-type particles in the different size fractions of the same sample. In extrapolating toward the end-member consisting of pure lath-type particles, the authors obtained ages of 100–110 Ma for the Albian–Aptian samples and 50–60 Ma for the Paleocene samples (Fig. 5). These values are close enough to the periods of deposition to consider that the recrystallization mechanism could have occurred soon after deposition.

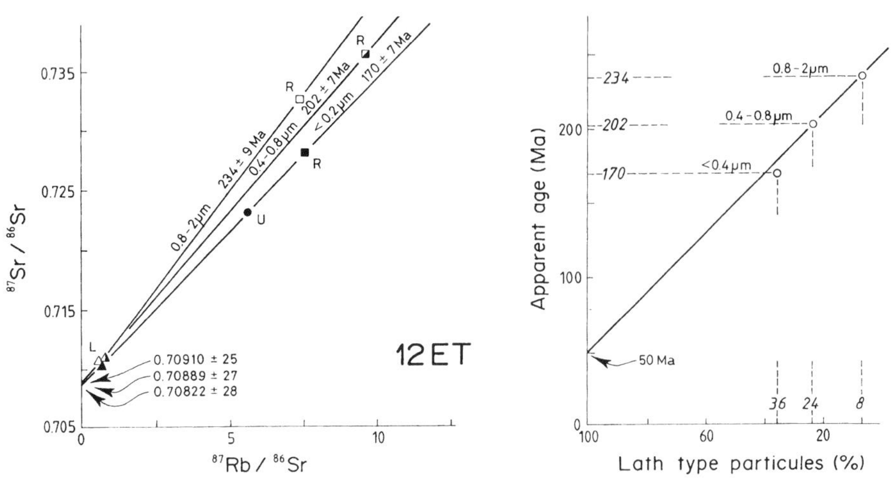

Fig. 5. Estimation of Rb—Sr ages of granulometric fractions in a black shale from the Atlantic Ocean as a function of the amounts of lath-type particles. (After Clauer et al. 1990)

Two additional results of this study are of prime importance: the initial $^{87}Sr/^{86}Sr$ ratios of the recrystallized smectites are systematically different and above that of contemporary seawater Sr, and the oxygen and hydrogen isotope compositions of the smectites are abnormally low for marine clay minerals. As the authors pointed out, this evidence requires that the recrystallization occurred in an environment insulated from seawater. Synsedimentary isotopic homogenization of clay minerals may, therefore, happen soon after deposition in an environment that is insulated from seawater. Consequently, Sr isotope measurement of a marine carbonate to determine the isotope composition of the immediate environment of the clays has to be checked by leaching of the clay fractions.

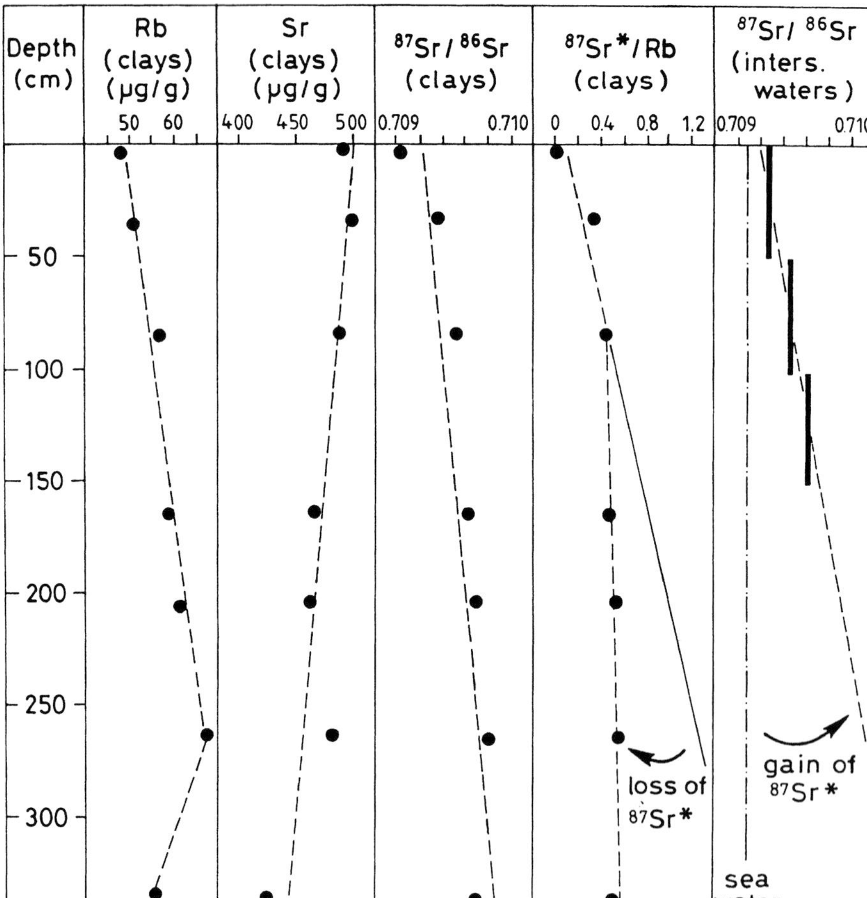

Fig. 6. $^{87}Sr/^{86}Sr$ ratios of deep-sea red clays and of the associated interstitial waters as a function of depth. (After Clauer 1982b)

The Rb–Sr studies of Clauer et al. (1975) and Clauer et al. (1982a) on smectites from deep-sea red clays have shown that these minerals might have interacted with their formation environment soon after deposition. Combination of the Rb–Sr isotopic data of the solid material and the interstitial fluids suggested that slight migration of radiogenic ^{87}Sr into the interstitial fluids may happen (Fig. 6). The implication of these results is that one cannot take for granted that the $^{87}Sr/^{86}Sr$ ratio of the formation environment is always identical to that of the contemporary seawater. In most studies, the difference between the Sr isotopic composition of the formation environment and of the seawater is probably small and acceptable, but this difference will systematically introduce an error by having a lower age when the environmental Sr isotopic composition is higher than that of the marine reference. A good illustration of this is given by a Rb–Sr isotopic study on Palaeozoic illites in which Gorokhov et al. (1994) differentiated two generations of authigenic illites forming in environments characterized by different $^{87}Sr/^{86}Sr$ ratios (Table 1). The oldest generation of syngenetic illite crystallized at about 530 Ma, whereas the youngest generation was late diagenetic and formed about 470 Ma ago. The Sr incorporated by each generation of clays yields different $^{87}Sr/^{86}Sr$ ratios, probably reflecting changed physical and chemical conditions during crystallization. The different initial $^{87}Sr/^{86}Sr$ ratios were found to be higher than the contemporary seawater ratio which was estimated to be at about 0.7091. This difference has been considered by the authors to reflect illite formation in restricted chemical systems. A subsequent comparison of the ages obtained by isochrons and those calculated conventionally by using the seawater Sr, outlines more or less pronounced differences depending directly on the Sr values obtained by regression. However, it might be noticed that if the isotopic ages of sedimentary minerals formed or modified in environments insulated from a marine environment and calculated by using the

Samples	Size of fraction	Number of points	Isochron age [MA]	Initial $^{87}Sr/^{86}Sr$	Convent. age [MA]
806/1	0.1-0.2	3	521	0.7136	543
806/8		3	534	0.7149	552
807/8		5	541	0.7133	560
77NC102	< 0.1	3	466	0.7134	484
77NC106		5	481	0.7109	498

The sizes of the fractions are in μm; the number of points is the number of analyses made for each isochron; the calculated age and the $^{87}Sr/^{86}Sr$ ratio were obtained by regression calculation; convent. age stands for conventional ages which were obtained with a marine Sr ratio (=0.7091); the ages were determined on the analytical point having the highest Rb/Sr and $^{87}Sr/^{86}Sr$ ratios.

Table 1. Rb—Sr ages of Estonian illites calculated by the isochron (after Gorokhov et al. 1994) and by the conventional technique with a contemporaneous marine $^{87}Sr/^{86}Sr$ ratio

marine $^{87}Sr/^{86}Sr$ of contemporaneous carbonates (Clauer 1982b), are in general different from the "true" ages, the differences remain "reasonable" in considering the analytical uncertainties.

Progressive isotopic equilibration for clay minerals with their formation environment, as outlined in several studies presented above, suggests that leaching of clay particles with diluted acid reagents should be routinely carried out to verify progressive equilibration, to characterize the environment in which the clay particles grew, but also to delineate occasional late chemical exchanges between early crystallized clay particles and lately intruded formational waters inducing significant age alterations (e.g. Clauer and Chaudhuri 1995).

3 Examples of Isotopic Dating of Clay Minerals

The examples of isotopic dating of clays picked from the literature and discussed below were chosen not only to illustrate that isotopic dating provides meaningful stratigraphic results, but also to emphasize that an approach combining mineralogical, morphological and multi-isotopic analyses on different size fractions of carefully selected samples is the ultimate way to generate reliable stratigraphic values.

3.1 Rb–Sr and K–Ar Dating

Odin and Hunziker (1982) presented Rb–Sr and K–Ar isotopic data for glauconitic fractions from Albian and Cenomanian units in Normandy (France), providing precise isotopic dates for these two stratigraphic units. Analysing eleven separated and purified mineral fractions of the Albian unit, the authors found a K–Ar isochron age of 98.6±2.1 Ma. They also studied 12 glauconitic samples of Cenomanian rocks by the K–Ar and the Rb–Sr methods and obtained dates of 93.0±1.4 and 93.5±1.6 Ma, respectively. The identity between the K–Ar and the Rb–Sr ages of the Cenomanian material is remarkable. The isotopic date of an associated carbonate plots near the initial of the Rb–Sr isochron, and it refines the age if taken into the calculation: the age becomes 92.9±1.4 Ma. The initial $^{87}Sr/^{86}Sr$ ratios of the different isochrons being identical to that of the contemporary marine Sr, suggest that isotopic homogenization occurred in an open chemical system.

Many Rb–Sr and K–Ar studies on non-glauconitic clay minerals of sedimentary sequences of Precambrian age from western Africa have also been carried out (Clauer 1976; Clauer et al. 1982b). The isotopic analyses for these clays provided a succession of Rb–Sr isochron ages from 998±34 to 595±45 Ma. The very high ages were found to be reasonable estimates for the time of deposition, as volcanic rocks located at the bottom of these sequences were dated at 1035 Ma in the Hoggar Mountains (Allègre and Caby 1972) and at 1090 Ma in the Senegal (Clauer, unpubl. data). These studies did also underline discrepancies between the K–Ar and Rb–Sr results, the K–Ar dates of the West African samples being systematically lower than the Rb–Sr dates by 30 to 10%. This behaviour which was also found in other Precambrian to Cambrian sediments (Clauer et al. 1993) without any explanation other than a very discrete loss of radiogenic ^{40}Ar, might become detectable, as suggested by the Ar diffusion code, when the process is protracted over tens or even hundreds of million years in shallow burial conditions.

3.2 Sm–Nd Dating

Beside the Sm–Nd dating attempt made by Stille and Clauer (1986) on shale whole-rock samples of Proterozoic age, Bros et al. (1992) studied the Sm–Nd isotopic systematics of several size fractions of two black-shale samples from a Lower Proterozoic sedimentary sequence of southwestern Gabon. The two clay fractions <0.2 µm and 0.2–0.4 µm of the two samples were leached with dilute hydrochloric acid (1 N) to analyse separately the amounts and isotopic compositions of the mobile REE located outside the mineral structures, and those of the REE trapped in the mineral structures (Fig. 7). The Sm–Nd results of the untreated aliquots, the leachates and residues for the two size fractions of each of the two samples, defined separate isochrons, one yielding a date of 2099±115 Ma with an initial $^{144}Nd/^{143}Nd$ ratio of about 0.5107 and the other giving a date of 2036±79 Ma with an initial $^{144}Nd/^{143}Nd$ ratio of about 0.5108. The authors interpreted these lines as isochrons and the ages as resulting from isotopic homogenizations that occurred during diagenetic evolution, probably related to the illitization process. In comparison to the Sm–Nd data, the K–Ar and the Rb–Sr dates were found to be lower with values at about 1.8 Ga. As the Rb–Sr and K–Ar ages were found to be similar, the authors considered that these isotopic systems did not suffer preferential diffusion losses of the radiogenic isotopes relative to the non-radiogenic ones, but were decoupled from the corresponding Sm–Nd isotopic system during a late diagenetic-to-hydrothermal activity, as found and described by Schaltegger et al. (1994) under similar conditions in Morocco.

Bros et al. (1992) also noticed that the analytical data point of an organic matter-enriched shale plotted in the region between the two isochrons defined above. Furthermore, the data point of a pure bitumen sample, which had the highest Sm/Nd ratio, also deviated slightly from either isochron. The line drawn through these two data points was parallel to and between the two previous isochrons, strongly suggesting that isotopic homogenization of the organic and inorganic materials occurred roughly at the same period, 2.07 Ga ago. The isotopic homogeneity among interstitial fluid (leachates), organic matter (kerogen) and clay minerals (residues) leads evidence to the reliability of the Sm–Nd isotopic method for the purpose of stratigraphic dating.

3.3 $^{40}Ar/^{39}Ar$ Dating

The $^{40}Ar/^{39}Ar$ technique is a recent extention of the $^{40}K/^{39}Ar$ technique (Merrihue and Turner 1966), but it was so far of limited use in studies of sedimentary rocks. It is based on the fact that ^{39}K decays to ^{39}Ar by fast-neutron irradiation in a reactor. Among the advantages of this technique relative to the K–Ar method is the simul-

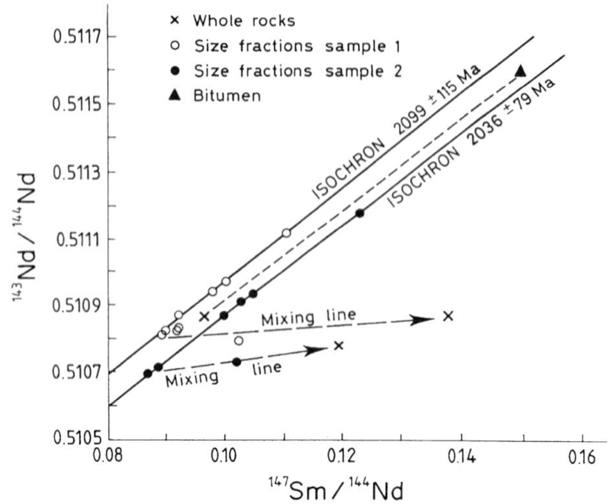

Fig. 7. Sm—Nd isochron diagram containing data points for clay fractions (*squares*) and whole rocks (*crosses*), and a single data point for bitumen (*triangle*) of Francevillian sedimentary rocks, Gabon. (After Bros et al. 1992)

taneous measurement of both radioactive parent and radiogenic daughter isotopes on the same aliquot, which avoids the uncertainty from sample heterogeneity and allows analysis by stepwise progressive heating. However, the analytical procedure is not without inconveniences and uncertainties including the recoil effect on the retention of Ar. Such effects can be especially significant when the particles are small, like clay particles, because the isotopes which are impacted by the neutrons are moved in the lattices and can be expelled out of the mineral structure.

Kunk and Brusewitz (1987) documented the problem of the ^{39}Ar recoil effect on an illite/smectite mixed-layer mineral placed in a quartz vial before irradiation. They could determine the amount of recoil, which has also been mentioned by Yanase et al. (1975), Brereton et al. (1976) and Foland et al. (1984) on glauconite-type minerals. However, loss of ^{39}Ar, which has been related to structural alterations attendant on irradiation, may not be the rule for all clay minerals, since Hunziker et al. (1986) and Reuter and Dallmeyer (1987a,b), for instance, reported cases of similar conventional K–Ar and ^{40}Ar/^{39}Ar ages for different size fractions of illite-type minerals. It is clear that the potential application of the ^{40}Ar/^{39}Ar technique to well-crystallized diagenetic illite for dating purpose needs understanding of the behaviour of large particles (in the 10 μm range) during irradiation. More about the potential merits of such a tool may be found in a review by Clauer and Chaudhuri (1992), but additional studies addressing the recoil-related loss of radiogenic ^{40}Ar by different types of clay minerals are warranted before any expectation of routine application of the ^{40}Ar/^{39}Ar method on sediments can materialize.

4 Isotopic Dating of Non-Clay Minerals

Salts have been studied by the various routine techniques, but the isotopic dates for such materials were often lower than the stratigraphic ages (e.g. Chaudhuri and Clauer 1992), langbeinite being the only salt phase able to yield reasonable K–Ar dates relative to depositional time (Oesterlé and Lippolt 1975). However, it should be recalled that these minerals are extremely sensitive to post-formational alteration related to migration of formational brines. Continuous diffusion of radiogenic isotopes has often been invoked to explain the anomalous ages for salts, in the knowledge that frequent dissolution–recrystallization processes would lead to similar results.

On the other hand, marine carbonate and phosphate materials appear very promising for geochronologic investigations. Considerable information exists about the use of isotopic compositions of Sr in marine carbonate rocks as a tool for stratigraphic dating. Pb isotope analyses of marine carbonate rocks have been applied more recently to determine stratigraphic ages.

4.1 Dating by Reference to the Secular Variation of Marine Sr

Sr is homogeneously distributed in the oceans because its residence time in seawater is much longer than the average time for complete mixing of these waters (Chester 1990). Having a chemical behaviour similar to that of Ca, Sr is easily incorporated in carbonate minerals, often by substitution, while Rb is present in nearly negligible amounts, so that the production of radiogenic ^{87}Sr from decay of ^{87}Rb is often negligible. Marine carbonate minerals thus inherit the Sr isotopic signature of the ocean water at the time of formation of the minerals. By analysing the Sr isotopic composition of palaeontologically well-dated marine fossils that have remained largely unaltered, seawater Sr has been well recognized for variations in its isotopic composition as a function of time. This trend in the Sr isotopic variation has been used to define the stratigraphic position of a carbonate rock by analysing its Sr isotopic composition. The idea of using the Sr isotopic composition of carbonates for dating sedimentary rocks goes to Wickman (1948) who thought that the isotopic composition of the marine Sr increased regularly with time. But the works of Peterman et al. (1970), of many others and later of Burke et al. (1982), have shown that the $^{87}Sr/^{86}Sr$ ratio evolves irregularly and not linearly as a function of time. The $^{87}Sr/^{86}Sr$ ratio of seawater depends on the mass balance between the continental and hydrothermal inputs to the oceans. The marine isotopic composition is regulated by the mantle Sr provided through hydrothermal activity in the oceanic crust, the continental Sr represented by riverine dissolved loads, input from aeolian dusts, flux from groundwaters, and the diagenetic Sr from recrystallized carbonate minerals in the ocean basins (Veizer 1989, 1992). Variations in these fluxes due to oceanic expansion, climatic changes, continental surface changes, etc. induced fluctuations of the marine $^{87}Sr/^{86}Sr$ ratio.

The secular variation of the isotopic composition of marine Sr, illustrated in Fig. 8, is theoretically easy to use, as it simply requires that the Sr isotopic composition of carbonates, phosphates, sulphates or oxides be matched with the value given in the curve for the time-dependent Sr isotopic variation. However, this method of dating has a large degree of uncertainty due to inherent weaknesses. Among the weaknesses is the fact that the Sr trapped by the primary mineral phases has its isotopic composition sytematically modified by recrystallization processes.

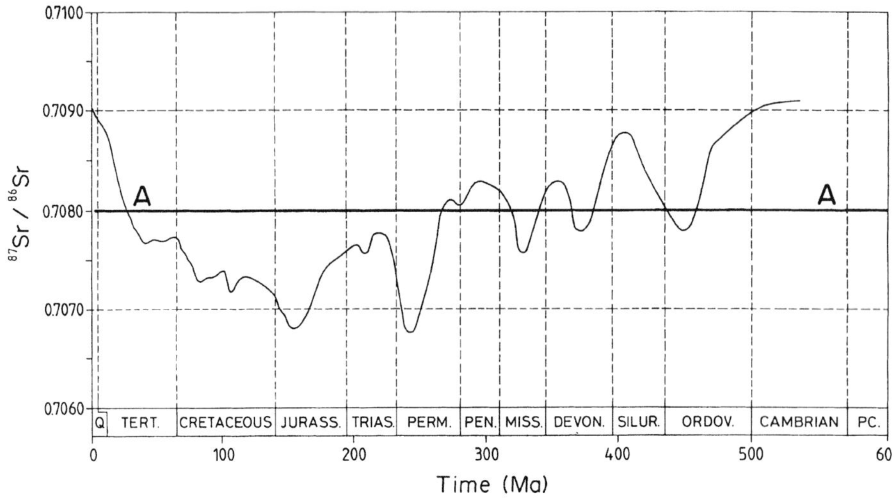

Fig. 8. Secular variation of the marine Sr during Paleozoic time (after Burke et al. 1982). The *horizontal line* at 0.7080 corresponds to the example discussed in the text

It should also be mentioned that the same isotopic value may belong to several periods. Figure 8 shows that an $^{87}Sr/^{86}Sr$ ratio of 0.708, for instance (line AA), indicates Ordovician, Devonian, Carboniferous, Permian and even Tertiary formation periods. Complementary information is, therefore, needed for material suitable for dating by this method. Furthermore, calculated ages will be precise only for the parts of the curve with very steep slopes, as in lower and middle Devonian, lower and middle Carboniferous, middle and upper Permian, lower Triassic and upper Tertiary times.

This method is especially well suited to dating primary phosphate minerals of Tertiary sequences precisely. Stille et al. (1994), for instance, found phosphate minerals without any evidence for secondary recrystallization and yielding mean ages between 19.1 and 17.2 Ma across a sediment a few metres thick. The use of such a result is obvious if one knows the thickness and the sequential order of the deposition of the studied sequence.

4.2 Pb–Pb Dating of Carbonates

Carbonate rocks did not represent a common target material for isotopic dating by the conventional methods until marine organisms were found to have concentrated U in their skeletal materials, which prompted U–Pb methods to be applied to ancient carbonate rocks. The first evaluation of dating carbonate rocks by the U–Pb and Pb–Pb methods was given by Doe (1970) who found such rocks to be unsuitable for isotopic dating. About two decades later, Moorbath et al. (1987) further explored the Pb–Pb dating of metamorphic carbonate rocks. Since then, several studies have been reported focusing on Pb-isotope analyses on marbles (Jahn

1988; Jahn et al. 1992; Taylor and Kalsbeek 1990) and non-metamorphic carbonates (Smith and Farquhar 1989; Smith et al. 1991; Jahn et al. 1990; DeWolf and Halliday 1991).

Moorbath et al. (1987) made various batch dissolutions of some of the samples and analysed both the leachates and the residues. The isotopic data points defined a line which inclined the authors to suggest that the isotopic date reflects the time of a metamorphic activity which is thought to have caused an isotopic equilibrium between the carbonate and silicate phases. By contrast, Smith and Farquhar (1989) and Jahn et al. (1990) obtained analytical scatters by using the same procedure. To understand the variation in the results, mineralogical results are essential, provided that information is available about the origin and diagenetic evolution of the rocks. The results should even be completed by petrographic observations, including cathodoluminescence, chemical data, and oxygen and carbon isotope geochemistry.

5 Conclusion

Isotopic dating for stratigraphic purpose is possible on authigenic mineral phases as long as these components form during sedimentation and do not undergo further changes inducing preferential gains or losses of radiogenic isotopes. Additionally, late diagenetic mineral components have also to be identified and separated from early formed. Glauconite- and 1M illite-type clay minerals, as well as palygorskite, are well suited if carefully separated from whole rocks, and characterized by using independent means to identify the formational process and to trace the chemical and isotopic compositions of the formational environment. This environment can be chemically reconstructed by leaching clay particles and analysing the leachates.

The Rb–Sr and Sm–Nd methods appear to be most promising. The K–Ar method is applicable to Phanerozoic materials which have not been buried to depths inducing loss of radiogenic ^{40}Ar; the values, in this case, are lower than those of the Rb–Sr and Sm–Nd methods, especially in studies on Proterozoic clay minerals.

Carbonate, phosphate and sulphate mineral phases formed during sedimentation time can also be dated by isotopic methods. The $^{87}Sr/^{86}Sr$ ratios of these marine components may, for instance, be projected on the path of the secular variation of the isotopic composition of the marine Sr to define the time of their crystallization. The Pb–Pb method applied to carbonate rocks or minerals seems also to have an interesting potential, but basic work is still needed before it can be used routinely. This is also the case for the $^{40}Ar/^{39}Ar$ method in which fundamentals have to be carefully controlled before a decision can be taken for routine use in stratigraphic dating.

References

Allègre CJ, Caby R (1972) Chronologie absolue du Précambrien de l'Ahaggar occidental. CR Acad Sci Paris 275, D:2095–2098

Amouric M, Parron C (1985) Structure and growth mechanism of glauconite as seen by high resolution transmission electron microscopy. Clays Clay Min 33:473–482

Baadsgaard H (1987) Rb–Sr and K–Ca isotope systematics in minerals from potassium horizons in the Prairie Evaporite formation, Saskatchewan, Canada. Chem Geol 66:1–15

Bath AH (1977) Experimental observation of exchange of Rb and Sr between clays and solution. 2nd Int Symp on Water-rock interaction, Strasbourg, France, IV, 244–249

Bernat M, Bieri RH, Koide M, Griffin JJ, Goldberg ED (1970) Uranium, thorium, potassium and argon in marine phillipsites. Geochim Cosmochim Acta 34:1053–1072

Birch GF, Willis JP, Rickard RS (1976) An electron microprobe study of glauconites from the continental margin off the west coast of S. Africa. Mar Geol 22:271–284

Blanco JA, Corrochano A, Montigny R, Thuizat R (1982) Sur l'âge du début de la sédimentation dans le bassin tertiaire du Duero (Espagne). Attribution au Paléocène par datation isotopique des alunites de l'Unité inférieure. CR Acad Sci Paris 295/II:259–262

Bofinger VM, Compston W, Vernon MJ (1968) The application of acid leaching to the Rb–Sr dating of a Middle Ordovician shale. Geochim Cosmochim Acta 32:823–833

Brereton NR, Hooker PT, Miller JA (1976) Some conventional potassium-argon and $^{40}Ar/^{39}Ar$ age studies on glauconite. Geol Mag 113:329–340

Brookins DG (1980) Geochronologic studies in the Grants mineral belt. New Mex Bur Mines Min Res Mem 27:87–98

Bros R, Stille P, Gauthier-Lafaye F, Weber F, Clauer N (1992) Sm–Nd isotopic dating of Proterozoic clay material. Example from Francevillian sedimentary series (Gabon). Earth Planet Sci Lett 113:207–218

Brueckner HK, Snyder WS (1979) Rb–Sr dating of chert: a potential chronological tool. Geol Soc Am Abstr Prog 11–2, San Jose, California, p 71

Burke WH, Denison RE, Hetherington EA, Koepnick RB, Nelson HF, Otto JB (1982) Variation of seawater $^{87}Sr/^{86}Sr$ throughout Phanerozoic time. Geology 10:516–519

Burns SJ, Haudernschild U, Matter A (1994) The strontium isotopic composition of carbonates from the late Precambrian (ca. 560–540 Ma) Huqf Group of Oman. Chem Geol Isot Geosci Sect 111:269–282

Chaudhuri S, Clauer N (1992) History of marine evaporites: constraints from radiogenic isotopes. In: Clauer N, Chaudhuri S (eds) Isotopic signatures and sedimentary records. Lecture Notes in Earth Sciences 43. Springer, Berlin Heidelberg New York, pp 177–198

Chester R (1990) Marine geochemistry. Unwin Hyman, London, 698 pp

Clauer N (1974) Utilisation de la méthode rubidium-strontium pour la datation d'une schistosité de sédiments peu métamorphisés: application au Précambrien II de la boutonnière de Bou Azzer-El Graara (Anti-Atlas). Earth Planet Sci Lett 22:404–412

Clauer N (1976) Géochimie isotopique du strontium des milieux sédimentaires. Application à la géochronologie de la couverture du craton ouest-africain. Sci Géol Mém (Strasb) 45:256 pp

Clauer N (1982a) Strontium isotopes of Tertiary phillipsites from the Southern Pacific: timing of the geochemical evolution. J Sediment Petrol 52:1003–1009

Clauer N (1982b) The rubidium-strontium method applied to sediments: certitudes and uncertainties. In: Odin GS (ed) Numerical dating in stratigraphy. Wiley, New York, pp 245–276

Clauer N, Chaudhuri S (1992) Indirect dating of sediment-hosted ore deposits: promises and problems. In: Clauer N, Chaudhuri S (eds) Isotopic signatures and sedimentary records. Lecture Notes in Earth Sciences 43. Springer, Berlin Heidelberg New York, pp 361–388

Clauer N, Chaudhuri S (1995) Clays in crustal environments. Isotopic dating and tracing. Springer, Berlin Heidelberg New York, 358 pp

Clauer N, Hoffert M, Grimaud D, Millot G (1975) Composition isotopique du strontium d'eaux interstitielles extraites de sédiments récents: un argument en faveur de l'homogénéisation isotopique des minéraux argileux. Geochim Cosmochim Acta 39:1579–1582

Clauer N, Hoffert M, Karpoff AM (1982a) The Rb–Sr isotope system as an index of origin and diagenetic evolution of southern Pacific red clays. Geochim Cosmochim Acta 46:2659–2664

Clauer N, Caby R, Jeannette D, Trompette R (1982b) Geochronology of sedimentary and metasedimentary Precambrian rocks of the West African craton. Precambrian Res 18:53–71

Clauer N, Giblin P, Lucas J (1984) Sr and Ar isotope studies of detrital smectites from the Atlantic Ocean (DSDP, Legs 43, 48, and 50). Isot Geosci 2:141–151

Clauer N, O'Neil JR, Bonnot-Courtois C, Holtzappfel T (1990) Morphological, chemical and isotopic evidence for an early diagenetic evolution of detrital smectite in marine sediments. Clays Clay Min 38:33–46

Clauer N, Keppens E, Stille P (1992a) Sr isotopic constraints on the process of glauconitization. Geology 20:133–136

Clauer N, Stille P, Keppens E, O'Neil JR (1992b) Le mécanisme de la glauconitisation: apports de la géochimie isotopique du strontium, du néodyme et de l'oxygène de glauconies récentes. CR Acad Sci Paris 315/II:321–327

Clauer N, Chaudhuri S, Kralik M, Bonnot-Courtois C (1993) Effects of experimental leaching on Rb–Sr and K–Ar isotopic systems and REE contents of diagenetic illite. Chem Geol 103:1–16

Clauer N, Srodon J, Francu J, Sucha W (1995) K–Ar dating of illite/smectite fundamental particles. Euroclay Conf, 19–25 Aug, Leuven, Belgium, 2 pp

Compston W, Pidgeon RT (1962) Rb–Sr dating of shales by the total-rock method. J Geophys Res 67:3493–3502

Cordani UG, Kawashita K, Thomas-Filho A (1978) Applicability of the Rb–Sr method to shales and related rocks. Am Assoc Petrol Geol, Spec Publ 6:93–117

Cordani UG, Thomaz-Filho A, Brito-Neves BB, Kawashita K (1985) On the applicability of the Rb–Sr method to argillaceous sedimentary rocks: some examples from Precambrian sequences of Brazil. Giorn Geol Ser 3,547:253–280

Cormier RF (1956) Rubidium-strontium ages of glauconite and their application to the construction of a Post-Precambrian time-scale. PhD Thesis, Massachusetts Institute of Technology, Cambridge, Massachusetts

Derry LA, Keto LS, Jacobsen SB, Knoll AH, Swett K (1989) Sr isotopic variations in Late Proterozoic carbonates from Svalbard and East Greenland. Geochim Cosmochim Acta 54:2331–2339

Derry LA, Brasier MD, Corfield RM, Rozanov AYu, Zhuravlev AYu (1996) Sr isotopes in Lower Cambrian carbonates from the Siberian Craton: a paleoenvironmental record during the "Cambrian explosion". Earth Planet Sci Lett (in press)

Dewolf CP, Halliday AN (1991) U–Pb dating of a remagnetized Paleozoic limestone. Geophys Res Lett 18:1445–1448

Doe BR (1970) Evaluation of U-Th-Pb whole-rock dating on Phanerozoic sedimentary rocks. Eclog Geol Helv 63:79–82

Faure G (1982) The marine-strontium geochronometer. In: Odin GS (ed) Numerical dating in stratigraphy. Wiley, New York, pp 73–79

Faure G (1986) Principles of isotope geology, 2nd edn. Wiley, New York, 589 pp

Foland KA, Linder JS, Laskowski TE, Grant NK (1984) $^{40}Ar/^{39}Ar$ dating of glauconites: measured ^{39}Ar recoil loss from well-crystallized specimens. Chem Geol Isot Geosci Sect 2:241–264

Gorokhov IM, Clauer N, Turchenko TL, Melnikov NN, Kutyavin EP, Pirrus E, Baskakov AV (1994) Rb–Sr systematics of Vendian-Cambrian claystones from east European platform: implications for a multi-stage illite evolution. Chem Geol 112:71–89

Hunziker JC, Frey M, Clauer N, Dallmeyer RD, Friedrichsen H, Flehmig W, Hochstrasser K, Roggwiller P, Schwander H (1986) The evolution of illite to muscovite: mineralogical and isotopic data from the Glarus Alps, Switzerland. Contrib Miner Petrol 92:157–180

Jahn BM (1988) Pb–Pb dating of young marbles from Taïwan. Nature 332:429–432

Jahn BM, Bertrand-Sarfati J, Morin N, Macé J (1990) Direct dating of stromatolitic carbonates from the Schmidtdrif Formation (Transvaal dolomite), South Africa, with implications on the age of the Ventersdorp Supergroup. Geology 18:1211–1214

Jahn BM, Chi WR, Yui TF (1992) A Late Permian formation of Taïwan (marbles from Chia-Li well no. 1): Pb–Pb isochron and Sr isotopic evidence, and its regional and geological significance. J Geol Soc China 35:193–218

Kaufman AJ, Jacobsen SB, Knoll AH (1994) The Vendian record of Sr- and C-isotopic variations in seawater/implications for tectonics and paleoclimate. Earth Planet Sci Lett 120:409–430

Keppens E, O'Neil JR (1984) Oxygen isotope variations in glauconies. Terra Cognita, Spec Issue:42

Kolodny Y, Luz B (1992) Isotope signatures in phosphate deposits: Formation and diagenetic history. In: Clauer N, Chaudhuri S (eds) Isotopic signatures and sedimentary records. Lecture Notes in Earth Sciences, 43. Springer, Berlin Heidelberg New York, pp 69–122

Kunk MJ, Brusewitz AM (1987) 39Ar recoil in an I/S clay from the Ordovician "Big Bentonite Bed" at Kinnekulle, Sweden. 21st Annu Meet North-Central Section, April 1987, St Paul, Minnesota, Geol Soc Am, Abstr with Prog, 19, p 230

Liewig N, Clauer N, Sommer F (1987) Rb–Sr and K–Ar dating of clay diagenesis in Jurassic sandstone reservoirs, North Sea. Am Assoc Petrol Geol Bull 71:1467–1474

Moorbath S, Taylor PN, Orpen JL, Treloar P, Wilson JF (1987) First direct radiometric dating of Archean stromatolitic limestone. Nature 326:865–867

Nägler T, Schäfer JL, Gebauer D (1992) A Sm–Nd isochron on pelites 1 Ga in excess of their depositional age and its possible significance. Geochim Cosmochim Acta 56:789–795

Nicolaysen LO (1961) Graphic interpretation of discordant age measurements on metamorphic rocks. Ann NY Acad Sci 91:198–206

Merrihue CM, Turner G (1966) Potassium–argon dating by activation with fast neutrons. J Geophys Res 71:2852–2857

Odin GS (1975) De glauconarium, origine, aetateque. Thèse Doc ès-Sci, Univ Paris VI, 280 pp

Odin GS (ed) (1982) Numerical dating in stratigraphy, 2 vols. Wiley, Chichester, 1094 pp

Odin GS, Bonhomme MG (1982) Argon behaviour in clays and glauconies during preheating experiments. In: Numerical Dating in Stratigraphy. Wiley, pp 333-349

Odin GS, Dodson MH (1982) Zero isotopic age of glauconies. In: Odin GS (ed) Numerical Dating in Stratigraphy. Wiley, pp 277–305

Odin GS, Hunziker JC (1982) Radiometric dating of the Albian–Cenomanian boundary. In: Odin GS (ed) Numerical Dating in Stratigraphy. Wiley, pp 537–556

Odin GS, Matter A (1981) De glauconarium origine. Sedimentology 28:611–641

Odin GS, Dodson MH, Hunziker JC, Kreuzer H (1979) Radiogenic argon in glauconies during their genesis. Bull Inform Int Geol Corr Prog, Project 133,6:7–8

Oesterlé FP, Lippolt HJ (1975) Isotopische Datierung der Langbeinitbildung in der Kalisalzlagerstätte des Fuldabeckens. Kali Steinsalz 11:391–398

Ohr M, Halliday AN, Peacor DR (1991) Sr and Nd isotopic evidence for punctuated clay diagenesis, Texas Gulf Coast. Earth Planet Sci Lett 105:110–126

Peterman ZE, Hedge CE, Tourtelot HA (1970) Isotopic composition of strontium in seawater throughout Phanerozoic time. Geochim Cosmochim Acta 34:105–120

Reuter A, Dallmeyer RD (1987a) $^{40}Ar/^{39}Ar$ age spectra of whole-rock and constituent grain-size fractions from anchizonal slates. Chem Geol 66:73–88

Reuter A, Dallmeyer RD (1987b) $^{40}Ar/^{39}Ar$ dating of cleavage formation in tuffs during anchizonal metamorphism. Contrib Miner Petrol 97:352–360

Schaltegger U, Stille P, Rais N, Piqué A, Clauer N (1994) Nd and Sr isotopic dating of diagenesis and low-grade metamorphism of argillaceous sediments. Geochim Cosmochim Acta 58:1471–1481

Shanin LL, Ivanov IB, Shipulin FK (1968) The possible use of alunite in K–Ar geochronology. Geokhimyia 1:109–111

Smith PE, Farquhar RM (1989) Direct dating of Phanerozoic sediments by the ^{238}U-^{206}Pb method. Nature 341:518–521

Smith PE, Farquhar RM, Hancock RG (1991) Direct radiometric age determination of carbonate diagenesis using U–Pb in secondary calcite. Earth Planet Sci Lett 105:474- 491

Stille P, Clauer N (1986) Sm–Nd isochron-age and provenance of the argillites of the Gunflint Iron Formation in Ontario, Canada. Geochim Cosmochim Acta 50:1141–1146

Stille P, Clauer N (1994) The process of glauconitization. Chemical and isotopic evidence. Contrib Mineral Petrol 117:253–262

Stille P, Gauthier-Lafaye F, Bros R (1993) The Nd isotope system as a tool for petroleum research and exploration. Geochim Cosmochim Acta 5:4521–4525

Stille P, Riggs S, Clauer N, Crowson R, Ames D, Snyder SW (1994) Sedimentation through one Miocene depositional cycle based on Sr and Nd isotopic analysis of phosphorite peloids: North Carolina continental shelf (Part 1). Mar Geol 117:253–273

Taylor PN, Kalsbeek F (1990) Dating the metamorphism of Precambrian marbles: examples from Proterozoic mobile belts in Greenland. Chem Geol 86:21–28

Veizer J (1989) Strontium isotopes in seawater through time. Annu Rev Earth Planet Sci 17:141- 167

Veizer J (1992) Depositional and diagenetic history of limestones: Stable and radiogenic isotopes. In: Clauer N, Chaudhuri S (eds) Isotopic signatures and sedimentary records. Lecture Notes in Earth Sciences, 43. Springer, Berlin Heidelberg New York, pp 13–48

Wasserburg GJ, Hayden RI, Jensen KJ (1956) Ar^{40}-K^{40} dating of igneous rocks and sediments. Geochim Cosmochim Acta 10:153–165

Wickman FE (1948) Isotope ratios – a clue to the age of certain marine sediments. J Geol 56:61- 66

Yanase Y, Wampler JM, Dooley RE (1975) Recoil-induced loss of ^{39}Ar from glauconite and other minerals. Trans Am Geophys Union 56:472

Zwingmann H (1995) Etude des conditions de mise en place des gaz naturels dans les réservoirs gréseux de la Rotliegende (Permien) en Allemagne. Aspects minéralogiques, géochimiques et isotopiques. Thèse, Univ Strasbourg, 189 pp

15 Concomitant Alteration of Clay Minerals and Organic Matter During Burial Diagenesis

BERNARD KÜBLER

1 Introduction

Georges Millot and his collaborators, in working on the genesis of clays during pedogenic or sedimentological evolutions, had no choice but to deal with diagenetic burial alteration to explain, in ancient sediments, the mineral assemblages which often constitute the source rocks of soils and sediments via erosion/alteration processes. Georges Millot claimed, in a famous shortcut, that any diagenetic alteration is characterized by a reduction of the number of argillaceous minerals to two types: micas and chlorites. This diagenetic evolution could even be delineated into stages in considering the clay mineral parageneses. On the basis of results obtained by Burst (1959), Weaver (1958) and Grim's «school», Georges Millot emphasized the importance of argillaceous minerals in petroleum prospection, as well as lithostratigraphic markers in azoic series and evolution markers during burial diagenesis. Thus, he contributed to the creation of a scientific community providing thousands of studies in many basins of different ages, across the five continents.

It became clear to Georges Millot that both syngenesis and diagenesis are only answers to dynamic geochemical environments. The research laboratory of the Société Nationale des Pétroles d'Aquitaine and of the French National Research Council at Strasbourg were, therefore, completed by equipment for the analysis of major and minor elements in sediments. The geochemical and mineralogic-crystallographic data basis was then used to understand how mineral phases appear or disappear, in relation with thermodynamic methods and chemical studies of interstitial solutions or those related to diagenetic changes.

If maturation of organic matter is induced by diagenetic processes whose stages could be progressively more constrained, both primary and secondary migrations are complex processes which have not yet been fully elucidated. For secondary migration, in addition to the fundamental parameters such as porosity and permeability, the driving factors are of diagenetic, but also of tectonic origin. In this case, partly because of investigations of Georges Millot's collaborators, attempts were made to determine the periods of fluid migrations, in applying the K/Ar or Rb/Sr methods on neoformed micas built in brine traps of oil reservoirs.

In fact, to establish clearly a relationship between maturation of hydrocarbons and alteration of minerals during burial diagenesis, it is essential to establish clearly the different evolutionary steps for each organic or mineral component and to determine precisely the specific parameters over the same rock sequences. As such detailed and combined studies are scarce, independent criteria, such as vitrinite reflectance, pyrochromatography, fluid inclusions or apatite fission tracks were sought to characterize the different steps independently. The aim of the present contribution is to reassess the maturation stages of organic matter with a tentative comparison of the equivalent alteration stages of the argillaceous phases.

2 Maturation Stages of Organic Matter

Since the work of Weaver (1960) and since Burst (1959,1969) correlated the disappearance of smectite with accumulation and migration of natural oils in sediments of the Gulf Coast area, special attention has been given to the comparison of mineral alteration and maturation of organic matter. However, the concepts have evolved a great deal, especially the significance of depth and temperature as reference variables. The analytical equipment available for organic geochemistry has also been considerably diversified and improved. Petrographic studies of dispersed organic matter by fluorescence, reflectance and coloration of spores and pollens, have completed the scope of the organic diagenetic markers. Pyrochromatography, by the «Rock-Eval» technique for instance (Espitalié et al. 1985), became one of the most widely used tools to distinguish the different maturation stages of organic matter, and to determine the petroleum potential of drilled sedimentary rock units.

In a number of studies, results from studies on organic geochemistry only proved to be powerful when combined with seismo-tectonic and seismo-stratigraphic interpretations (Cramez-Dias and Kübler 1982). Interpretation of the seismic results improved the understanding of the ages of source rocks in terms of sequential stratigraphy and their location in global tectonic models of various types of basins and margins, each having its own thermal, burial, extensional, compressional and hydrocabron migrational histories. All these facts have been combined by Cramez (pers. comm.) to suggest the concept of «petroleum systems». Application of this concept to mineral alterations would be extremely useful for understanding the distribution of argillaceous minerals and their diagenetic alterations.

The notion of diagenetic markers for clay minerals was a strictly static concept, but the approach became much more dynamic because of the need to understand the processes governing primary migration, decompaction of undercompacted (overpressured) zones and secondary migration, all of which processes are closely associated with mineral alterations. During the pioneering period, the organic markers used to establish relationships with the alteration of clays are those of the "potential oil window", that is to say reflectance of organic matter, pyrochromatography and, in some cases, chemistry of adsorbed gases.

2.1 The Oil Window

Pusey III (1973) grouped under the term «oil window» any economic accumulation of petroleum occurring in an oil field at limited depths and geothermal gradients, in a temperature range of 65.5–150 °C (Fig. 1). By extension, the same term was used for the stage during which liquid hydrocarbons are generated and are defined by the quantity of organic extracts normalized in organic coal contents (Le Tran et al. 1974; Fig. 2). In fact, this definition does not correspond to the real oil window, but to a potential oil window whose definition is not exclusively based upon the quantity of organic extracts but on the nature of the hydrocarbons, the parity and the amount of the n-alkanes, the number of cycles of the cyclo-alkanes, and the number of carbons in the aromatics (Tissot and Welte 1978; Fig. 3).

The organic geochemists decided later to divide the domain of diagenesis defined by the geologists into three different stages: (1) a diagenetic one occurring between the surface and the beginning of the potential oil window, also called the immature stage; (2) a catagenetic one being the stage at which liquid hydrocarbons are formed, also called the mature stage or potential oil window; and (3) a metagenetic one corresponding to the stage of methane or dry gas formation, also called the super-mature or sterile stage. The notion of the potential oil window (POW) is

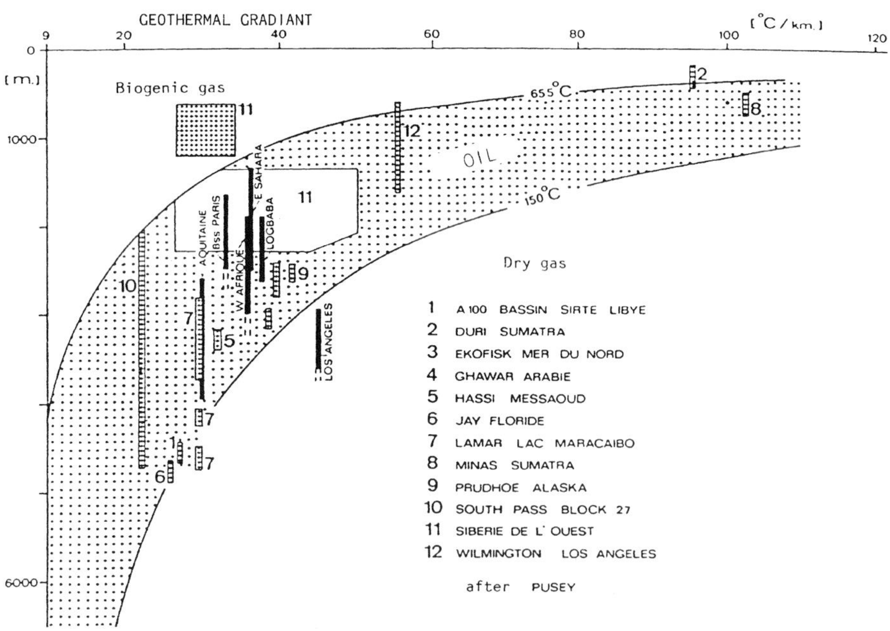

Fig. 1. The concept of the oil window according to Pusey III (1973, in Kübler 1980). Depth of the oil window depends on temperature, in other words, on the geothermal gradient. Dates from Aquitaine basin, Paris basin, Logbaba basin, Los Angeles basin, and central (Sahara) and western Africa have been added to the original graphic

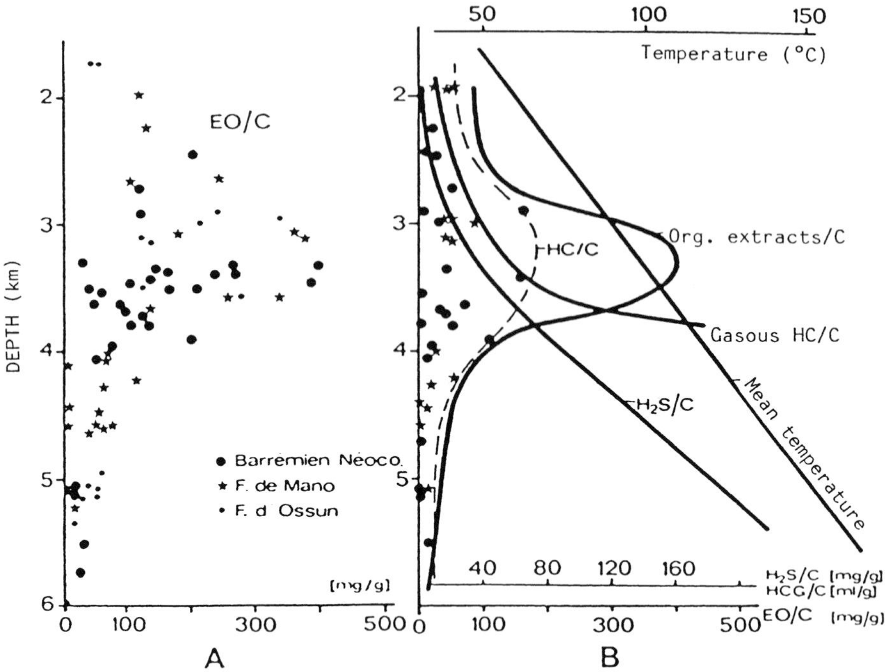

Fig. 2. The concept of the oil window compared to the example of the Aquitaine basin (according to Le Tran et al. 1974). The *dashed curve* is only defined by about 15 experimental points, the others plotting inside the curved area. The production of H_2S is a diagenetic process which accompanies that of the gases; it should also be found in the fluid inclusions

still a widely considered aspect in petroleum prospection. It represents one among the six conditions required to succeed in oil exploration and which are: a source rock, a maturation of the POW stage, a reservoir, an impermeable caprock, a tectonic or stratigraphic trap, and a primary or secondary migration at the appropriate moment.

In early research, most scientists based the POW on depth alone, i.e. on an adequate burial to reach the required temperature range. However, the geothermal gradient was soon taken into consideration in order to adjust depths and durations during which the source rocks have to be exposed to undergo the accepted temperatures for the POW. The empirical relationship published by Connan (1974) relates temperature and duration based on POWs determined by the organic extracts:

$$\log t = 3014 \times 1/T - 6.498$$

where t is expressed in millions of years and T in degrees Kelvin (Fig. 4). According to this equation, whose coefficients might have different values (Connan 1974), the start of the POW stage could be reached in 12 million years at a temperature of 120 °C, or in 400 million years at 60 °C. This approach had the merit of attracting attention to the fact that the quantity of energy required, also termed the «Parmentier principle» (Kübler 1980), was the cardinal variable for oil to reach the maturation stage. However, application of these values to Pleisto-Pliocene and Cretaceous sequences of the Gulf Coast area studied by Dow (1978), did not provide very convincing results (Kübler 1984). Since Connan's first equation does not take into account geodynamic aspects of basin histories, it was rapidly improved by inclusion of «iso-thermo-chrono-lines» (Cornelius 1975; Fig. 5). This latter concept takes into account the exposure time to the required temperatures. However, to avoid their direct use, Cornelius (1975) calibrated the curves on the basis of vitrinite reflectance (Ro).

Since Lopatin (1971), it has been widely accepted that vitrinite reflectance of coals is an excellent temperature indicator integrating the geodynamics of the basin, i.e. the burial history, the geothermal gradient and the residence time split into successive temperature increments of 10 °C. Vitrinite reflectance increases exponentially with temperature and linearly with time. Wapples (1981) took Lopatin's basic equations and defined a time temperature index (TTI), i.e. the sum of the maturities acquired at each temperature interval, which he applied to the POW. This TTI integrates the depth, the geothermal gradient, i.e. the temperatures and the residence periods calculated from burial curves. This is a highly pragmatic approach useful at a first glance and which has, in addition, the advantage of including the tectonic history of the sedimentary basins. In fact, basins often experience successive extensional and compressional phases and the tectonic structures which are objectives of the petroleum companies, are frequently displaced by tec-

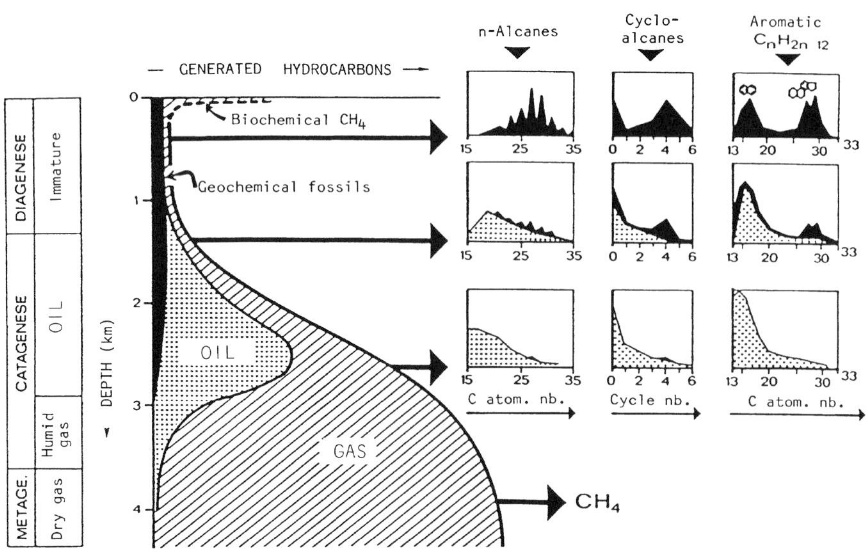

Fig. 3. Diagram of the formation of hydrocarbons according to Tissot and Welte (1978). The depths are only indicative. Catagenesis begins when liquid hydrocarbons are produced, metagenesis when dry gases are within the methane zone

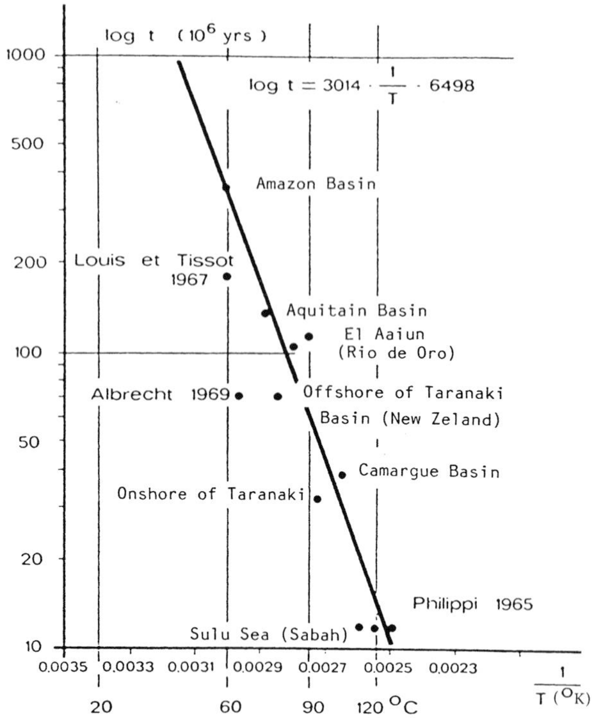

Fig. 4. Influence of time duration on the stage at which liquid hydrocarbons are generated (Connan 1974). The ages are plotted vertically on a logarithmic scale and the temperatures horizontally in °C and in °K

tonic activities from maturation depths to shallower burial. Mineral alterations would certainly gain from being compared to the TTI.

The range of energy required for the production of liquid hydrocarbons is different according to the nature of the kerogen. For type I kerogens (Green River Shale of the Uinta Basin), depth and Ro (vitrinite reflectance) are higher than those of type III (see references in Kübler et al. 1979a). Espitalié (1986) defined these differences on the basis of a comparison between temperatures of Rock-Eval pyrolysis, organic extracts and Ro reflectance. If alteration of smectite is related to the generation of liquid hydrocarbons, the kerogen type of the diagenetic host rocks should be kept in mind. Introduction of kinetic parameters in the genetic models of liquid hydrocarbons was suggested by Tissot (1969). Completed by Tissot et al. (1975), this constitutes a large sector of the THEMIS (Doligez et al. 1986) and OPTIM (Ungerer et al. 1986) models. It would be useful for understanding the alteration of clays to take into account fluid migration resulting from these simulations. Since the POW depends on all mentioned parameters, claiming the importance of a given depth or a given temperature is no longer determining, as was the case when these comparisons were first made.

Mineral alterations, especially disappearance of smectite, only gain full significance if they are closely compared to all parameters integrating burial history, geothermal gradient and exposure time – all of these being useful to define the POW.

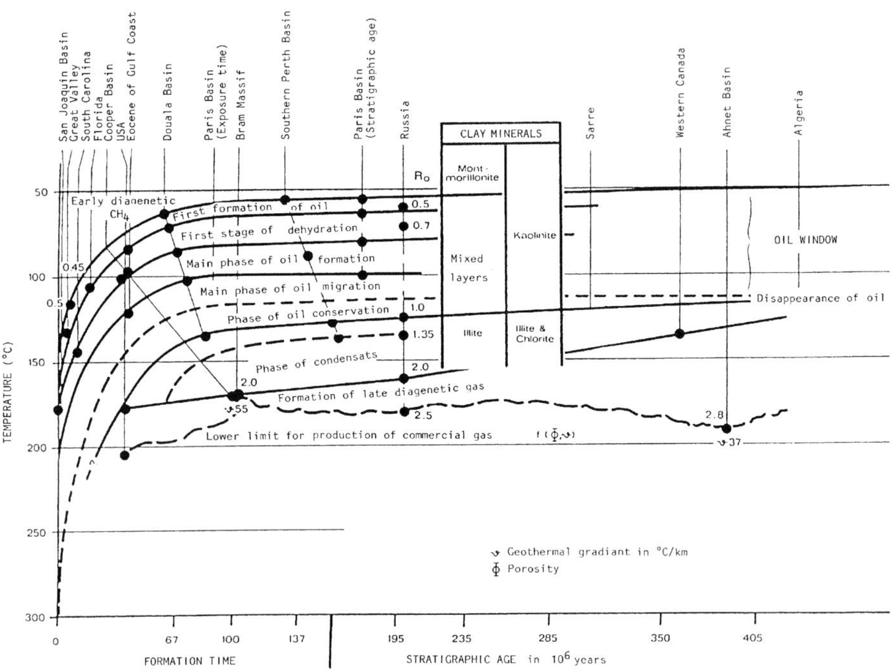

Fig. 5. Diagram of the thermo-iso-chrono-lines by Cornelius (1975) integrating formation times and temperatures as far as 150 million years. The various stages of hydrocarbon formation are correlated with vitrinite reflectances and mineral alterations

2.2 The Vitrinite Reflectance

Vitrinite is a maceral of coals whose vitrinite reflectance (Ro) increases with coalification, coal ranks or, in other terms, the maturation of the global humic matter. In both peats and soils, during the very first alteration stage of organic matter, fungi and bacteria amorphise the ligneous and cellulosic debris into humin which is a gel partially transformed first into huminite and later into vitrinite. Vitrinite is, thus, an amorphous cement, a linking phase which results from gelification of terrestrial (continental) plants, or of aquatic plants.

Vitrinite reflectance increases from 0.2% in brown coal or lignites to more than 5% in anthracites. This reflectance was first calibrated by Teichmüller (1971) for coals on the basis of volatile substances and fixed carbon, which are the routine methods used by coal industries to evaluate the quality of coal. These calibrations are very useful because coal ranks were only expressed in terms of volatile substances, or of fixed carbon before reflectance was applied. Because of Teichmüller's curves, it is possible to translate the former values into Ro values. A detailed analysis of the calibrations reveals that the result is not a curve, but rather a succession of several areas whose centres of gravity form a generalized curve-shaped evolution with more or less linear segments (Kübler 1984).

In the calibration of vitrinite reflectance vs. volatile components, the peats and lignites form two units, probably two types of organic matter characterized by extremely large variations of the volatiles without increase of the Ro (Fig. 6). The Ro values of the first group are low, between 0.1 and 0.3%, and the volatiles vary from 68 to 60%. The variation of the volatiles is larger in the second group, from 63 to 40% for a Ro between 0.38 and 0.52%. A third group includes coals with a volatile content of 44 to 32% and a Ro between 0.5 and 1%. There are no quantifiable relationships between volatiles and Ro for the three groups. However, when Ro varies from 0.7 to 2.3%, the volatiles decrease from 48 to 10%, according to a linear relationship with a rate of 0.0567% Ro pro percent of volatiles. Between 2.5 and 5%, Ro increases significantly for a small reduction in the remaining volatiles, the gradient being 0.520% Ro pro percent of volatiles.

According to the careful calibration by Teichmüller (1971) on the basis of measurements of some 500 coal samples, the only reliable relationship between vitrinite reflectance and volatiles begins at a Ro of 0.7% and terminates at a Ro of 2.3% Ro. As it will be mentioned below, the limits usually set for the potential oil window vary between 0.5–0.6% and 1.3–1.4% Ro. The POW is included, to a large extent, in the linear domain of Teichmüller's calibration. The increase in reflectance is attributed to the aromatic cores losing their hetero-atomics due to calorific energy, temperature and time, which allows a «condensation» into polyaromatic structural units (Oberlin et al. 1980) which become progressively oriented according to the perpendicular planes along the c axis of graphitoid structures. For the groups with or without relationships between Ro and volatiles, the difference is due to the original composition of the organic matter (Rouzaud 1984). There are as many types of peats and lignites as there are environments governing deposits or climates. This can certainly be addressed by the variable contents in C, H, O and N

Fig. 6. Reinterpretation of the data of Teichmüller (1971). The correlation of the Ro reflectance with the volatile components on pure-dry material outlines several areas. From *top* to *bottom*, the areas of water expulsion and alcohol function, of decarboxylation, of expulsion of gaseous hydrocarbons, especially between 0.6 to 2.5 Ro and of expulsion of dry methane and probably of hydrogen, follow each other

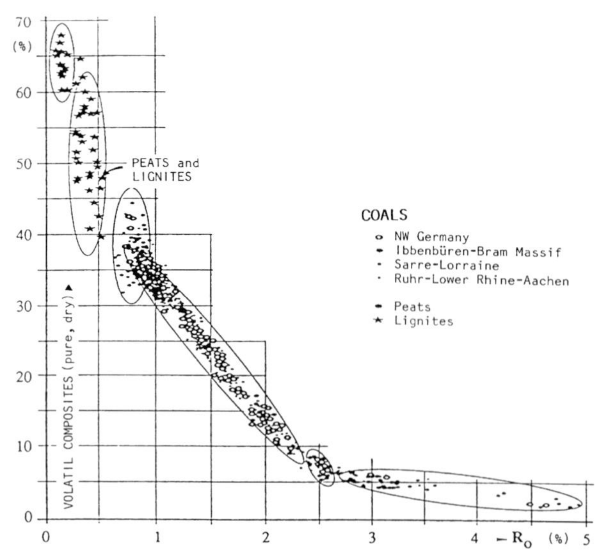

in the composition of the huminites (precursor of vitrinite). In the first stages, energy is consumed for the rupture of the alcohol and the carboxyl functions. The orientation of the aromatic cores is weak, and the reduction of sulphur and its precipitation into pyrite is frequent at this stage.

When Ro ranges between 0.7 and 2.3%, the increase in orientation which is a purely mechanical process, is linearly related to the loss of volatiles which is a chemical process. In the following step, the significant increase in orientation is more or less proportional in linear terms to a small loss in volatiles, the mechanical process dominating the chemical reaction. The participation of these two types of reaction – chemical and physical – led Durand et al. (1986) to determine that there were no pragmatic models linking reflectance, time and temperature by mathematical expressions. Among the different models, that of Lopatin (1971) had such an impact in the study of the diagenesis of coals that it is worth recalling the procedure. In relating the equation of decrease in radioactivity which is widely used in thermodynamics, to the decrease in the number of initial reactants (N°) in relation to the number of reacting products (N) in following form:

$$\ln(N^\circ/N)=kt.$$

Lopatin applied the Arrhenius principle to calculate k, the constant of the reaction rate for a given temperature:

$$\mathrm{k}=\exp A(-E/RT).$$

Then, like Tissot (1969), Johns and Shimoyama (1972), Connan (1974), Karweil (1975), Tissot and Welte (1978), he linked A, the frequency factor (also termed Arrhenius or impact factor), to E the activation energy with the equation:

$$\ln(N^\circ/N)=At\exp(-E/RT).$$

Since K varies with temperature, Lopatin introduced an empirical thermal coefficient of reaction activity, in admitting that each increase of 10 °C doubles the reaction speed. Then, without reference to actual activation energies, he supposed that the energy increases by 20.9 kJ (5 kcal/mol) with each temperature interval of 10 °C. This allowed him to introduce the notion of kinetics by adding a thermal coefficient to the velocity reaction for an empirical calculation as follows:

$$\mathrm{g}=k(T)/k/T+10).$$

To control the whole of these considerably simplified hypotheses, Lopatin introduced t, a «corrected residence time» calculated as follows:

$$t=g\times t.$$

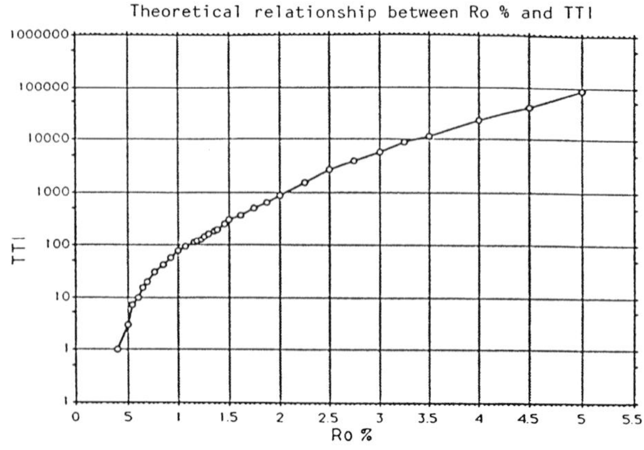

Fig. 7. Correlation between the time temperature index (TTI) of Wapples (1981) and vitrinite reflectance. The curve is not continuous but shows a succession of segments like that of Teichmüller (1971)

The correlation coefficient between this corrected residence time and the vitrinite reflectance measured on the Münsterland coals is 0.999 (Lopatin 1971; Fig. 7).

It should be noted that Lopatin's correlations are only valid for a vitrinite reflectance of 1.1–5% and only for coals. It is surprising that they are not influenced by the important change in the slope of Ro at 2.4–2.5% (Fig. 6). If one accepts Lopatin's demonstration, the physical process of orientation, which is responsible for the increase in Ro (Oberlin et al. 1980; Rouzaud et al. 1984), has to be considered as the only consequence of the chemical reactions. The good correlation between Ro and H/C ratios found by Burnham and Sweeney (1989) confirms this process (Fig. 8). However, the correlation using McCartney and Ergun's equation (1967) is:

$$\%Ro = 15.64\exp[-3.6(H/C)]$$

and it completes the correlation by taking into account the oxygen content, especially in areas of weak reflectance:

$$\%Ro = 12\exp[-3.3(H/C)] - (O/C).$$

This is only valid for vitrinite reflectance above 1%, which then agrees with Lopatin's empirical correlations.

Even when correcting the oxygen content, the correlation is not satisfactory for Ro values below 0.5%. This implies that, in the first maturation stages of coal, the precursor of vitrinite loses functions other than the carboxyle or methyl functions, which is not taken into account by the empirical equations mentioned above. In the very early stages of the reaction, the water and alcohol functions are lost and Burnham and Seeney (1989) decided to include the following four independent reactions in their model:

$$\text{Precursor} - \text{residual vitrinite} + H_2O \qquad (1)$$

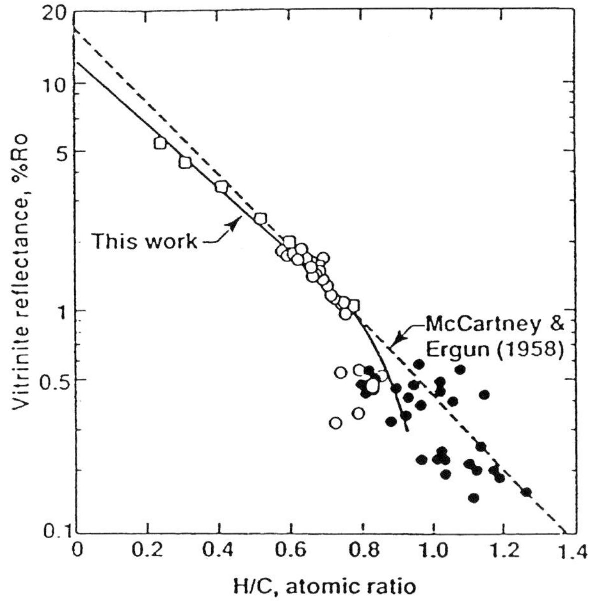

Fig. 8. Confirmation of Burnham and Sweeney's (1989) modelling by analytical data which take into account the H/C and O/C relationships in the stages of weak reflectance. The correlation only becomes valid for a reflectance above 1%

$$\text{Precursor–residual vitrinite} + CO_2 \quad (2)$$

$$\text{Precursor–residual vitrinite} + CH_n \quad (3)$$

$$\text{Precursor–residual vitrinite} + CH_4 \quad (4)$$

The term vitrinite used by the authors has been replaced by that of precursor because the real precursors are humin and then huminite in the very first evolution stages. These four simultaneous reactions, even if the third does not play a very important role in coals, should be taken into account for the mineral alterations in the coal seams. They determine the composition of the interstitial water and the redox potential.

Since the published kinetic constants were insufficient, Burnham and Sweeney (1989) adjusted the VITRIMAT model to geological dates, to results of pyrochromatography and to results of mass spectrometry following pyrolysis (Burnham et al. 1989). The application to several selected examples allowed the demonstration that the spectrum of activation energies by «species» ranges from 38 to 52 kgcal/mol for H_2O, from 42 to 54 kcal/mol for CO_2, from 46 to 58 kcal/mol for CH_n and from 50 to 74 kgcal/mol for CH_4, for an Arrhenius factor of 1×10^{13}/s (2×10^{13}/s for CH_n). The low-activation energy values of Karweil's or Lopatin's equations have not been confirmed, nor have those of Barker (1989). According to Sweeney and Burnham (1990), such low values are not chemically realistic and result from application of

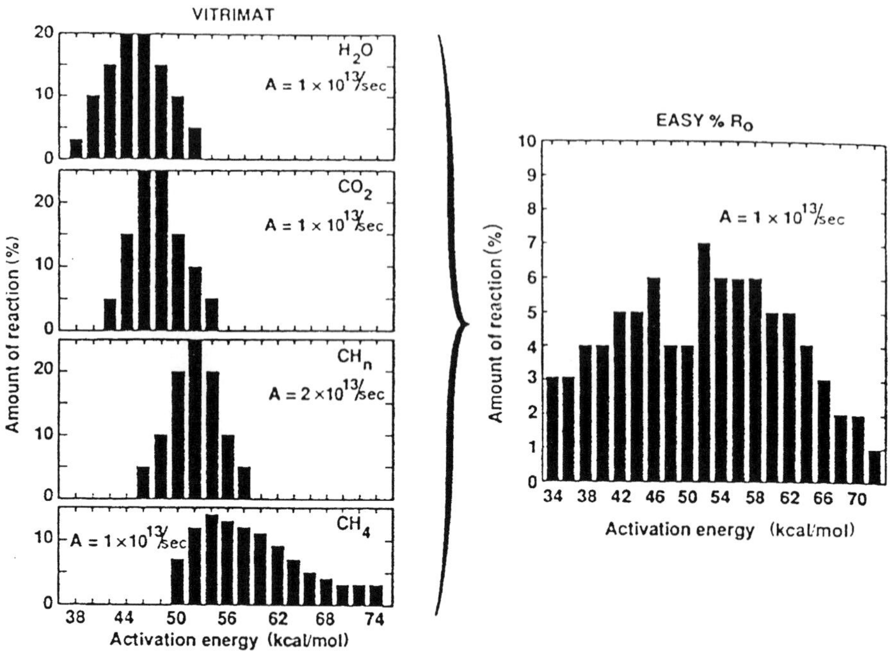

Fig. 9. Histograms of energy activation frequency according to Burnham and Sweeney (1989; VITRIMAT), and Sweeney and Burnham (1990; EASY% Ro) taking into account the energies necessary to liberate water and alcohol functions (H_2O), for liberation of carboxyl functions (CO_2) and hydrocarbons CH_n and CH_4. The apparent activation energies for the liberation of hydrogen are not shown

Arrhenius' reaction with a single activation energy. However, there is concordance between the dates of Burnham and Sweeney (1989) and those of Ungerer et al. (1986), at least for CH_n and CH_4. In the EASY%Ro model which is a simplified alternative of VITRIMAT, Sweeney and Burnham (1990) chose the same spectrum of activation energies and maintained the Arrhenius constant at 1×10^{13} kcal/mol (Fig. 9). Karweil (1975) obtained apparent activation energies of 35.2 kJ/mol (8.4 kcal/mol) for two measurements of coal samples in the Ruhr, whilst this activation energy is 100.5 kJ/mol (24 kcal/mol) for the formation of benzene from peat (Galway 1969; Lopatin 1971 in Connan 1974). According to pyrochromatography and calibration by means of the OPTIM program (Ungerer et al. 1986), the spectrum of energies averages 240 kJ/mol (58 kcal/mol) for Tertiary coals in the Mahakam delta. This spectrum ranges from 217 (52) to 334 kJ/mol (80 kcal/mol) for an Arrhenius factor of 10^{13} to 10^{16}/s. The low values obtained by Karweil (1975) and Lopatin (1971) are excluded from this spectrum which, however, does not mean that they do not exist. The three parallel reactions H_2O, CO_2 and CH_4 have not been taken into account in these approximations.

In summary, reflectance of vitrinite has its limitations as is the case for any other criterion applied to the zoneography of diagenesis. Even with coals, which constitute the best lithology for reliable measurements of reflectance on well-identified vitrinites, the authors were able to distinguish three types of vitrinite: an A type (gel), a B type (textured) and a C type (oxidation tracks), all being chemically different with variable degrees of reflectance (Alpern 1980; Durand et al. 1984). If one considers the difficulties of distinction and the problem of anistrophy, it is clear that reflectance values should not be uncritically accepted. As far as retromorphosis is concerned, one should no longer admit that reflectance works like a maximum calorimeter. Heroux (pers. comm.) observed a decrease in reflectance (and in the illite-crystallinity index) under the influence of mineralizing circulations of Mississippi Valley-type fluids. Barker (1986) even proposed to discard vitrinite reflectance. The results illustrate, at least, the importance of fluid advection and outline the research work necessary to understand diagenetic mineral alterations.

In correlations between reflectance of coal vitrinites and mineral alterations, it would be extremely useful to check on the thermal history, and to compare the measured and the calculated Ro for simulation models, e.g. EASY%Ro or Basin Mode.

2.3 The Dispersed Organic Matter

In studying the zoneography of coals (ranks) and that of liquid hydrocarbons, Hood et al. (1975) extracted oil components and liquid hydrocarbons from sequences of carbonaceous veins in which vitrinite could be measured according to coal petrographic techniques. From a compilation made by Teichmüller (1971), it was deduced that the POWs range from 0.5% to 1.3–1.4% Ro. If the comparison had been a rigorous one, i.e. vitrinite in the coal and organic extracts in the other lithologies, the progress of Ro would not have been regular and would have provided several gradients (Kübler 1984). However, the limitations of such a comparison are more important only for a more fundamental point of view, because it is rare to observe oil-prone series interstratified with coal veins. Thus the search for vitrinite is more often carried out on concentrates of dispersed organic matter. This not only induces technical problems, such as how to evaluate the influence of chemical treatments, of polishing, etc. on the dates, but above all it raises difficulties in properly characterizing the «real» vitrinite, even for well trained petrographists (Alpern 1980; Durand et al. 1986). More important is the fact that Bostick and Foster (1975), using Timofeev and Bogoliyubova's (1970) results, demonstrated that at identical stages of maturation, Ro is always higher in coals than in other lithologies such as limestone, sandstone, shale or clay. Buiskool Toxopeus (1983) compared two series of measurements, one on coals and the other on clays, and those for the clays were always lower and evolved more slowly with depth than those for the coals. At approximately 3000 m, the difference between the Ro of coal (=0.80) and the Ro of clays (=0.45%) is more than 40% (Fig. 10). If one chooses vitrinite reflectance as an indicator for maturation of organic matter or for reconstruction of the thermal history of any basin, it is essential to identify precisely the type of lithology in which vitrinite evolved, at least in the early stages of diagenesis. Consequently, it is no longer possible to compare mineral evolution with that of vitrinite without this

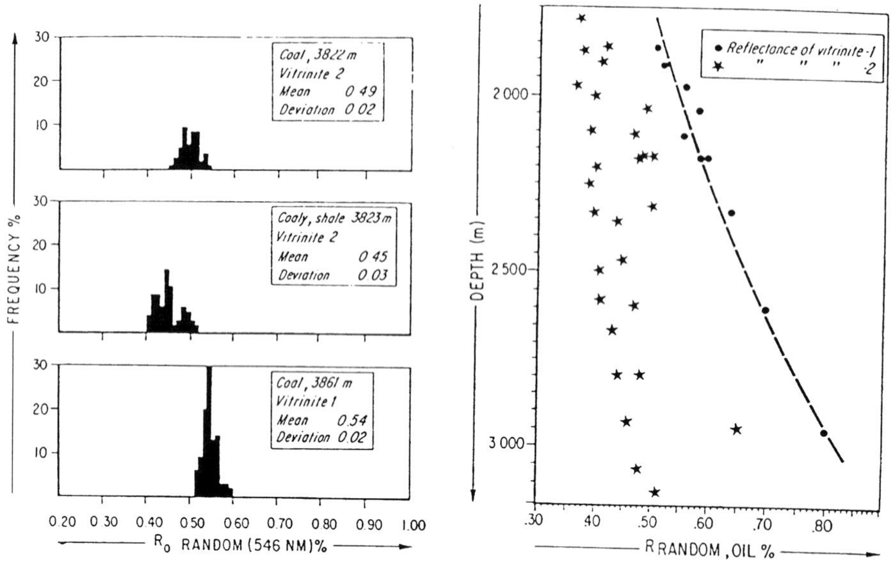

Fig. 10. Influence of lithology on vitrinite reflectance according to Buiskool Toxopeus (1983). At identical stages of alteration, the shales (*right*) have an average weaker Ro than coals (*left*). The progression of Ro with depth is slower in shales than in coals. At 3000 m, the difference reaches 40%

precision. The differential evolution between mineral and organic alterations (Robert 1985) has to be analysed in the light of these differences.

The difficulties of calibration between POW and Ro do not arise from vitrinite alone, but also from the types of kerogens. In their compilation, Kübler et al. (1979a) established that the maximum of POW determined by organic extracts, oily components and hydrocarbons, corresponds to Ro=0.8% for kerogen of type III (Hood and Gutjahr 1972), to Ro=0.9% for kerogen of type II to III (Logbaba; Tissot and Welte 1978), and to Ro=1% for kerogen of type I (Green River Shale; Tissot and Welte 1978). If Ro is taken as a reference, it has to be assumed that more energy is required to generate liquid hydrocarbons in a type I kerogen than in a type III kerogen. This takes the same trend as the apparent activation energy spectra by Tissot and Espitalié (1975), whose mode is at 70 kcal/mol (293 kJ/mol) for type I and at 50 kcal/mol (210 kJ/mol) for type II. However, the mode is proportionally weaker and the spectrum is designed more for type III. This means that generation begins at lower and ends at higher energies, which is confirmed by the T_{max} of pyrochromatography (Espitalié 1986). However, vitrinite can only be used as a reference if it is measured in the coals or possibly from concentrates of type III kerogen. It is extremely rare and poorly defined in type I (Green River shales) and difficult to use in some type II kerogens such as those of the Toarcian of the Paris Basin (Alpern and Cheymol 1978). It is, therefore, essential for POW–Ro calibration to take into account the kerogen types, which is of some difficulty. It is no longer possible to compare the results of organic geochemistry and those of petrography of the organic matter without stating the lithologies to which they belong.

Despite these fundamental restrictions, Wapples' (1980) Time Temperature Index (TTI) has been widely used for reconstructing the thermal history of basins and to define POWs. This TTI is based on Lopatin's empirical calculations and it was calibrated more than 400 times on Ro from throughout the world. It was also compared to Staplin's (1969) Thermal Alteration Index (TAI) and to other data from organic geochemistry such as the Carbon Preference Index (CPI) of n-alkanes. If, as pointed out by Tissot and Welte (1984), Burnham and Sweeney (1989) and Durand et al. (1986), this TTI is only an empirical approach and by no means a physico-chemical approach, the activation energy for maturation of organic matter is approximately 42 kJ/mol (10 kcal/mol; Barker 1983), since the reaction is only an initial stage and the stabilization period spans between 1 and 10 million years. The maturation of organic matter can easily be modelled in using only the maximum temperature or in taking into account the temperature and the duration, as proposed by Lopatin. The two models may be used to predict the palaeotemperatures with a similar degree of precision (Barker 1989). These remarks are important because it is possible to explain the formation of these «early» hydrocarbonated gases with activation energies as low as 10 kcal/mol. These latter play the role of reducing agents in the first stages of diagenesis and affect the balance of Fe^{2+} and Fe^{3+}. The lack of precision of the TTI for the different stages of diagenesis is not greater than the lack of precision in the burial curve, residence periods, or that of the definition of types of kerogen. We shall keep the TTI as a reference for comparison with mineral alterations.

3 Mineral Alterations

Two major directions have been explored to identify the stages of mineral diagenesis: disappearance of the swelling smectite layers in the illite/smectite mixed-layer phases and neoformation of index minerals such as regular mixed layers, pyrophyllite, paragonite or phengite.

3.1 The Disappearance of Swelling Layers

The disappearance of smectite swelling layers in the illite/smectite phases relative to burial was described long ago (Grim 1953, 1958; Powers 1959; Milne and Earley 1958; Burst 1959; Weaver 1960; Füchtbauer and Goldschmidt 1963; Millot 1964). When Millot (1964) published his synthesis on the *Geology of Clays*, the term montmorillonite was generally used for minerals consisting only of 2:1 swelling layers. Since then, this term has been replaced by that of smectite which, according to recommendations by the AIPEA Commission for Terminology, covers both dioctahedral and trioctahedral types. The disappearance of the swelling layers was related, at a very early stage, to the search for oil (Weaver 1960). Since the 1970s, however, many have attempted to understand the relationship between mineral

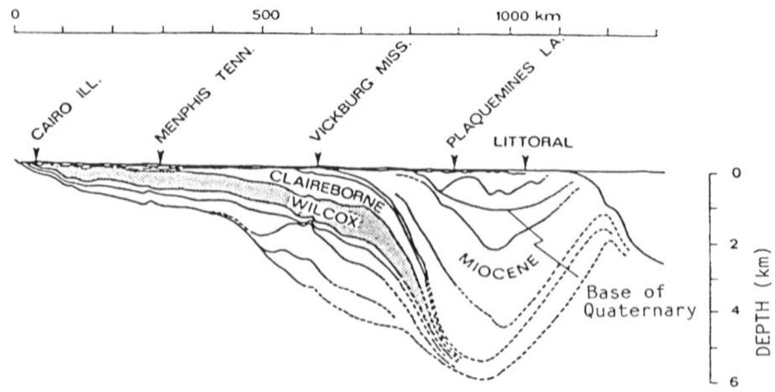

Fig. 11. The unity rule for diagenetic alterations. Unity of time: deposition of an identical lithofacies of the Wilcox Formation; unity of place: the same formation in continuity throughout the basin; unity of action: progressive subsidence in the Gulf Coast. (According to Burst 1969)

alteration in argillaceous minerals with swelling layers and the processes which may lead to economically interesting accumulations of liquid hydrocarbons. The zone where smectite swelling layers in illite/smectite particles are modified into illite layers is often described as the zone of thermal illitization of smectite, or the smectite disappearance zone.

In analysing the mineral alterations in the Eocene Wilcox Formation, Burst (1969) detected a reduction of the swelling layers in three successive stages, taking into account burial ranging from several hundred to more than 4000 m (Fig. 11). These three stages were: a stage of stability without alteration, a stage of illitization with 80 to 20% of swelling layers, and a deep stage in which the remaining swelling layers are very slowly altered into illite (Fig. 12). For Burst, the first stable stage corresponds to expulsion of gravity water from large pores solely by compaction. However, alteration of the swelling layers is not progressive, as Powers (1959) claimed, but relatively rapid at the top of the sequence and relatively progressive at the bottom. The upper rapid illitization corresponds to the second stage of dehydration and the slow reduction of the swelling layers with depth to the third stage of

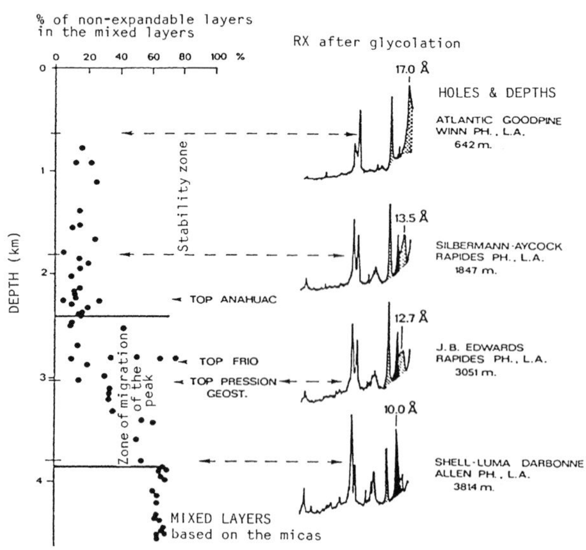

Fig. 12. Diagenetic alteration in the Gulf Coast area according to Burst (1969). Note the *stable zone* without smectite alteration in the shallow part; the transition zone with a reduction in the smectite amount and migration of the major peak towards that of the micas in the *intermediate part*; the *mixed-layer zone* with the XRD peaks centered on that of the micas in the *lower part*

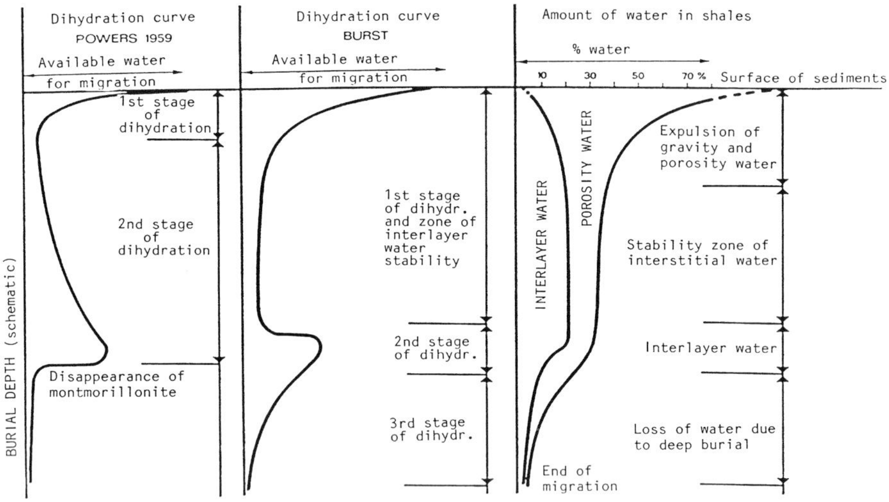

Fig. 13. Stages of dehydration of shales according to depth. First stage characterized by removal of porosity water; second stage with redistribution of fluids; third stage with dehydration of the smectitic layers. (After Burst 1969)

dehydration (Fig. 13). Since the water loss from this alteration is between 25 to 30%, on the basis of a calculation of the reduction of the unit cell volume, it is tempting to identify it as the necessary driving force for the expulsion of drops of liquid hydrocarbons from source rocks which represents the highly controversial primary migration and to compare the depth of this mineral alteration with the depth of oil accumulation. This comparison was so successful that it induced numerous analyses and publications. In order to distinguish the beginning and end of the transition zone, a crystallographic question must be addressed.

3.2 Definition of theTransition Zone

Dioctahedral smectites, such as montmorillonite or beidellite in the strict sense of the term, are no longer considered well-defined crystalline structures such as kaolinite, micas or chlorites, but are considered as mixed layers (Brindley 1980). The notion of lattice electric charge can no longer be used as a distinctive criterion since the definition of «transformation smectite» (Robert 1972) which results from an efficient pedogenesis under a graminaceous lawn, no longer exists (Pochon 1978). The average grain size of smectites is very small with a mean at 0.2 µm (i.e. 200 nm or 2000 Å). This mode is a mechanical aggregate of «micro-flakes», also termed tactoids, or a coherent diffraction volume made up of 3, 5, 7 or more layers.

The beginning of this third stage, termed diagenetic energy isopleth, increases in the deeper zones with temperature, which appears to contradict Parmentier's principle. However, if one considers that in the Gulf Coast sediments, the geothermal gradients are extremely low in the Plio–Pleistocene and normal in the Eocene and the contiguous Cretaceous period, a limited depth and a low temperature corrected over a long period agree with a greater depth and higher temperature, as long as the duration is shorter in the latter case. This is then no longer an isopleth in the sense defined by Burst, but a thermo-iso-chrono-line in the sense of Cornelius. The consequence is that the reflexion at 17 Å on an X-ray diagram after saturation with ethylene glycol, can no longer be used as a criterion for the definition of smectite. In fact, for three-layer crystallites, the reflection remains in this position as long as at least 60% of non-swelling layers are present (Brindley and Brown 1980). The nature of the original swelling minerals which will be subjected to burial, is not necessarily of smectite-type with 100% of swelling layers. In the palaeogeothermal reconstructions of the Gulf Coast region, it is generally admitted that the starting material already was a mixed layer with 70 to 80% of swelling layers (Freed and Peacor 1992). This must be linked to the fact that the diverging margin of the major palaeodelta of the Mississippi River was fed with smectites probably of beidellite type or with «transformation smectites» which are products of continental weathering. Only bentonites of volcanoclastic origin produce smectites with 100% swelling layers in diagenetic conditions.

In addition, the methods of calculating the amount of swelling layers in mixed layers have been improved. The 10–14 Å calibration curve of Weaver (1956) for an infinite crystallite, the 10–17.5 Å calibration curve, which is highly sensitive to the number of layers per tactoid (MacEwan et al., in Brown 1961), or that used by Burst (1969) between [003] smectite and [002] illite XRD peaks, needed both the use of probabilities (p) in establishing the succession of the sheets within swelling layers (S) and non-swelling layers (I) (p(SS), p(SI), p(IS)) and the use of the «Reichweite» (R) which, when zero, describes a series of randomly organized sheets. The R1 type corresponds to a regular interstratification (notation: ISISIS) and R3 to a probability that an S layer appears after three I sheets (notation: ISII). With burial, the ordering shifts from R0 to R1 (previous allevardite stage, currently rectorite stage) and R3 types (Kalkberg; Reynolds and Hower 1970; Perry and Hower 1970; Srodon and Eberl 1987). Among all methods of measurement, that of Reynolds (in Moore and Reynolds 1989) seems to be the most appropriate (Fig. 14). The angle of diffraction, the d space of mixed sheets formed by the harmonic peaks 2 and 3 of smectites and 1 and 2 of micas is less sensitive to the number of sheets per tactoid. There is no solution of continuity between the two end members at 100% or 0% of swelling layers saturated with ethylene-glycol as for 1–1 composite lines. According to Reynolds (in Moore and Reynolds 1989), the difference between the 1(I)–2(S) and 2(I)–3(S) composite lines decreases with increase in swelling layers according to the equation (Rolli 1991):

$$\%\ \text{«illite»} = -4.694 \times \Delta^2 + 91.4504 \times \Delta - 448.0472$$

$$\%\ \text{swelling layers} = 4.694 \times \Delta^2 - 91.4504 \times \Delta + 448.0472$$

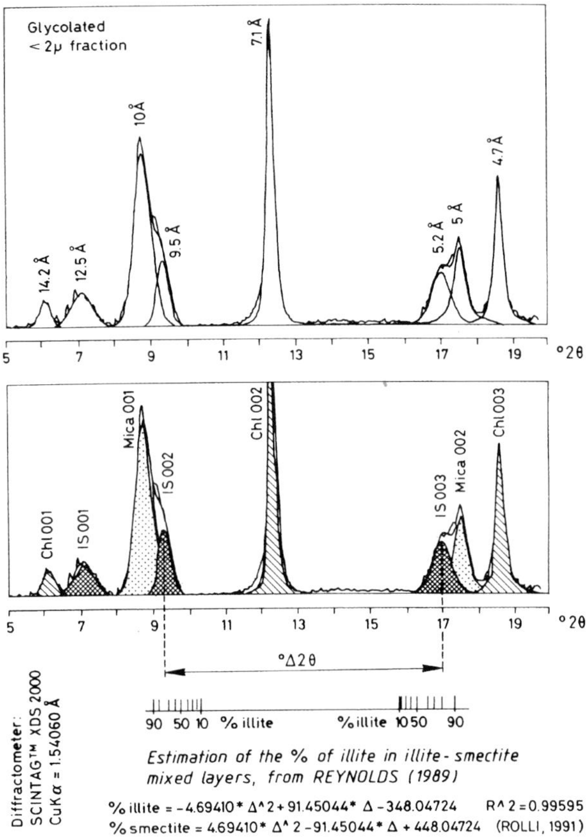

Fig. 14. Assessment of the amounts of swelling layers according to Reynolds (1980) by decomposition, and to Pearson VII by comparison of XRD peaks of detritic micas and mixed layers

This method can be applied successfully to very fine <0.6 μm and <0.2 μm fractions, and even to smaller ones. In these fractions, the detritic micas are generally absent and the asymmetry of the peaks [001] and [002] is nonexistant. However, this is the case for the <2 μm fractions, but the asymmetry of the first and second peaks is no longer an argument against the application of Reynolds' method since, via functions such as those of Pearson VII (Howard and Preston 1989), the individualization of the peaks of mixed layers by deconvolution has become reliable.

The range of a mineral transition can only be compared if the same method of measurement has been applied, whether that of Burst (1969), Reynolds (1980) or Reynolds (in Moore and Reynolds 1989). A comparison of Srodon's results and those of Reynolds emphasizes very minor differences in the evaluation of the amounts of non-swelling layers (Anceau and Adatte pers. comm.). However, disappearance of the XRD peak at 17 or 14 Å which was believed to be typical for smectites in earlier studies (Kübler 1966; Dunoyer de Segonzac 1969; Monnier 1982) shall be examined. The position of the diffraction angle remains at 17 Å, as long as the amount of layers is less than 60%. It is, therefore, impossible to extrapolate the former profiles with those for which a rate of swelling layers has been actually measured. A simulation by NEWMODE (Reynolds in Moore and Reynolds, 1989)

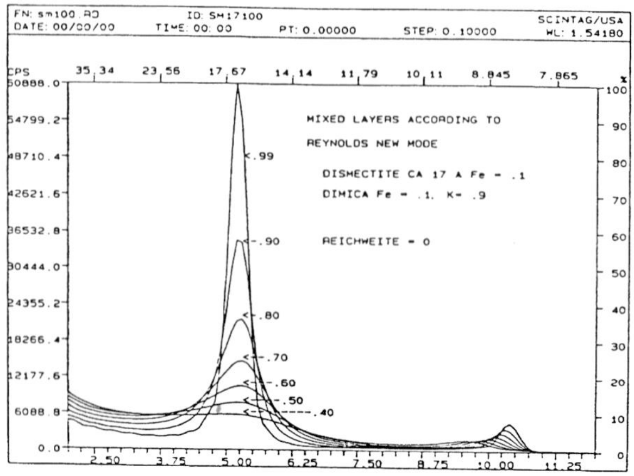

Fig. 15. Decrease in the intensity of the XRD peak at 17 Å by an increase in the amount of non-swelling layers by a modelled «solid random solution» (R=0; after Reynolds 1989; NEWMODE)

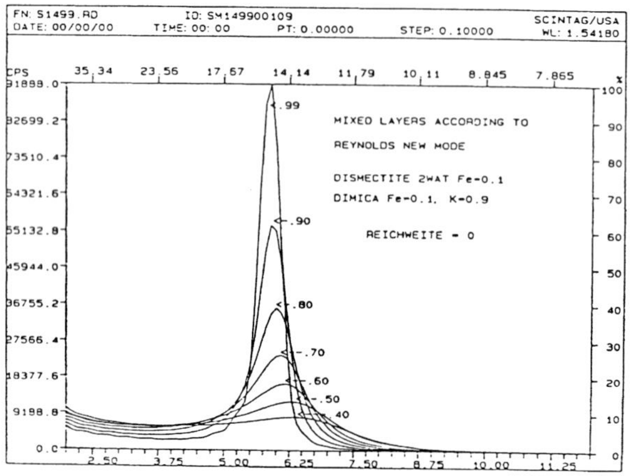

Fig. 16. Decrease in peak intensity characteristic for a two water layering saturated in Ca by a modelled increasing of the non-swelling layers. (After Reynolds 1989; NEWMODE)

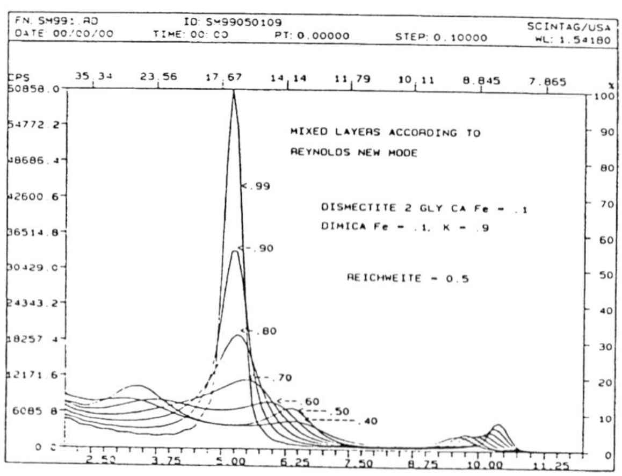

Fig. 17. Effect of Reichweite increase on the shift of the 17 Å XRD peak on the increase in the amount of non-swelling layers. The decrease in the intensity of the 17 Å peak and its peak, modelled by NEWMODE, provides a better image of the recorded effects

Fig. 18. Calibration of «smectite distribution» measured according to the intensity of the 17 Å peaks (after Monnier 1982) and decrease in the amounts of swelling layers according to Pytte and Reynolds (1989). The 17 Å peak disappears for mixed layers containing 60—50% of swelling layers

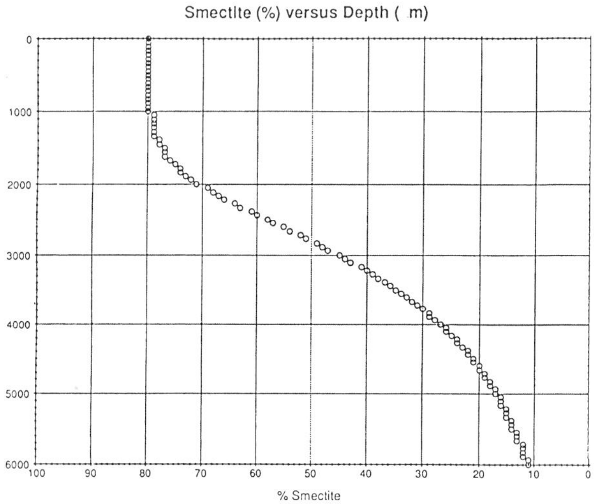

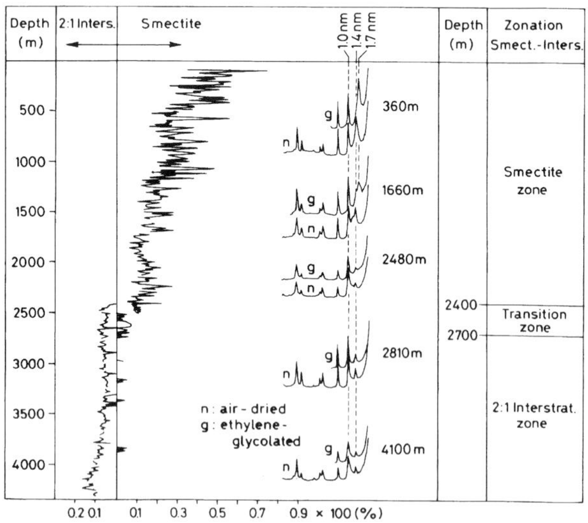

Smectite disappearance (right-hand axis) and 2:1 mixed layers (left-hand axis) in Linden 1 well

with decreasing amounts of swelling layers saturated with ethylene glycol, shows a rapid decrease in the intensity of the peak at 17 Å with a very small shift towards the high angles (Fig. 15).

Comparison between simulations and experimental diffractograms leads towards the conclusion that the disappearance of smectite claimed by Monnier (1982) corresponds to less than 60% of the swelling layers. For diffractograms generated by Kübler (1966) and Dunoyer de Segonzac (1969) of non-saturated <2 µm fractions, the simulation also showed a strong decrease and a small shift of the peaks (Fig. 16), the limit of smectite disappearance being at about 60 to 50% of the

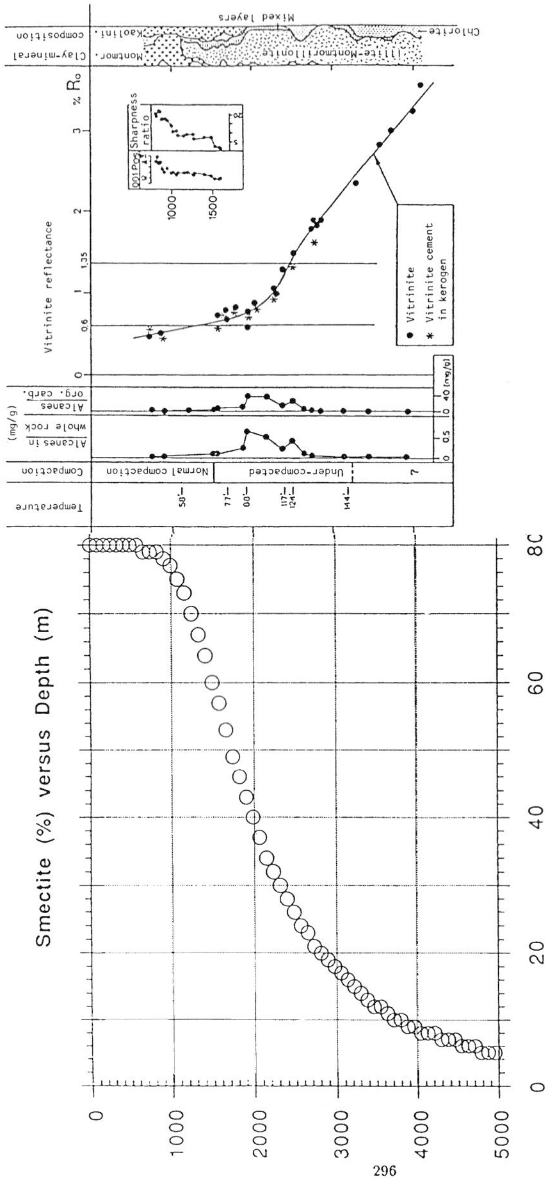

Fig. 19. Calibration of «smectite distribution» according to Dunoyer de Segonzac (1969), sharpness ratio and shift of the 14 Å peak with decrease in the rate of swelling layers modelled by Pytte and Reynolds (1989)

swelling layers. These two simulations were made using a zero «Reichweite». For a Reichweite of 0.5, the shift of the peak is more significative and a superstructure appears towards the low angles. This behaviour has been frequently observed and may have led to the recognition of a trend towards an allevardite structure, now termed rectorite (Fig. 17). On the basis of these simulations, the disappearance of smectite seems not to be strictly indicated by the disappearance of the first peak at 17 or 14 Å as claimed by the three above-mentioned authors, but rather by the reduction of intensity of the 10–17 Å interstratification at a 50 to 60% content of swelling layers. This disappearance corresponds to the mean point of the transition curves as defined by the measurement of the proportion of swelling layers; the interstratification of the non-swelling layers beginning at shallower depth and lower temperature, as visualized in Pytte and Reynolds' (1989) model.

In the Linden drill hole, Monnier's (1980) «transition zone» corresponds to a rate of interstratification between 60 and 50% of the swelling layers (Fig. 18), when based on a geothermal gradient of 34 °C/km, a sedimentation rate of 1.65 mm/year and a composition of the initial mixed layer containing 80% of swelling layers. It should be noted that the relative intensity of reflection of the mineral, which was defined as a smectite by Monnier, decreases constantly until a depth of 2400 m, agreeing with the decrease in intensity predicted by the NEWMODE code. The same control on Logbaba drilling samples confirmed «disappearance of montmorillonite» for a rate of interstratification of 50% of non-swelling layers (Dunoyer de Segonzac 1969), which has been modelled according to Pytte and Reynolds' model

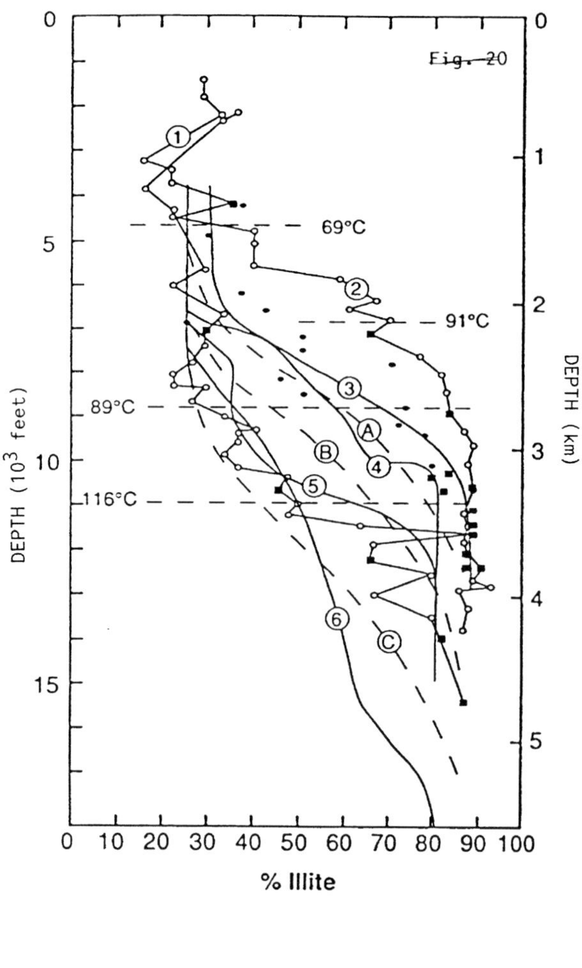

Fig. 20. Compilation of the smectite—illite transition zones in connection with depth: *1* Hidalgo County, temperatures 69 and 81 °C: *2* Brazoria, temperatures 89 and 116 °C (Freed and Peacor 1989); *3* North Sea, Viking Graben (Pearson et al 1982); *4* Gulf Coast (Perry and Hower 1982); *5* Gulf Coast (Aronson and Hower 1976); *6* Gulf Coast (Perry and Hower 1982); *A*, *B* and *C* correpond to Pytte and Reynolds' (1989) model

Fig. 21. Smectite—illite transition according to temperature. (After Srodon and Eberl 1987)

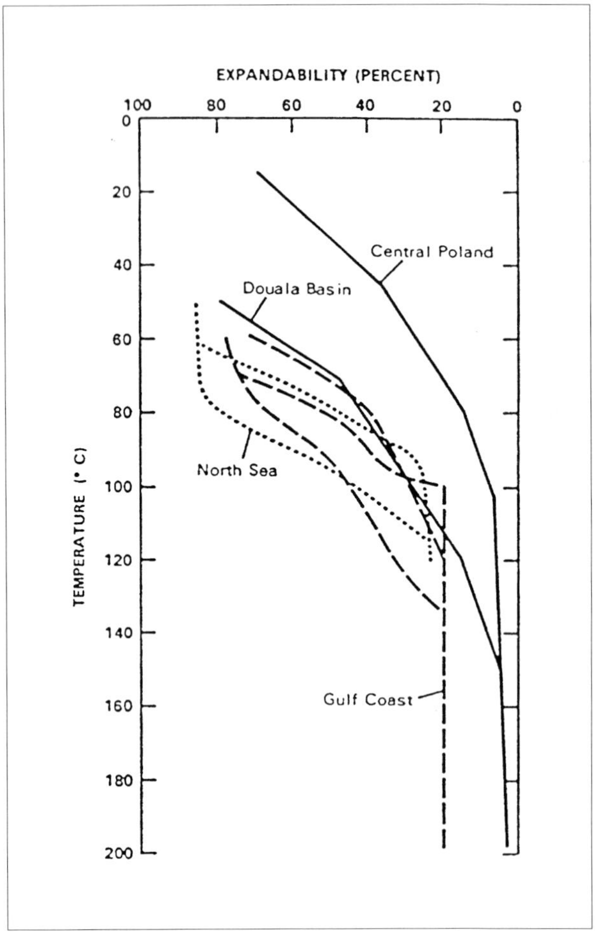

(Fig. 19). This method also allows more accurate comparison with former data based on the measurement of the first smectite peak alone, and current data integrating the proportion of swelling layers.

3.3 Relationship between Transition Zone, Depth and Temperature

The impression has frequently been given that the transition zone of smectite illitization is confined to a limited depth interval (Kisch 1987). In fact, the extent of this zone is variable, as shown by compilation of the curves published by Perry and Hower (1970), Aronson and Hower (1976), Pearson et al. (1982), Shoonmaker et al. (1986) and Freed and Peacor (1989; Fig. 20). The depth interval may reach 3000 m (Fig. 21), the start being usually between 1000 and 2000 m. It is generally accepted that the beginning and the thickness depend upon the geothermal gradients (Srodon and Eberl 1987; Weaver 1989; Chamley 1989). The geothermal gradient is 25 °C/km for curve 6 in Fig. 20, 40 °C/km in southwest Texas (Boles and Franks 1979), and it varies between 22 and 43 °C/km in the Frio Formation (Freed 1980). The contrast is even more marked when the curves are reported according to temperature (Fig. 21). The difference of gradient in the illitization of smectite is not due to the difference in geothermal gradients for Srodon and Eberl (1979), but to age; the duration of exposure favouring illitization. However, anomalies cannot be explained only by these two external factors. Changes in the deposit environment and the nature of detritic clays are also partly responsible for the mentioned anomalies (Kübler 1984). However, the factor which appears to be of some importance is overpressure (undercompaction).

3.4 Relationship Between Smectite/Illite Transition, Overpressure and Organic Maturation

Overpressure which is also termed undercompaction, may be considered as a dynamic process. To occur, it needs a rate of water expulsion from rock pores to be lower than the rate of burial. These conditions take place in basins with important subsidence, because of intercalated clays or clay–silt whose hydraulic permeabilities are approximately 10^{-8} to 10^{-10} m/s. If the pore pressure exceeds confinement pressure or lithostatic pressure, mud volcanoes are formed. Such fluid overpressures are not stable over geological periods. On the other hand, undercompacted zones due to occurrence of permeability seals or barriers such as saliferous or black shales are much more stable. They acquire the properties of permeability barriers only after a given compaction which depends on the clay content, the sedimentation rate and the geothermal gradient. For example, in the divergent margin of the Labrador offshore, the seal only became efficient at a depth of 2500 m when the shales reached a density of 2.1 (Cramez-Diaz and Kübler 1982; Kübler 1984). According to Powers (1959), who was first to note a concordance between the transition zone and abnormal pressures, dehydration due to the smectite–illite mineralogical transformation would appear to be one of the causes for the creation of undercompaction. Comparison of published examples shows that abnormal pressures and the transition zone are concordant at Karlsefni (Fig. 22), whereas abnormal pressures precede the transition zone by about 800 m in the Panonian basin (Fig. 23; Francu et al. 1989), In Burst's studies, undercompaction begins only in the middle of the Gulf Coast transition zone (Fig. 12), while it begins at the bottom of the transition zone in the Logbaba boreholes (Dunoyer de Segonzac 1969; Kübler 1984; Fig. 19). In the Mahakam delta, the undercompacted zones were found in much deeper parts of the basin than the transition zone.

If one admits, as did Powers (1967) and Burst (1969), that the fluid volume released by mineral alteration induces undercompacted zones, since there is no close relation between depth of mineral transition and that of undercompaction, one should accept that the latter has probably a different origin or that the zone decompacted as a result of tectonic movements. This hypothesis cannot be confirmed due to lack of elements for comparison such as deferred-well loggings or pressure gradients. If one admits, as is often the case, that tectonic structurations did take place after maturation of the organic matter, then it can be admitted that

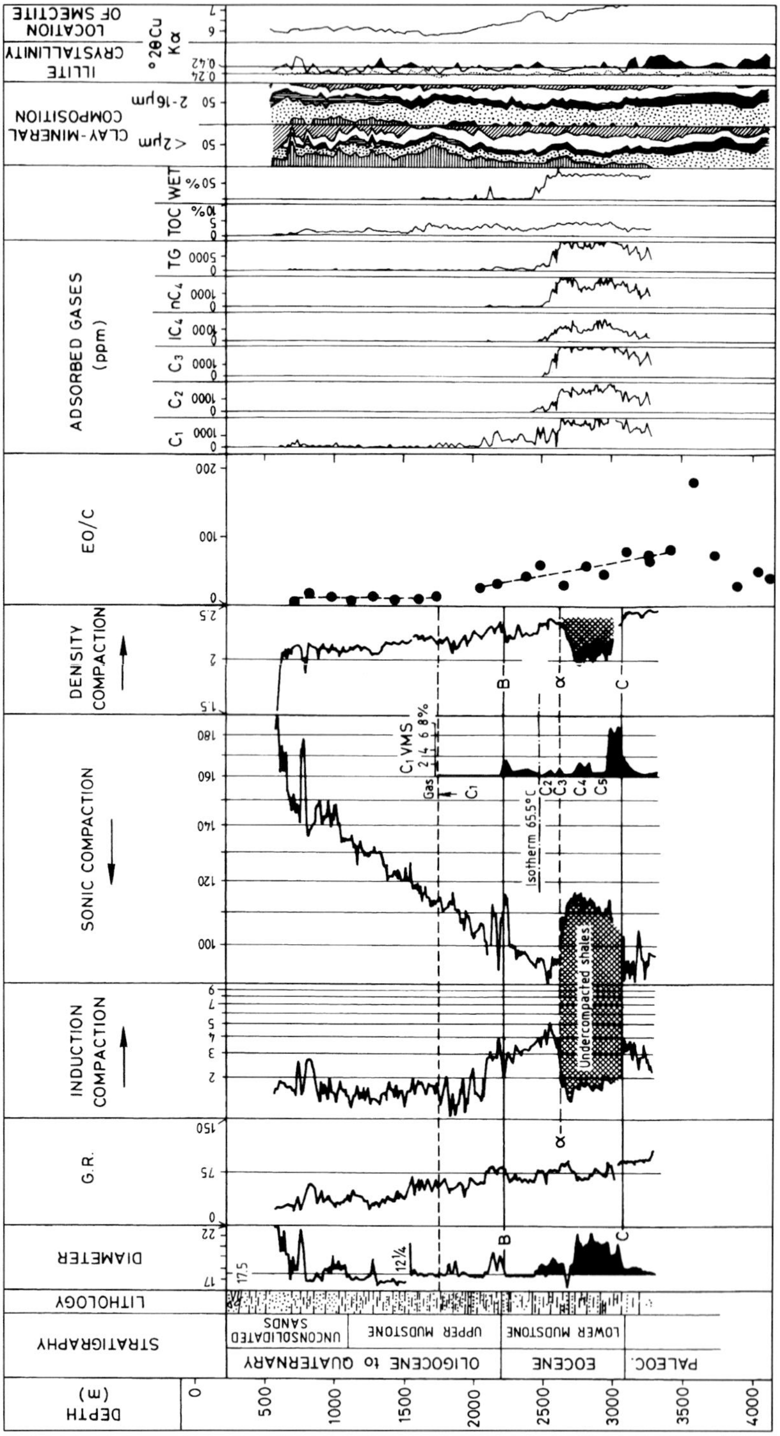

Fig. 22. Example of Karlsefni, off-shore Labrador. Comparison between delayed diagraphies, gamma-ray, induction, sonic and density, and organic extracts, adsorbed gas, mineralogical composition of the 2—16 µm and <2 µm clay fractions, and the shift of the 14 Å peak. (After Heroux et al. 1981; Kubler 1984)

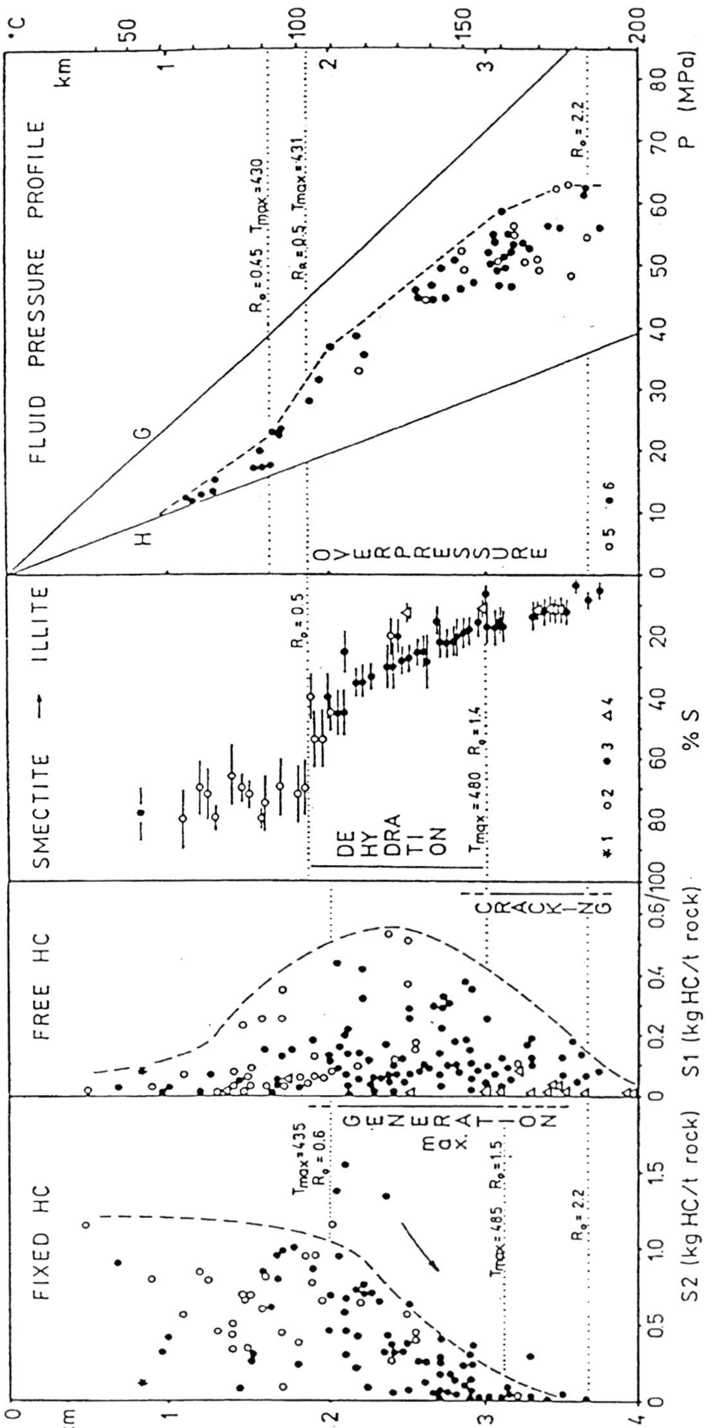

Fig. 23. Comparison between the S_2 and S_1 values (Rock-Eval), the smectite—illite transition and the pressure profiles in an intra-arc panonic basin. (After Francu et al. 1989)

undercompacted zones, which were uplifted via these tectonic activities to shallower depths, can blow up by reduction of the lithostatic confinement pressure. The illite–smectite transition has a sealing effect for Freed and Peacor (1989), in reducing the permeability and increasing to a similar extent the fluid retention capacity. This mechanism is well known in gritty reservoirs, where hairy or platy illites (Sommer 1978) crystallized in the aqueous phase in relation to invasion of water in the reservoirs, reduce the permeability to a point at which production of crude oil is blocked. It would then appear that creation of overpressures is not a process due to compaction or mineral alterations alone, but also to generation of gas and hydrocarbons. It is possible to distinguish undercompacted zones which have been subsequently decompacted, in considering argillaceous layers and in admitting that creation of mixed layers has a regular tendency, since formation of rectorite is favoured by fluid overpressures. However, comparisons are still needed to confirm this hypothesis.

To evaluate the relationships between mineral and organic transformations, compaction and temperature, it would be appropriate to examine a study case for which it is possible to determine sedimentation and history and for which tectonic activity only played an extremely minor role. The divergent margin of Labrador offshore is probably such a case. Opening took place during Cretaceous after a «rifting» episode, and the divergent margin was permanently active until the present time (Cramez-Diaz and Kübler 1982). It is, therefore, possible to follow the initial diagenetic stages continuously from Palaeocene to Pleistocene until the POW. The clay mineral assemblage is detrital in these margino–coastal or turbiditic detritic sediments, as revealed by the fraction between 2 and 16 µm which consists of well-crystallized micas and chlorite, kaolinite and mixed layers. In the <2 µm fraction, smectite occurs until a depth of 1750 m and below, the [001] XRD peak shifts from 14 to 11 Å until 3200 m (Heroux et al. 1981). This smectite–illite transition zone is marked by a reduction in the relative intensity of the 14 Å peak, as already seen in Logbaba drillings.

The inductolog provides information about the salinity of the interstitial waters. According to the induction, the resistivity remains constant (<2 ohm m^2/m), while the speed of the «sonic» increases regularly from 160 to 110 µs/foot. These two deferred diagraphies give a precise image of Burst's first dehydration phase: free water of the large pores is expelled by continuous compaction, without changing the chemical composition of the interstitial water. It is of interest to notice that concomitantly neither adsorbed hydrocarbon gases, nor organic extracts, nor vitrinite reflectance increase. Kübler (1984) concluded from this that during the initial stage of compaction, the water and the transformation products of the organic matter, which could have been generated by temperature increase, also return to the ocean. Some intercalated clay-type sediments begin to act as permeability barriers only at 1800 m.

This is recorded by an increase in methane, ethane and in the adsorbed gases. In the clays, the peak at 14 Å begins to shift towards 11 Å and the relative smectite content tends to decrease progressively. According to the inductolog, resistivity increased to 5 ohm m^2/m. According to the filtering capacity often attributed to clays, the resistivity should, on the contrary, decrease by relative enrichment of interstitial water in cations. We can only attribute the increase in resistivity to the increase in gases which first reach a plateau at 500 µl/l and then a second one at nearly 1000 µl/l for methane. The organic extracts also begin to increase at 2000 m. At 2600 m, boreholes penetrate an undercompacted zone marked by a striking reduction in resistivity, sonic speed and density, but also by a rapid increase in the adsorbed gases, whose total exceeds 10.000 µl/l. However, neither the progression in organic extract, nor that of the reflectance capacity is affected by this undercompaction, whereas the peak at 14 Å continues to shift and the relative smectite content continues to decrease.

At Karlsefni, the undercompacted zone has no influence on the decrease in swelling layers in the clay minerals, which confirms the results from the Mahakam Delta Basin. Moreover, and contrary to the proposal by Rumeau and Sourisse (1972), undercompaction does not slow down mineral transformations, or at least the illitization of swelling layers in the case of a divergent margin, with continuous clay–silt–gravel sedimentation from Palaeocene to present time. Interstratified argillaceous units are needed because they are sensitive to a given compaction to become a permeability barrier by reduction of free porosity water. Overpressure is favoured by the thermocatalytic generation of methane, ethane and their upper equivalents. The circulation of fluids is usually restricted in undercompacted zones and yet this has no influence on the progression of illitization in the swelling layers, nor on that of organic extracts and vitrinite reflectance. In the Karlsefni borehole, the proportion of about 80% of non-swelling layers is reached at the beginning of the POW. However, kerogen being mainly continental, more than an average type III, it is not certain that it could generate oil.

Comparison with the numerous studies of the Gulf Coast area is probably not reasonable. In fact, many fields explored in the Gulf Coast area could testify that the undercompacted zones were destroyed, at least in part, favouring primary and secondary migrations as shown by Asakawa and Fujita (1979 in Kübler 1984). Nonetheless, given the current state of knowledge and the available results, it is impossible to assume a strict relationship between disappearance of smectite at the beginning of the illitization zone of the swelling layers and the beginning of the oil window. The question remains open, and the example of the Slovakian sector of the Panonian Basin is a good illustration.

Francu et al. (1989) compared fixed and free hydrocarbon contents of Rock-Eval pyrochromatography in the transition zone as identified by Burst, or the smectite illitization zone, and the profile of pressures which define the undercompacted zone. The kerogen is of type III and even of type IV; the continental component of the organic matter being, therefore, highly dominant with more gaseous than oil-prone potential. The results clearly confirm that undercompaction does not block mineral transformations, especially the illitization of swelling layers. Undercompaction begins before and continues after the POW framed by reflectances of 0.5 and 1.4%. Moreover, these values are considered by the authors as the limits of the transition zone reaching a thickness of 1100 m despite a geothermic gradient of

45°/km, which is relatively high but normal for a type A intra-arc subduction basin (Cramez, pers. comm.).

3.5 Stability–Instability of Smectite

Temperature is not a criterion for smectite stability. It is well known that under laboratory conditions, it is possible to heat a smectite to 700 °C without destroying the mineral structure, nor altering the physical properties. To do so, it is just needed to increase and decrease the temperature very slowly. This temperature of 700 °C would correspond to the pyroxene hornfels or to the granulite metamorphic facies. In the micritic clay massifs of the Doldenhorn nappe which contain up to 99% of carbonates, Burkhard (1988) found extremely well crystallized smectites together with epimetamorphic micas. These smectites suffered temperatures of 300 to 350 °C during at least 10 million years. This is an excellent example of thermal stability. Organic matter is rare in such micrites and would be of type II. The instability of smectites is, therefore, not directly linked to temperature, but mainly to chemical conditions.

Millot (1964) already described the framework for the transformation process. For Dunoyer de Segonzac (1969), illitization of smectite is definitely dependent on the availability of K and Al and a possible migration of Al from octahedral to tetrahedral positions, to create the necessary charge defect for K to neutralize the layer. In this case, the transformation process is strictly a chemical reaction and the chemical composition of the water necessary for this transformation to take place can be easily evaluated by thermodynamic considerations (Fritz 1981). It is, therefore, surprising that temperatures for both the beginning and the end of the transformation of swelling layers can be defined, as it is then necessary to ensure that temperature also influences other parameters, whose consequences are to induce smectite transformation.

These parameters are kinetics and permeability. As far as kinetics are concerned, it is clear, as for organic matter transformation, that an increase in temperature increases the reaction rate. For permeability, the processes are complex. When the compaction phase starts, gravitic water returns to the ocean without changing in composition as in Karlsefni (Kübler 1984), and no smectite transformation is observed. However, once the first seal appears, illitization of the swelling layers begins. The creation of seals is due to compaction which reduces the permeability of the interbedded clay or clay–silt units. In the Panonic basin, the current pressure to form a seal is at 12 MPa (130 kg cm^{-2}), the beginning of the transition zone being at 31 MPa (330 kg cm^{-2}), whereas the hydrostatic pressure should be at 18.5 MPa (200 kg cm^{-2}) for this depth. In Karlsefni, the beginning of illitization coincides with the appearance of the first seal at 28 MPa (302 kg cm^{-2}), whilst the continuous undercompacted zone only begins at 34 MPa (367 kg cm^{-2}). The pressures are higher in Karlsefni, because the tectonics have not raised the structures as they have done in the Panonic basin. However, if this permeability becomes too low, the core of the bentonite layer does not have a degree of illitization as high as that of the rims (Altaner et al. 1984). If permeability is too high in the first dehydration phase, all the transformation products return to the ocean and no illitization of the swelling layers is observed. In order for this to begin, permeability has to decrease, but if it becomes to low, even after the Cretaceous period, illitization may be incomplete.

4 Conclusions

In the first stages of diagenesis, mineral and organic transformations cannot be dissociated. However, correct interpretation of their interactions requires an evaluation of all information usually gathered for petroleum prospection: seismic dates to reconstruct the sedimentary and tectonic history, deferred electric or nuclear well loggings to simulate decompacted sedimentation, chemistry of interstitial palaeowaters, argillosity, gas content and desorbable gas profiles of the sediments. Only then can petrography, geochemistry of organic matter, mineralogy of clays and mineral geochemistry be considered to constrain the incertitudes. However, despite the already considerable amount of available information which allows cross-checkings, non-hydrocarbon gas profiles are still missing to help both the organic geochemists and mineralogists.

It is also interesting to recall that the concept of fluid migration, which has generally been admitted for a long time to explain hydrocarbon accumulations of gas and oil fields, is almost never suggested to explain the inorganic transformations of diagenetic origin. As fluid migrations depend on the shape and dynamic of the sedimentary basins under consideration, as well as on their tectono-sedimentary evolution, it might be considered that forthcoming studies on diagenesis will also evaluate hydrodynamic aspects together with the usual parameters to end with a definition of integrated diagenetic systems.

Acknowledgements. I would like to thank the editors for having conceived this book dedicated to Georges Millot and for having achieved the project. I also sincerely thank the reviewers for their help in improving the script.

References

Alpern B (1980) Pétrographie du kérogène. In: Durand B (ed) Kerogen, insoluble organic matter from sedimentary rocks. Editions Technip, Paris, pp 39–384

Alpern B, Cheymol D (1978) Réflectance et fluorescence des organoclastes du Toarcien du bassin de Paris en fonction de la profondeur et de la température. Rev Inst Fr Pétrol 33/4:515–535

Altaner SP, Hower J, Whitney G, Aronson JL (1984) Model for K-bentonite formation: evidence from zoned K-bentonites in the disturbed belt, Montana. Geology 12:412–415

Arronson JL, Hower J (1976) Mechanism of burial metamorphism of argillaceous sediment. 2. Radiogenic argon evidence. Geol Soc Am Bull 87:738–744

Asakawa T, Fujïta Y (1979) Organic metamorphism and hydrocarbon generation in sedimentary basins of Japan. Proc Seminar on Generation and maturation of hydrocarbon in sedimentary basins, Manila, 1977, pp 142–162

Barker CE (1983) Influence of time on metamorphism of sedimentary organic matter in liquid dominated geothermal systems, western North America. Geology 11:384–388

Barker CE (1989) Temperature and time in the thermal maturation of sedimentary organic matter. In: Naeser ND, MacCulloh TH (eds) Thermal history of sedimentary basins. Methods and case history. Springer, Berlin Heidelberg New York, pp 75–98

Boles JR, Franks SG (1979) Clay diagenesis in Wilcox Sandstones of southwest Texas. J Sediment Petrol 49:55–70

Bostick NH, Foster JM (1975) Comparison of vitrinite reflectance in coal seams and in kerogen of sandstones, shales and limestones in the same part of a sedimentary section. Coll Int Pétro Matière Organique des Sédiments, CNRS, Paris, 15–17 Sept 1973, pp 14–25

Brindley GW, Brown G (eds) (1980) Crystal structure of clay minerals and their X-ray identification. Monograph 5. Mineralogical Soc, London, 495 pp

Brown G (ed) (1961) The X-ray identification and crystal structures of clay minerals. Mineralogical Soc, London, 544 pp

Buiskool Toxopeus JMA (1983) Selection criteria for the use of vitrinite reflectance as a maturity tool. In: Brooks J (ed) Petroleum geochemistry and exploration of Europe. Blackwell, London, pp 295–307

Burkhard M (1988) L'Helvétique de la bordure occidentale du massif de l'Aar (évolution tectonique et métamorphique). Eclog Geol Helv 81:63–114

Burnham AK, Sweeney JJ (1989) A chemical kinetic model of vitrinite maturation and reflectance. Geochim Cosmochim Acta 53:2649–2657

Burnham AK, Oh MS, Crawford RW, Samoun AM (1989) Pyrolysis of the Argonne premium coals: activation energy distributions and related chemistry. Energy Fuels 3:42- 55

Burst JF (1959) Post diagenetic clay mineral environmental relationship in the Gulf Coast Eocene. 6th Natl Conf 1957, Berkeley, Clays Clay Min, pp 327–341

Burst JF (1969) Diagenesis of Gulf Coast clayey sediments and its possible relation to petroleum migration. Bull Am Assoc Petrol Geol 53:73–93

Chamley H (ed) (1989) Clay sedimentology. Springer, Berlin Heidelberg New York, 623 pp

Connan J (1974) Time-temperature relationship in oil genesis. Am Assoc Petrol Geol Bull 58:2516–2521

Cornelius CD (1975) Geothermal aspects of hydrocarbon exploration in the North-Sea Area. Nor Geol Unders 316:29–67

Cramez-Dias C, Kübler B (1982) Les gaz adsorbés (C1 à C4) dans huit puits de l'offshore du Labrador. Impact de la sismographie sur l'interprétation de la maturation, de la migration et de la genèse des hydrocarbures. Notes Mém, Total (CFP) 17:22–33

Doligez B, Bessis F, Burrus J, Ungerer P, Chenet PY (1986) Integrated numerical simulation of the sedimentation, heat transfer, hydrocarbon formation and fluid migration in a sedimentary basin: The THEMIS model. In: Burrus J (ed) Thermal modeling in sedimentary basins. Editions Technip, Paris, p 173

Dow WG (1978) Petroleum source beds on continental slopes and rises. Am Assoc Petrol Geol Bull 62:1584–1606

Dunoyer de Segonzac G (1969) Les minéraux argileux dans la diagenèse. Passage au métamorphisme. Mém Serv Carte Géol Als Lorr (Strasb) 29:320 pp

Durand B, Oudin JL (1979) Exemple de migration des hydrocarbures dans une série déltaïque: le delta de Mahakam. Proc 10th World Petr Congr Bucarest, Romania, 1, pp 1–9

Durand B, Ungerer P, Chiarelli A, Oudin JL (1984) Modélisation de la migration de l'huile. Application à deux exemples de bassins sédimentaires. Proc 11th World Petr Congr, London, pp 1–13

Durand B, Alpern B, Pittion JL, Pradier B (1986) Reflectance of vitrinite as a control of the thermal history of sediments. In: Burrus J (ed) Thermal modeling in sedimentary basins. IFP Research Conf on Exploration, Editions Technip, Paris, pp 441–474

Espitalié J (1986) Use of Tmax as a maturation index for diffrent types of organic matter. Comparison with vitrinite reflectance. In: Burrus J (ed) Thermal modeling in sedimentary basins. IFP Research Conf on Exploration. Editions Technip, Paris, pp 475–496

Espitalié J, Deroo G, Marquis F (1985) La pyrolyse Rock-Eval et ses applications. Rev Inst Fr Pétrol 40:563–579

Francu J, Rudinec R, Simanec V (1989) Hydrocarbon generation zone in the East Slovak Neogen Basin: model and geochemical evidence. Geol Carpathica, 40:355–384

Freed RL (1980) Shale mineralogy and burial diagenesis in four geopressured wells, Hidalgo and Brazoria counties, Texas. In: Loucks RG, Richmann DL, Milliken KL (eds) Factor controlling reservoir quality. In: Tertiary sandstones and their significance to geopressured geothermal productions, Division of the Geothermal Energy, US Department of Energy Contract No DOE/ET/2711–1, Appendix A, pp 111–172

Freed RL, Peacor DR (1989) Geopressured shale and sealing effect of smectite to illite transition. Am Assoc Petrol Geol Bull 73:1223–1232

Freed RL, Peacor DR (1992) Diagenesis and the formation of authigenic illite-rich I/S crystals in Gulf Coast Shales: TEM study of clay separates. J Sediment Petrol 62:220–234

Fritz B (1981) Etude thermodynamique et simulation des réactions hydrothermales et diagénétiques. Sci Géol Mém (Strasb) 65:179 pp

Füchtbauer H, Goldschmidt H (1963) Beobachtungen zur Ton-Mineral Diagenese. 1st Int Clay Conf, Göttingen, vol 1, pp 99–111

Galway AK (1969) Heterogeneous reaction in petroleum genesis and maturation. Nature 223:1257–1260

Grim RE (1953) Clay mineralogy. McGraw-Hill, London, 384 pp

Grim RE (1958) Concept of diagenesis in argillaceous sediments. Am Assoc Petrol Geol Bull 42:246–253

Heroux Y, Chagnon A, Bertrand A (1979) Compilation and correlation of major thermal maturation parameters. Am Assoc Petrol Geol Bull 63:2128–2144

Heroux Y, Bertrand R, Chagnon A, Connan J, Pittion JL, Kübler B (1981) Evolution thermique et potentiel pétroligène par l'étude des kérogènes, des extraits organiques des gaz adsorbés, des argiles du sondage Karlsefni H-13 (Offshore Labrador Canada). Can J Earth Sci 18:1856–1877

Hood A, Gutjahr CCM (1972) Organic metamorphism and the generation of petroleum. Pap Annu Meet Geol Soc Am, Minneapolis 1972

Hood A, Gutjahr CCM, Heacock CL (1975) Organic metamorphism and the generation of petroleum. Bull Am Assoc Petrol Geol 59:986–996

Howard SA, Preston KD (1989) Profile fitting of powder diffraction patterns. Rev Mineral 20:217–272

Hower J, Eslinger EV, Hower ME, Perry EA (1976) Mechanism of burial metamorphism of argillaceous sediment. 1. Mineralogical and chemical evidence. Am Assoc Petrol Geol Bull 87:725–737

Johns WD, Shimoyama A (1972) Clay minerals and petroleum forming reactions during burial and diagenesis. Am Assoc Petrol Geol Bull 56:2160–2167

Karweil J (1975) The determination of paleotemperature from the optical reflectance of coaly particles in sediments. Coll Int Pétro Matière Organique des Sédiments, CNRS, Paris 1973, pp 195–203

Kisch HJ (1987) Correlation between indicators of very low-grade metamorphism. In: Frey M (ed) Low temperature metamorphism. Blackwell, Oxford, pp 227–300

Kübler B (1966) La cristallinité de l'illite et les zones tout à fait supérieures du métamorphisme. In: Coll sur les étages tectoniques. A la Baconnière, Neuchâtel, pp 105–122

Kübler B (1979) Adsorbed gases, C1 to C4 in relation to petroleum exploration. In: Generation and maturation of hydrocarbons in sedimentary basins. Proc Seminar, Manila, Tech Publ 6, pp 81–97

Kübler B (1980) Les premiers stades de la diagenèse organique et de la diagenèse minérale. 2. Zonéographie par les transformations minéralogiques, comparaison avec la réflectance de la vitrinite, les extraits organiques et les gaz adsorbés. Bull Ver Schweiz Petrol Geol Ing 46:110–122

Kübler B (1984) Les indicateurs des transformations physiques et chimiques dans la diagenèse. Température et calorimétrie. In: Lagache M (ed) Thermométrie et barométrie géologique. Soc Fr Minér Crist, Paris, pp 489–596

Kübler B, Betrix MA, Monnier F (1979a) Les premiers stades de la diagenèse minérale, une tentative d'équivalence. 1. Zonéographie par la maturation de la matière organique. Bull Ver Schweiz Petrol Geol Ing 45(108):1–22

Kübler B, Pittion JL, Heroux Y, Charollais J, Weidmann M (1979b) Sur le pouvoir réflecteur de la vitrinite dans quelques roches du Jura, de la molasse et des nappes préalpines, helvétiques et penniques (Suisse occidentale et Haute-Savoie). Eclog Geol Helv 72:347–373

Kübler B, Cramez-Diaz C, Bertrand R, Desjardins M (1982) Les hydrocarbures gazeux adsorbés dans les sédiments: leur utilisation en exploration pétrolière. Notes Mém, Total (CFP) 13:50 pp

Le Tran K, Connan J, De la Pasture BM (1974) Diagenesis of organic matter and occurrence of hydrocarbon and hydrogen sulfide in the SW Aquitaine Basin (France). Bull Centre Rech Pau-SNPA 8:111–137

Lopatin NV (1971) Temperature and geologic time as factors in coalification. Akad Nauk URSS, Serv Geol Izbestiya 3:96–106 (Translation by NW Bostick)

McCartney JT, Ergun S (1967) Optical properties of coals and graphite. US Bur Mines Bull 641:1–49

Millot G (1964) Géologie des Argiles. Masson, Paris, 499 pp

Milne IH, Early JW (1958) Effect of source and environment on clay minerals. Am Assoc Petrol Geol Bull 42:328–338

Monnier F (1982) Thermal diagenesis in the Swiss molasse Basin: implications for oil generation. Can J Earth Sci 19:328–342

Moore DM, Reynolds RC Jr (1989) X-ray diffraction and the identification and analysis of clay minerals. Oxford Univ Press, Oxford, 332 pp

Oberlin A, Boulimier JL, Villey M (1980) Electron microscopy of kerogen microstructure selected criteria for determining the evolution path and evolution stage of kerogen. In: Durand B (ed) Kerogen. Technip, Paris, pp 191–241

Pearson MJ, Watkins D, Small JS (1982) Clay diagenesis and organic maturation in northern sea sediments. In: van Olphen H, Veniale F (eds) Developments in sedimentology 35. Proc Int Clay Conf, Bologna, 1981. Elsevier, Amsterdam, pp 665–675

Perry EA Jr, Hower J (1970) Burial diagenesis of Gulf Coast pelitic sediments. Clays Clay Min 18:165–177

Pochon M (1978) Origine et évolution des sols du Haut Jura Suisse. Mém Soc Helv Sci Nat 90:190 pp

Powers MC (1959) Clay diagenesis in the Chesapeake Bay area. Clays Clay Min, Natl Acad Sci Nat Res Council, Washington, DC, Publ 327, pp 68–80

Pusey WC III (1973) How to evaluate potential gas and oil source rock. World Oil 176:71–75

Pytte AM, Reynolds RC Jr (1989) The kinetics of the smectite to illite reaction in contact metamorphism shales. In: Naeser ND, McCulloh TH (eds) The thermal history of sedimentary basins. Springer, Berlin Heidelberg New York, pp 133–140

Reynolds RC Jr (1980) Interstratified clay minerals. In: Brindley GW, Brown G (eds) Crystal structures of the clay minerals and their X-ray identification. Monograph 5. Minerological Soc, London, 495 pp

Reynolds RC Jr, Hower J (1970) The nature of interlaying in mixed-layer illite–montmorillonite. Clays Clay Min 18:25–36

Robert M (1972) Transformation expérimentale des micas en vermiculite ou smectites. Propriétés des smectites de transformation. Bull Groupe Fr Argiles 24:137–151

Robert P (1985) Histoire géothermique et diagenèse organique. Centre Rech Explor-Prod Elf–Aquitaine Mém 8, Pau, France, 345 pp

Rolli M (1991) Appréciation des taux de couches gonflantes selon Reynolds (1980) par désommation selon Pearson VII, des pics des micas détritiques et des interstratifiés. Cahiers de l'Institut de Géologie de Neuchâtel, Suisse, 6 pp

Rouzaud JN (1984) Relation entre la microstructure et les propriétés des matériaux carbonés. Application à la caractérisation des charbons. Thèse, Univ Orléans, 138 pp

Rouzaud JN, Oberlin A (1984) Contribution of high resolution transmission electron microscopy (TEM) to organic materials characterization and interpretation of their reflectance. In: Durand B (ed) Thermal phenomena in sedimentary basins. Editions Technip, Paris, pp 127–134

Rumeau JL, Sourisse C (1972) Compaction, diagenèse et migration dans les sédiments argileux. Bull Centre Rech Pau-SNPA 6:313–346

Schoonmaker J, Fred T, Mackenzie J (1986) Tectonic implications of Illite/Smectite diagenesis, Barbados accretionnary prism. Clays Clay Min 34:465–472

Sommer F (1978) Diagenesis of Jurassic sandstones in the Viking graben. J Geol Soc 135:63- 67

Srodon J (1984) X-ray powder diffraction identification of illitic materials. Clays Clay Min 32:337–349

Srodon J, Eberl DD (1987) Illite. In: Bailey SW (ed) Micas. Rev Mineral 13, Miner Soc Am, Washington, DC, 584 pp

Staplin F (1969) Sedimentary organic matter. Organic metamorphism and oil gas occurrence. Bull Can Petrol Geol 1711:47–66

Sweeney JJ, Burnham AK (1990) Evaluation of a simple model of vitrinite reflectance based on chemical kinetics. Bull Am Assoc Petrol Geol 74:1559–1570

Teichmüller M (1971) Anwendung kohlenpetrographischer Methoden bei der Erdöl–Erdgasprospektion. Erdöl Kohle 24:69–76

Timofeev PP, Bogoliyubova LI (1970) Poastsedimentsionnyye izmenenia organicheskogo veshchestva zavisimosti ot lithologicheskikh tipov porod i fatisial'nykh uslovity ikh nakolenyia (Post-sedimentational changes of organic matter in relation to lithology and depositional facies). In: Vassoyevitch NB (ed) Organicheskoye veschestrvo sovremennykh i iskopayemikh osadkov. Nauka Press, Moscow, pp 169–190

Tissot BP (1969) Premières données sur les mécanismes et cinétique de la transformation du pétrole dans les sédiments, simulation d'un schéma réactionnel sur ordinateur. Rev Inst Fr Pétrol 24:470–501

Tissot BP, Espitalié J (1975) L'évolution thermique de la matière organique des sédiments: application d'une simulation mathématique. Rev Inst Fr Pétrol 30:734–777

Tissot BP, Welte DH (1978) Petroleum formation and occurrence. A new approach to oil and gas exploration. Springer, Berlin Heidelberg New York, 538 pp

Tissot BP, Welte DH (1984) Petroleum formation and occurrence. Springer, Berlin Heidelberg New York, 699 pp

Tissot BP, Pelet R, Roucaché J, Combaz A (1975) Utilisation des alcanes comme fossiles géochimiques indicateurs des environnements géologiques. In: Campos R, Goni J (eds) Proc 7th Int Meet on Organic geochemistry, Madrid, 1975, pp 117–154

Ungerer P, Espitalié J, Marquis F, Durand B (1986) Use of kinetic models of organic matter evolution for the reconstruction of paleotemperatures. Application to the case of the Gironville Well (France). In: Burrus J (ed) Thermal modeling in sedimentary basins. IFP Research Conf on Exploration, Editions Technip, Paris, pp 531–546

Wapples DW (1980) Time and temperature in petroleum formation: application of Lopatin's method to petroleum exploration. Am Assoc Petrol Geol Bull 64:916–926

Wapples DW (1981) Organic Geochemistry for exploration geologists. Burgess, p 103

Weaver CE (1956) The distribution and identification of mixed-layer clays in sedimentary rocks. Am Mineral 41:202–221

Weaver CE (1956) Mixed-layers in sedimentary rocks. Am Mineral 41:202–221

Weaver CE (1958) The effect and geologic significance of potassium «fixation» by expandable clay minerals derived from muscovite, biotite, chlorite and volcanic material. Am Mineral 43:839-861

Weaver CE (1960) Possible use of clay minerals in the search for oil. Am Assoc Petrol Geol Bull 44:1505–1518

Weaver CE (1989) Clays, muds, and shales. Developments in sedimentology 44. Elsevier, Amsterdam, 819 pp

Index

Symbols

A

B

C

D

E

J

K

L

M

R

S

T

U

V

W

Y

Z